Vitamin C

Volume III

Author

C. Alan B. Clemetson

Professor
Tulane University School of Medicine
New Orleans, Louisiana

CRC Press
Taylor & Francis Group
Boca Raton London New York

CRC Press is an imprint of the
Taylor & Francis Group, an informa business

First published 1989 by CRC Press
Taylor & Francis Group
6000 Broken Sound Parkway NW, Suite 300
Boca Raton, FL 33487-2742

Reissued 2018 by CRC Press

Library of Congress Cataloging-in-Publication Data

Clemetson, C. Alan B.
    Vitamin C / C Alan B. Clemetson.
        p.   cm.
    Includes bibliographies and index.
    ISBN 0-8493-4841-2 (v. 1)
    ISBN 0-8493-4842-0 (v. 2)
    ISBN 0-8493-4843-9 (v. 3)
    1. Vitamin C deficiency.   2. Vitamin C--Metabolism.   I. Title.
RC627.S36C53 1989
616.3'9--dc19   88-14735

A Library of Congress record exists under LC control number: 88014735

ISBN 13: 978-1-315-89848-3 (hbk)
ISBN 13: 978-1-351-07758-3 (ebk)

Visit the Taylor & Francis Web site at http://www.taylorandfrancis.com and the
CRC Press Web site at http://www.crcpress.com

FOREWORD

While frank scurvy is rare nowadays, subclinical vitamin C deficiency is common and is now known to be associated with elevated blood histamine levels, which rapidly return to normal when ascorbic acid is administered. Epidemiological and experimental evidence suggests that our common metabolic defect, the inability to synthesize ascorbic acid from simple sugars, may be largely responsible for the development of subendothelial hemorrhage, thrombosis, atheroma, and degenerative vascular disease. This book is more concerned with factors affecting ascorbic acid metabolism, such as aging, smoking, infection, trauma, surgery, hormone administration, heavy metals, pregnancy, hemolysis, ionizing radiation, aspirin, alcohol, and other drugs which cause a disturbance of ascorbic acid metabolism and may thereby lead to vascular disease, than it is with simple dietary deficiency of ascorbic acid. The clinical, pathological, and chemical changes observed in ascorbic acid deficiency are discussed in detail; several diseases and disorders associated with abnormalities of ascorbic acid metabolism are described. Possible toxic effects resulting from the oxidation of ascorbic acid are noted, and reasons for the use of D-catechin or other chelating fiber to prevent or minimize the release of ascorbate free radical are detailed.

PREFACE

About 60 years ago, and before the isolation of ascorbic acid, Mme. L. Randoin (1923)* found the number of research studies on the antiscorbutic vitamin so great as to make it impossible for her to review them all.

"Jai maintenant à parler des recherches de toute nature faites sur le facteur antiscorbutique. La tâche est bien ingrate, car le nombre de ces recherches est si grand qu'il m'est évidemment impossible de les passer toutes en revue et, au surplus, elles présentent, par défaut de convergence, de telles lacunes, qu'il est vraiment difficile d'en donner une idée d'ensemble."

Now I must speak of all kinds of studies of the antiscorbutic factor. It is a thankless task because the number of research studies is so great that it is clearly impossible to review them all; moreover, by failure of agreement, they present such gaps that it is truly difficult to present a consistent thesis.

Today, the profusion of the literature on this subject is even more overwhelming. It is growing so fast that it is impossible to do justice to all the work that had been done in this field. Moreover, having written 57 chapters in 36 months, it is inevitable that the chapters written first will not be as up-to-date as those written last.

Undoubtedly, some important works have been omitted, either because they have escaped my notice or because they were written in a language that I cannot read. Any workers whose contributions have been omitted must accept my assurance that it was not by intent.

It is hoped that this book presents a consistent thesis and that its main message is clear. It does not so much concern the amount of vitamin C in the diet, as it does the factors affecting ascorbic acid metabolism, the diseases that may result from abnormalities of ascorbic acid metabolism, and some suggestions as to what we may be able to do to prevent them.

Although the title of this book is *Vitamin C,* it could equally well have been entitled *Vitamin C, Heavy Metals, and Chelating Fiber.*

C. Alan B. Clemetson, M.D.
Pineville, Louisiana
February, 1987

THE AUTHOR

C. Alan B. Clemetson, M.D., was born in England. He attended the King's School, Canterbury, Magdalen College, Oxford, and Oxford University School of Medicine, graduating as a physician (B.M., B.Ch.) in 1948. He is an obstetrician and gynecologist, with fellowships in British, Canadian, and American colleges (F.R.C.O.G., F.R.C.S.C., and F.A.C.O.G.), but he has devoted most of his life to research and has published papers on many diverse subjects.

His career has included academic positions at London University, the University of Saskatchewan, the University of California at San Francisco, the State University of New York, and at Tulane University in Louisiana, where he is currently Professor at the School of Medicine.

He has challenged many conventional ideas and believes that, "certainty of knowledge is the antithesis of progress." Thus, every statement in this book is backed by reference to experiments and observations in the literature; contrary findings are cited, weighed, and given due credence.

VITAMIN C

Volume I

Vitamin C Deficiency
Classical Scurvy: A Historical Review
Chronic Subclinical Ascorbic Acid Deficiency
Factors Affecting the Economy of Ascorbic Acid
Inadequate Ascorbic Acid Intake
Smoking
Aging
Sex
Menstrual Cycle, Estrus Cycle, Ovulation
Infection
Trauma, Surgery, and Burns
Heavy Metals, Water Supplies: Copper, Iron, Manganese,
Mercury, and Cobalt
Bioflavonoids
Dietary Protein
Hormone Administration: Birth Control Pills
Pregnancy
Hemolysis
Stress and the Pituitary-Adrenal System
Lack of Sleep
Time of Day
Season
Achlorhydria
Ionizing Radiation
Aspirin and Salicylates
Alcohol
Other Factors Affecting Ascorbic Acid Needs

Volume II

Clinical and Pathological Findings in Ascorbic Acid Deficiency
Vascular Changes
Diabetes Mellitus
Anemia
Defective Wound Healing
Bone Changes
Joint Lesions
Dental and Periodontal Changes
Atherosclerosis
Mental Depression
Amyloid
Venous Thrombosis
Decreased Resistance to Infection
Liver, Bile, and Gallstones

Volume III

Chemical Changes Associated with Vitamin C Deficiency
Histamine Metabolism
Proline and Lysine Metabolism
Carbohydrate Metabolism
Folic Acid Metabolism
Cholesterol Metabolism
Tyrosine and Phenylalanine Metabolism
Tryptophan Metabolism
Adrenal Corticoid Metabolism
Uric Acid Clearance
Clinical Conditions Associated with Disorders of Ascorbic Acid Metabolism
Rheumatic Fever
Menorrhagia
Wound Dehiscence
Habitual Abortion
Abruptio Placentae
Prematurity and Premature Rupture of the Fetal Membranes
Megaloblastic Anemia of Infancy, Pregnancy, and Steatorrhea
Gastrointestinal Ulcers and Hemorrhage
Ocular Lesions
Cerebral Hemorrhage and Thrombosis
Coronary Thrombosis and Myocardial Infarction

TABLE OF CONTENTS

CHEMICAL CHANGES ASSOCIATED WITH VITAMIN C DEFICIENCY

Chemical Changes Associated with Vitamin C Deficiency

Chapter 1

HISTAMINE METABOLISM

I. INTRODUCTION

Since the original observation of Parrot and Richet (1945) that scurvy increases the sensitivity of guinea pigs to histamine, several workers, including Dawson and West (1965), Dawson et al. (1965), and Lewis and Nicholls (1973), have investigated the relationship between ascorbic acid and histamine metabolism. During the same period other workers, including Sayers and Sayers (1947), Hicks (1965), Csaba and Toth (1966), Kovacs and Suffiad (1968), Freeman (1968), Muszbek et al. (1969), Porter et al. (1970), Latif and Sultan (1971), Reilly and Schayer (1972), Hirose et al. (1976), and Bruce et al. (1976), were working on the related problem of histamine and the adrenocortical system in rats, mice, chicks, guinea pigs, dogs, and asthmatic human subjects.

II. GUINEA PIGS

It was the elegant studies of Subramanian et al. (1973, 1974), Chatterjee et al. (1975a, b), and Subramanian (1978) which clarified the relationship between blood histamine and ascorbic acid levels. These workers demonstrated a progressive rise in the blood histamine levels of guinea pigs on a scorbutogenic diet (Figure 1A). It is interesting to note that the histamine level began to rise very early, on the third day of the diet, when the blood ascorbic acid level had only just fallen below 1 mg/100 ml. Moreover, after 2 weeks on the diet, when the blood histamine level was markedly elevated, it could be returned rapidly to normal by the administration of a single dose of ascorbic acid (5 mg/100 g of guinea pig). They obtained very similar results when studying urinary histamine (Figure 1B) and studying tissue histamine levels in guinea pigs on a vitamin C-deficient diet (Figure 2).

Subramanian et al. (1973) found that neither ascorbic acid nor dehydroascorbic acid (DHAA) and H_2O_2, the products of oxidation of ascorbic acid, were able to break down histamine. However, histamine is broken down when it is added to a system in which ascorbic acid is allowed to undergo oxidation in the presence of a catalyst such as Cu^{++} or tissue homogenates. Chatterjee et al. (1975b) reported that in a model system of ascorbic acid (5 mmol), Cu^{++} (0.05 mmol as $CuSO_4$, $5H_2O$), and histamine (1 mmol), in a total volume of 12.5 ml 0.05 M sodium phosphate buffer, pH 7.2, incubated 37°C for 4 h, the histamine is completely broken down to aspartic acid. Hydantoin acetic acid was identified as an intermediate in this conversion (Figure 3).

The various intermediates identified at different intervals of incubation were as follows: 0 h, histamine; 2 h, histamine and hydantoin-5-acetic acid; 4 h, aspartic acid. Aspartic acid was identified even after 24 h of incubation, indicating that it was the end product of incubation. These workers suggest that the probable intermediate in the conversion of histamine to hydantoin acetic acid would be 2,4-dihydroxyimidazole acetic acid, believing it to be another example of an ascorbic acid-mediated hydroxylation of a heterocyclic compound.

Studying the effects of stress, Chatterjee et al. (1975b) observed that the administration of antibiotics, vaccines, heat, cold, and even pregnancy caused an elevation of the blood histamine levels of guinea pigs (Table 1), but ascorbic acid supplements brought their histamine levels down to normal, thus blocking this effect of stress.

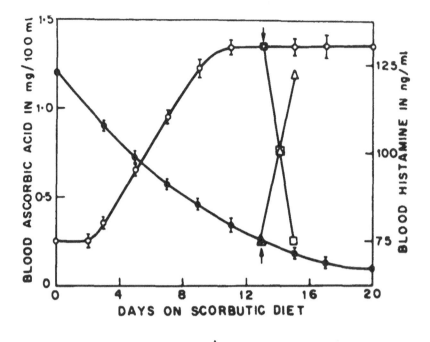

A

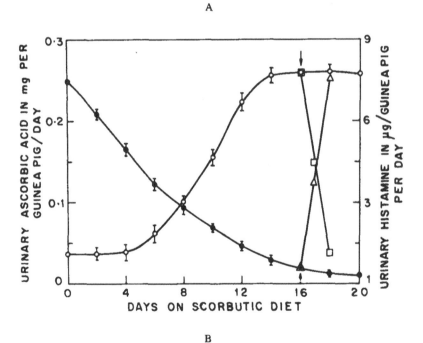

B

FIGURE 1.   (A) Ascorbic acid and histamine levels in blood from guinea pigs fed an ascorbic acid-free diet. ●–●, ascorbic acid; ○–○, histamine. ↑, ↓, administration of a single dose of ascorbic acid, 5 mg/100 g body weight guinea pig; △–△, □–□, subsequent ascorbic acid and histamine levels. (B) Ascorbic acid and histamine levels in urine from guinea pigs fed an ascorbic acid-free diet. Symbols ↑, ↓, as in Figure 1A. In control guinea pigs fed an ascorbic acid-free diet plus a daily oral dose of ascorbic acid 1 mg/100 g guinea pig, the basal value of ascorbic acid and histamine in urine did not change within the experimental period. (From Chatterjee, I. B., Das Gupta, S., Majumder, A. K., Nandi, B. K., and Subramanian, N. [1975a], *J. Physiol. (London)*, 251, 271. With permission.)

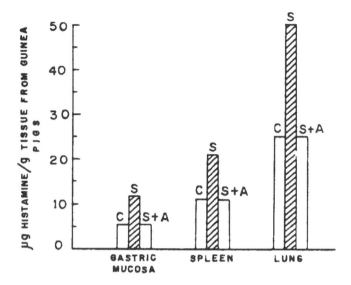

FIGURE 2. Effect of a single administration of ascorbic acid on the tissue histamine contents of guinea pigs fed an ascorbic acid-free diet for 16 d. The values were obtained after killing of the animals, on the 17th day. C, control; S, scorbutic diet; S + A, scorbutic diet after administration of ascorbic acid. (From Chatterjee, I. B., Majumder, A. K., Nandi, B. K., and Subramanian, N. [1975b], *Ann. N.Y. Acad. Sci.*, 258, 24. With permission.)

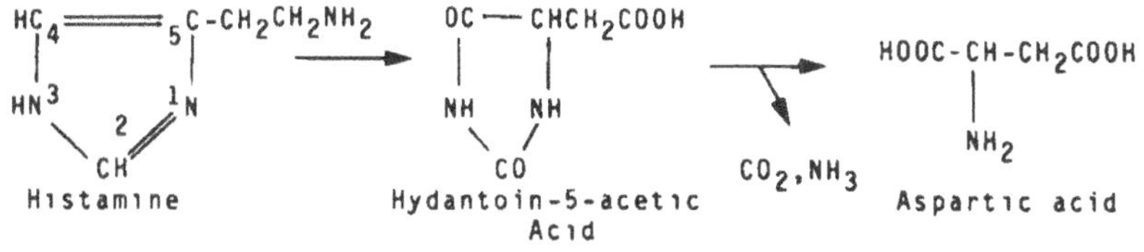

FIGURE 3. Scheme proposed by Chatterjee (1975b) whereby histamine is broken down when it is added to a system in which ascorbic acid is allowed to undergo oxidation in the presence of a catalyst, such as Cu^{++} or tissue homogenates.

III. HUMANS

The simple plan of analyzing the same blood samples, both for histamine and for ascorbic acid, was adopted by the Department of Obstetrics and Gynecology at the Methodist Hospital of Brooklyn, NY, and has provided much useful information. Blood samples were drawn from pregnant women attending the prenatal clinic and from healthy men and women volunteers, including physicians, nurses, laboratory technicians and medical students, before they had taken any vitamin C-containing food, in the morning.

The results of analysis of 437 such human blood samples for histamine by the spectro-fluorometric method of Shore et al. (1959) and for reduced ascorbic acid by the method of Roe (1954) were reported by Clemetson (1980) and are shown in Table 2. When the mean blood histamine level for each 0.1-mg/100 ml ascorbate group is plotted on a graph, as shown in Figure 4, it becomes evident that plasma ascorbic acid levels below 1.0 mg/100 ml are associated with increasing whole blood histamine levels. Moreover, this increase becomes highly significant ($p < 0.001$) for plasma ascorbate levels below 0.7 mg/100 ml. Indeed, the blood histamine level rises exponentially as the plasma ascorbate falls.

Table 1
EFFECT OF ASCORBIC ACID ON BLOOD HISTAMINE
LEVELS IN GUINEA PIGS UNDER STRESS CONDITIONS

	Blood histamine level (ng/ml)[a]	
Treatment	Without ascorbic acid	With ascorbic acid (5 mg/100 g/d)
None	75 ± 5	70 ± 2
Drugs, b.d.		
Penicillin, streptomycin	120 ± 2	75 ± 5
Chloramphenicol	110 ± 2	76 ± 4
Tetracycline	102 ± 2	78 ± 2
Vaccines and toxoids, single dose		
Triple antigen	118 ± 2	78 ± 2
Tetanus, TABC	110 ± 2	75 ± 5
Cholera	105 ± 2	76 ± 4
Physical stress		
Heat (39 ± 1°C)	160 ± 5	75 ± 5
Cold (6 ± 1°C)	135 ± 5	76 ± 4
Pregnancy (50—55 d)	120 ± 5	75 ± 5

Note: The blood histamine levels of guinea pigs were increased significantly after drug treatment or by different forms of stress, but ascorbic acid administration brought down the levels to normal values.

[a]   48 h after treatment.

From Chatterjee, I. B., Majumder, A. K., Nandi, B. K., and Subramanian, N. (1975b), *Ann. N.Y. Acad. Sci.,* 258, 24. With permission.

There is a considerable scatter of individual values, as shown in Figure 5 where the results of analysis of 240 blood samples from pregnant women are plotted as points on a graph, but the upward trend of the histamine levels of the women with the lower ascorbate levels is clearly seen.

Oral administration of ascorbic acid (1 g daily for 3 d) to 11 selected volunteers led to an elevation of the plasma ascorbic acid level and a reduction of the whole blood histamine level in every instance, as shown in Table 3 and Figure 6.

A. Pregnancy

Originally it was thought that the mean blood histamine level for each ascorbate group was lower in pregnant than in nonpregnant women, as suggested by Clemetson (1980). This apparent difference was attributed to the placental histaminase activity described by Ahlmark (1944) and by Kapeller-Adler (1949, 1951), but a subsequent study by Clemetson and Cafaro (1981), comparing the results of analysis of 541 blood samples from pregnant women attending the prenatal clinic and 84 blood samples from women attending the post-partum clinic at the same hospital, 6 weeks after childbirth (Table 4), did not show any difference between the blood histamine levels of pregnant and nonpregnant women with similar ascorbate levels.

B. Age

There was also the thought that older people might have higher blood histamine levels, and this may be true, but analysis of our data by age does not provide any evidence of this among women with similar ascorbic acid levels. In Table 5, all blood samples from healthy pregnant women, 6 weeks post-partum women, and other nonpregnant women, not taking

Table 2

RELATIONSHIP BETWEEN PLASMA ASCORBIC ACID AND BLOOD HISTAMINE[a]

	Plasma ascorbic acid mg/100 ml	Number of samples	Mean whole blood histamine ng/ml ± SEM	Statistical significance of differences	Mean age of group years
A	0 00—0.19	14	59 1 ± 10 1	A vs. B $p < 0.001$	27 2
B	0.20—0 39	35	36.5 ± 6 2	B vs C $p < 0$ 001	28.9
C	0.40—0.59	61	27.9 ± 1 4	C vs D $p < 0$ 001	28.1
D	0.60—0.79	101	21 0 ± 1.0	D vs. E N/S	26 8
E	0 80—0 99	99	19 2 ± 1.0	E vs. F N/S	27 6
F	1 00—1 19	61	16.8 ± 0.9	F vs G N/S	30.7
G	1.20—1 39	30	18.1 ± 2.4	G vs. H N/S	28.5
H	1.40—1.59	17	17.1 ± 1.7		31 8
I	1.60—1.79	5	13.8		27 2
J	1 80—1 99	5	16.0		31.8
K	2.00—2 19	6	12.0		29.0
L	2.20—2.39	1	13.0		31.0
M	2.40—2.59	2	17 0		36 0

Note. Analysis of human blood samples, both for whole blood histamine and for plasma reduced ascorbic acid. The blood histamine levels were found to be significantly higher in people with low ascorbic acid levels

[a] Results of analysis of 437 blood samples are divided into even-numbered 0.2-mg/100 ml plasma ascorbic acid groups

From Clemetson, C. A. B. (1980), *J Nutr.*, 110, 662 © American Institute of Nutrition. With permission.

birth control pills or other medications, have been combined to give a total of 840 blood samples from women; no women in labor and no sick women have been included. They have been divided into three age groups — younger than 25 years, 25 to 35 years, and over 35 years. Also, each group has been subdivided into 16 plasma ascorbic acid (AA)* groups. The blood histamine levels are seen to be higher in the women with low ascorbic acid levels, in all three age groups, but age does not seem to affect the blood histamine level for a given ascorbate level in this study. It is possible that a study including older age groups or more people with low ascorbate levels might show an effect of age.

C. Sex

We do not as yet have enough blood samples from healthy, unstressed men to make a definite statement as to whether men differ from women in the relationship of blood histamine to ascorbic acid, but we have no reason to believe that they would.

D. Subsequent Work

Sharma et al. (1981) have provided some anachronous data for the blood histamine and whole blood total ascorbic acid (TAA)** levels of women in late pregnancy. This may be due to their modification of the chemical methods; they used trichloracetic acid instead of perchloric acid in their analysis for histamine.

Bates et al. (1983), studying lactating women in the Republic of The Gambia, observed very high blood histamine levels (mean 106.1 ± S.E. 4.9 ng/ml). There was a trend towards

*   AA — ascorbic acid, reduced form.
** TAA — total ascorbic acid, reduced and oxidized forms.

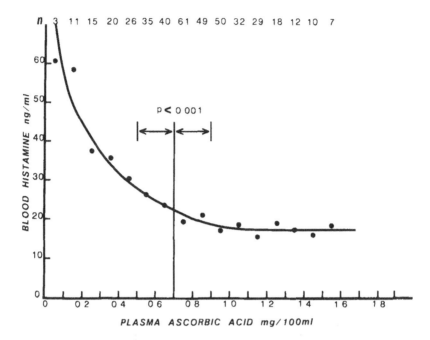

FIGURE 4.   The same data as in Table 2 are shown here in 0.1 mg/100 ml plasma ascorbic acid groups, in the range from 0 to 1.6 mg/100 ml The numbers along the top of the graph indicate the number of blood samples which fell in each group. It is apparent that the mean blood histamine levels began to rise when the ascorbic acid level fell below 1.0 mg/100 ml. The increase in the blood histamine level became highly significant when the ascorbic acid level fell below 0.7 mg/100 ml. This included 150 of the 437, or 34%, of the men and women studied

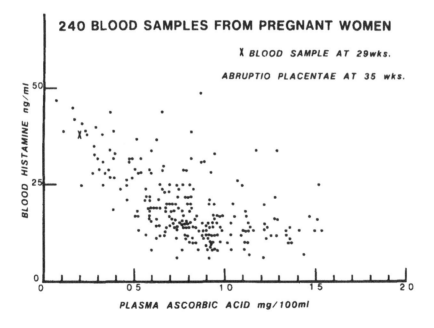

FIGURE 5.   Although there is considerable scatter, it is quite evident that the women with low plasma ascorbic acid levels had a tendency towards high whole blood histamine levels. The problem of abruptio placentae, mentioned here, is explored in greater depth in a later section of this book. (From Clemetson, C. A. B. [1980], *J. Nutr.*, 110, 662. © American Institute of Nutrition With permission.)

Table 3
THE EFFECT OF ORAL ASCORBIC ACID SUPPLEMENTS ON THE WHOLE BLOOD HISTAMINE LEVELS OF SELECTED VOLUNTEERS

Initials and condition of volunteer	Date	Sex	Age	Smoker (S) or Non Smoker (NS)	Before ascorbic acid supplement		After oral ascorbic acid 1 g daily for 3 d	
					Plasma ascorbic acid mg/100 ml	Duplicate whole blood histamine levels ng/ml	Plasma ascorbic acid mg/100 ml	Duplicate whole blood histamine levels ng/ml
S. G. F. (M.D.)								
After day and night on duty	7/28/78	M	31	S	0.14	180 & 180		
After night of sleep	7/31/78				0.20	80 & 84		
After night of sleep	8/4/78						1.15	15 & 18
R. A. M. (M.D.)								
After day and night on duty	8/15/78	M	29	S	0.57	28 & 29		
After night of sleep	8/16/78				0.65	29 & 31		
After night of sleep	8/23/78						1.54	22 & 25
R. S. (Medical student)								
After night of sleep	8/15/78	M	28	NS	0.31	24 & 22	1.71	14 & 16
After night of sleep	8/23/78							
R. S. (M.D.)								
After night of sleep	8/7/78	F	31	NS	0.86	34 & 33	1.82	20 & 21
After day and night on duty	8/21/78				0.95	45 & 46		
After day and night on duty	9/19/78							
E. T. (Registered Nurse)								
After night on duty	3/19/79	F	23	S	0.85	27 & 28	1.92	10 & 10
After night on duty	3/23/79							
C. F. (Ward Clerk)								
After night of sleep	3/19/79	F	34	NS	0.73	24 & 24	2.18	6 & 7
After night of sleep	3/23/79							
R. S. (Medical student)								
After night of sleep	3/20/79	M	23	S	0.55	26 & 26	1.56	11 & 12
After night of sleep	3/26/79							

Table 3 (continued)
THE EFFECT OF ORAL ASCORBIC ACID SUPPLEMENTS ON THE WHOLE BLOOD HISTAMINE LEVELS OF SELECTED VOLUNTEERS

Initials and condition of volunteer	Date	Sex	Age	Smoker (S) or Non Smoker (NS)	Before ascorbic acid supplement		After oral ascorbic acid 1 g daily for 3 d	
					Plasma ascorbic acid mg/100 ml	Duplicate whole blood histamine levels ng/ml	Plasma ascorbic acid mg/100 ml	Duplicate whole blood histamine levels ng/ml
A. F. (Medical student)								
After night of sleep	3/27/79	M	28	NS	0.54	22 & 22		
After night of sleep	4/3/79						1.23	7 & 8
J. M. (Subintern)								
After night of sleep	3/30/79	M	28	S	0.63	23 & 26		
After night of sleep	4/6/79						1.46	10 & 9
G. S. (Subintern)								
After night of sleep	3/30/79	M	44	NS	0.43	28 & 28		
After night of sleep	4/6/79						1.87	11 & 7
E. H. (Patient with purpura simplex)								
After night of sleep	4/5/79	F	24	S	0.44	28 & 27		
After night of sleep	4/9/79						1.41	13 & 13

Note: Morning blood samples were obtained form 11 selected volunteers both before and on the day after oral loading with ascorbic acid, 500 mg twice daily for 3 d. Analysis showed a reduction of the blood histamine level in all 11 subjects, as illustrated in Figure 4. Volunteers were selected for this study solely on the basis of having elevated blood histamine levels.

From Clemetson, C. A. B. (1980), *J. Nutr.*, 110, 662. © American Institute of Nutrition. With permission.

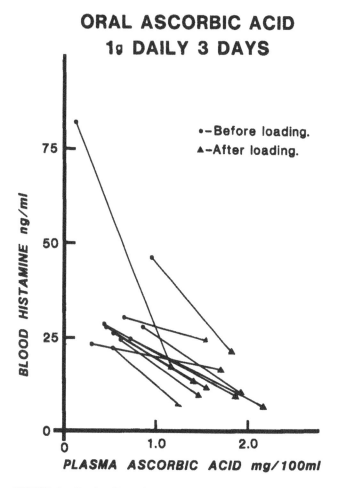

ORAL ASCORBIC ACID
1g DAILY 3 DAYS

•–Before loading.
▲–After loading.

BLOOD HISTAMINE ng/ml

PLASMA ASCORBIC ACID mg/100ml

FIGURE 6. Graphic illustration of the results of the oral ascorbic acid loading experiment, the details of which are shown in Table 3.

lower levels with increasing vitamin C supplements, but this failed to reach statistical significance. A combination of treatment for malaria and other parasitic diseases and vitamin C supplements much greater than 24, 47, or 69 mg/d would almost certainly have caused a marked reduction in the blood histamine levels of these women.

The finding, by Enwonwu and Okolie (1983), of increased brain histamine levels, both in protein-deficient and in ascorbate-deficient monkeys, as shown in Table 7 of Chapter 12, Volume I, is a very interesting observation. It may be largely responsible for the apathy which is common to both of these nutritional deficiencies.

Nakano and Suzuki (1984) studied the effects of immobilization stress on the tissue levels of histamine and ascorbic acid in rats. Stress caused an increase in the histamine levels of the blood, lungs, kidneys, and stomach, but caused a marked decrease in both the histamine and ascorbic acid levels of the adrenal glands. The same stress caused a marked increase in the serum and total body ascorbic acid levels. This was largely due to an increase in the liver content of this vitamin, presumably due to synthesis. It is our misfortune that we cannot react to stress in the same way because of our inability to synthesize ascorbic acid.

Mohsenin and DuBois (1987) reported that, " the airway response to aerosolized histamine was enhanced in scorbutic guinea pigs and animals partially deficient in vitamin C." Moreover, Chatham et al. (1987) have observed that ascorbic acid and vitamin E, when given concurrently, appear to ameliorate ozone-induced bronchoconstriction in normal human subjects.

Table 4
WHOLE BLOOD HISTAMINE LEVELS OF PREGNANT AND NONPREGNANT WOMEN

Plasma ascorbic acid (mg/100 ml)	Ante partum samples			Post partum samples		
	n	Blood histamine (ng/ml)	SD	n	Blood histamine (ng/ml)	SD
0 4—0.499	20	22 58	±4.7			
0.5—0.599	44	19.98	±6.2	4	16 4	
0.6—0.699	85	16.50	±3 8	18	16.33	±4.3
0.7—0 799	112	15.17	±4.4	14	15 25	±3.3
0 8—0 899	103	12.71	±4.0	22	11 61	±3.0
0.9—0.999	101	11 20	±4 2	18	11.25	±3.0
1.0—1 099	41	9.55	±3.7	8	6.56	±2.7
1.1—1 199	35	9.91	±3.6	4	7.5	
	Average age 24 6 years			Average age 24 8 years		

Note: Earlier work reported by Clemetson (1980) suggested that pregnant women might have lower blood histamine levels than nonpregnant women. However, this table, showing the results of a study conducted in the same laboratory, comparing 541 blood samples from women attending a prenatal clinic and from 84 women attending a 6-week post-partum clinic at the same hospital, shows no such difference.

Table 5
AGE, HISTAMINE, AND ASCORBIC ACID

Ascorbic acid (mg/100 ml)	Less than 25 years old			25 To 35 years old			Over 35 years old		
	n	Mean blood histamine (ng/ml)	SD	n	Mean blood histamine (ng/ml)	SD	n	Mean blood histamine (ng/ml)	SD
0—0.09				1	47				
0.1—0.19	2	40		2	42				
0.2—0.29	3	29 3		5	37 0		2	28.5	
0.3—0.39	14	26.6	±5.6	6	30.3	±5.6	2	29.0	
0.4—0.49	16	23.2	±5.2	12	25.7	±7.6	6	35.4	±17.9
0.5—0.59	35	20.3	±7.5	20	24.2	±10.4	4	27.4	
0.6—0.69	62	17.6	±5.1	46	18.5	±8.3	8	23.6	±9.5
0.7—0.79	69	15.9	±4.4	58	15.6	±5.4	11	17.1	±9.6
0.8—0.89	81	14.2	±7.0	43	15.5	±7.7	27	13.9	±5.6
0.9—0.99	57	11.7	±6.2	55	12.2	±6.0	13	13.5	±5.9
1.0—1.09	29	11.2	±5.8	22	12.2	±8.6	12	13.4	±6.5
1.1—1.19	22	9.9	±3.5	19	13.4	±8.5	6	9.9	±3.7
1.2—1.29	13	11.9	±5.5	11	11.6	±12.7	7	9.4	±4.4
1.3—1.39	9	10.4	±3.1	10	9.6	±1.9			
1.4—1.49	6	6.5		10	10.6	±3.2			
1.5—1.59	1	16.0		2	19.0		1	12.0	
Totals	419			322			99		

Note: Here the blood histamine levels of 840 healthy pregnant and nonpregnant women, not in labor and not taking birth control pills or other medications, are tabulated. They have been divided into 3 age groups and subdivided into 16 plasma ascorbate groups. The blood histamine levels were higher in the women with low plasma ascorbate levels, but age did not show any effect on the histamine level for each ascorbate group.

IV. IMPLICATIONS

Using electron microscopy to study the endothelial changes produced by ascorbic acid deficiency in the aorta of the guinea pig, Gore et al. (1965) observed a separation of the endothelial cells. Gapping of the intercellular junctions could be seen, as well as a distinct depletion of collagen between the endothelium and the internal elastic lamina. Since sub-endothelial collagen is sparse and inconstant in the capillary channels from which scorbutic bleeding originates, they concluded that endothelial cell disjunction must be the essential structural basis for the occurrence of hemorrhage in scurvy. While conceding that weakness of the perivascular collagen may contribute to the dilation and increased porosity of the small blood vessels, they were nevertheless struck by the similarity of the changes they observed in scurvy and the widening of intercellular junctions with leakage of leukocytes and erythrocytes, reported by Majno and Palade (1961), following the administration of serotonin or histamine. They did not know, at that time, that the blood histamine level was markedly increased in scurvy, but referring to their obsevations in scurvy, they stated, "The prompt and reversible production of the same changes by pharmacological stimuli or by inflammation clearly must have some other explanation." Indeed, the explanation may well be that the loss of endothelial integrity in scurvy is largely due to the action of histamine.

It is especially interesting to note that the blood histamine level starts to rise when the plasma ascorbic acid level falls below 1 mg/100 ml in human beings, just as it does when the ascorbic acid level falls below 1 mg/100 ml in guinea pigs. This is the first time that there has been any evidence of a metabolic change arising from suboptimal vitamin C nutrition. Previously it was thought by many that vitamin C deficiency caused only one problem, namely scurvy, when AA was completely absent from the diet and had virtually disappeared from the body.

Now it is clear that about one third of an apparently healthy group of people in New York City has a suboptimal vitamin C status (plasma AA <0.7 mg/100 ml) which is associated with an increased blood histamine level.

If it is confirmed that atherosclerosis begins with endothelial damage due to histamine, as suggested by Owens and Hollis (1979) and by De Forrest and Hollis (1980), studying rabbit aortas, and that antihistamines protect against the development of atherosclerosis, as shown by Harman (1962) and Hollander et al. (1974), then the subendothelial cholesterol deposition may be a secondary phenomenon. We may think of the cholesterol deposition as an aberrant form of wound healing in arteries. If so, it is the histamine damage that must be prevented and dietary ascorbic acid is undoubtedly the simplest and safest antihistamine available.

REFERENCES

Ahlmark, A. (1944), Studies on the histaminolytic power of plasma with special reference to pregnancy, *Acta Physiol. Scand.*, 9 (Suppl. XXVIII), 1.

Bates, C. J., Prentice, A. M., Prentice, A., Lamb, W. H. and Whitehead, R. G. (1983), The effect of vitamin C supplementation on lactating women in Keneba, a West African rural community, *Int. J. Vitam. Nutr. Res.*, 53, 68.

Bruce, C., Weatherstone, R., Seaton, A., and Taylor, W. H. (1976), Histamine levels in plasma, blood and urine in severe asthma, and the effect of corticosteroid treatment, *Thorax*, 41, 724.

Chatham, M. D., Eppler, J. H., Jr., Sauder, L. R., Green, D., and Kulle, T. J. (1987), Evaluation of the effects of vitamin C on ozone-induced bronchoconstriction in normal subjects, *Ann. N.Y. Acad. Sci.*, 498, 269.

Chatterjee, I. B., Das Gupta, S., Majumder, A. K., Nandi, B. K. and Subramanian, N. (1975a), Effect of ascorbic acid on histamine metabolism in scorbutic guinea-pigs, *J. Physiol. (London)*, 251, 271.

Chatterjee, I. B., Majumder, A. K., Nandi, B. K., and Subramanian, N. (1975b), Synthesis and some major functions of vitamin C in animals, *Ann N.Y Acad Sci*, 258, 24

Clemetson, C. A. B. (1980), Histamine and ascorbic acid in human blood, *J Nutr*, 110, 662.

Clemetson, C. A. B. and Cafaro, V. (1981), Comparison of blood histamine and plasma ascorbic acid levels in pregnant and post-partum women, unpublished work

Csaba, B. and Toth, S. (1966), The effect of ascorbic acid on anaphylactic shock in dogs, *J. Pharm. Pharmacol.*, 18, 325

Dawson, W., Maudsley, D. V., and West, G. B. (1965), Histamine formation in guinea pigs, *J Physiol (London)*, 181, 801

Dawson, W. and West, G. B. (1965), The influence of ascorbic acid on histamine metabolism in guinea pigs, *Br. J Pharmacol*, 24, 725.

De Forrest, J. M. and Hollis, T. M. (1980), Relationship between low intensity shear stress, aortic histamine formation, and aortic albumin uptake, *Exp Mol. Pathol.*, 32, 217.

Enwonwu, C. O. and Okolie, E. E. (1983), Differential effects of protein malnutrition and ascorbic acid deficiency on histidine metabolism in the brains of infant nonhuman primates, *J. Neurochem.*, 41, 230.

Freeman, B. M. (1968), Depletion of ascorbic acid from the adrenal of the intact embryo of gallus domseticus by adrenocorticotrophic hormone or histamine, *Comp. Biochem Physiol.*, 24, 905

Gore, I., Fujinami, T., and Shirahama, T. (1965), Endothelial changes produced by ascorbic acid deficiency in guinea pigs, *Arch Pathol.*, 80, 371.

Harman, D. (1962), Atherosclerosis inhibiting effect of an anti-histamine drug. Chlorpheniramine, *Circ. Res.*, II, 277.

Hicks, R. (1965), Some effects of corticosteroids on tissue histamine levels in the guinea-pig, *Br. J. Pharmacol*, 25, 664.

Hirose, T., Matsumoto, I., and Suzuki, T. (1976), Adrenal cortical secretory responses to histamine and cyanide in dogs with hypothalamic lesions, *Neuroendocrinology*, 21, 304.

Hollander, W., Kramsch, D. M., Franzblau, C., Paddock, J., and Colombo, M. A. (1974), Suppression of atheromatous plaque formation by antiproliferative and anti-inflammatory drugs, *Circ. Res.*, 34, 131

Kapeller-Adler, R. (1949), Histamine metabolism in pregnancy, *Lancet*, 2, 745

Kapeller-Adler, R. (1951), A new volumetric method for the determination of histaminase activity in biological fluids, *Biochem. J.*, 48, 99.

Kovacs, E. M. and Suffiad, K. (1968), Histamine release by cortisone induced hyperglycaemia, *Br. J. Pharmacol. Chemother.*, 32, 262.

Latif, E. A. A. A. and Sultan, I. H. (1971), Cortisone withdrawal rebound phenomena and histamine, *Ann. Allergy*, 29, 19.

Lewis, A. J. and Nicholls, P. J. (1973), Histamine metabolism and sensitivity in scorbutic guinea pigs, *Pharmacol. Res Commun.*, 5, 131.

Majno, G. and Palade, G. E. (1961), Studies on inflammation. I. The effect of histamine and serotonin on vascular permeability An electron microscopic study, *J. Biophys. Cytol.*, 11, 571.

Mohsenin, V. and DuBois, A. B. (1987), Vitamin C and airways, *Ann. N.Y. Acad. Sci.*, 498, 259.

Muszbek, L., Csaba, B., Kassay, L., and Kovacs, K. (1969), Effect of cortisone treatment and adrenalectomy on histamine metabolism in rat tissue, *Acta Physiol. Acad. Sci Hung.*, 36, 67

Nakano, K. and Suzuki, S. (1984), Stress-induced change in tissue levels of ascorbic acid and histamine in rats, *J. Nutr.*, 114, 1602.

Owens, G. K. and Hollis, T. M. (1979), Relationship between inhibition of aortic histamine formation, aortic albumin permeability and atherosclerosis, *Atherosclerosis*, 34, 365.

Parrot, J.-L. and Richet, G. (1945), Accroissement de la sensibilité a l'histamine chez le cobaye soumis a un régime scorbutigène. Reduction de cet accroissement par l'administration de vitamine P, *C. R. Soc. Biol.*, 139, 1072.

Porter, J. F., Young, J. A., and Mitchell, R. G. (1970), Effects of corticosteroids on distribution of histamine in the blood of asthmatic children, *Arch. Dis. Child.*, 45, 54.

Reilly, M. A. and Schayer, R. W. (1972), Effect of glucocorticoids on histamine metabolism in mice, *Br. J. Pharmacol.*, 45, 463.

Roe, J. H. (1954), Photometric indophenol method for the determination of ascorbic acid, *Methods of Biochemical Analysis*, Vol. 1, Glick, D., Ed., Interscience, New York, 121.

Sayers, G. and Sayers, M. A. (1947), Regulation of pituitary adrenocorticotrophic activity during the response of the rat to acute stress, *Endocrinology*, 40, 265.

Sharma, S. C., Molloy, A., Walzman, M., and Bonnar, J. (1981), Levels of total ascorbic acid and histamine in the blood of women during the 3rd trimester of normal pregnancy, *Int. J. Vitam. Nutr. Res.*, 51, 266.

Shore, P. A., Burkhalter, A., and Cohn, V. H. (1959), A method for the fluorometric assay of histamine in tissues, *J. Pharmacol. Exp. Ther.*, 127, 182.

Subramanian, N. (1978), Histamine degradative potential of ascorbic acid: considerations and evaluations, *Agents Actions,* 8, 484

Subramanian, N., Nandi, B. K., Majumder, A. K., and Chatterjee, I. B. (1973), Role of L-ascorbic acid on detoxification of histamine, *Biochem. Pharmacol.,* 22, 1671.

Subramanian, N., Nandi, B. K., Majumder, A. K., and Chatterjee, I. B. (1974), Effect of ascorbic acid on detoxification of histamine in rats and guinea pigs under drug treated conditions, *Biochem Pharmacol ,* 23, 637.

Chapter 2

PROLINE AND LYSINE METABOLISM

Ascorbic acid is essential for the synthesis of collagen, a protein characterized by its relatively large content of hydroxyproline. While mature extracellular collagen consists of long strands of amino acids in the form of triple-helical chains joined by link protein, the collagen precursor formed by the ribosomes of the fibroblasts is believed to consist of shorter water-soluble polypeptides called "procollagen". This substance is identical in amino acid composition to collagen, except that it does not contain any hydroxyproline or hydroxylysine, but is rich in proline and lysine. The hydroxylation of proline and lysine takes place at a later stage and it seems to be this step that requires ascorbic acid. Actually hydroxylation of proline and lysine can occur to some extent in the absence of ascorbic acid, but both prolyl and lysyl hydroxylases are markedly potentiated by ascorbic acid.

It was the work of Stetten and Schoenheimer (1944) and Stetten (1949) which established that collagen hydroxyproline arises from proline rather than from free hydroxyproline. Similarly Sinex et al. (1959) showed that lysine rather than hydroxylysine is the major precursor or collagen hydroxylysine.

Histological evidence of defective collagen formation in scurvy was reported by Wolbach and Howe (1926). Studying the organization of blood clots in wounds of the thigh muscles of scorbutic guinea pigs, Wolbach (1933) observed a prompt migration of fibroblasts from adjacent tissues into the blood clot and division of those cells, but no collagen or reticulum formation could be seen in absolute scorbutus. However, the fibroblasts showed vacuoles, and Wolbach postulated that soluble protein was discharged from these vesicles at the cell surface. This liquid was rapidly converted into collagen when vitamin C was provided by the feeding of orange juice. In normal guinea pigs, collagen fibrils formed in immediate contact with the fibroblasts and extended in the direction of the long axis of these cells. During rescue from scurvy, "The first material having the staining properties of collagen which appeared around the cells was homogenous, resembling lightly stained amyloid and therefore presumably not of great density. It was best observed where fibroblasts were grouped closely together and between fibroblasts and fibrin strands. This homogenous state was of short duration, though best seen on the second day when collagen deposition was rapid.

"Next in sequence was the appearance of delicate fibrils coincidently with the appearance of reticulum, or argyrophil fibrils. With isolated, rounded or ovoid cells the direction of the fibrillary collagen, as well as that of the argyrophil fibrils, was concentric to the cell body; if the cells were processed the fibrils paralleled the cell processes and the fibroglia fibrils.

"The conclusion is unavoidable that the earliest formed collagen fibrils are identical with reticulum."

Wolbach made the following conclusions:

1. Fibrin and other preformed materials do not contribute to collagen formation in repair by organization.
2. Collagen and reticulum represent physical differences of the same material.
3. Collagen is the product of secretory activity of fibroblasts, and its alignment and distribution are determined by the shape of the cell and its processes, including fibroglia fibrils.

Early attempts at chemical estimation of collagen formation by Robertson (1952) led to conflicting results. The first chemical evidence that ascorbic acid is involved in the synthesis

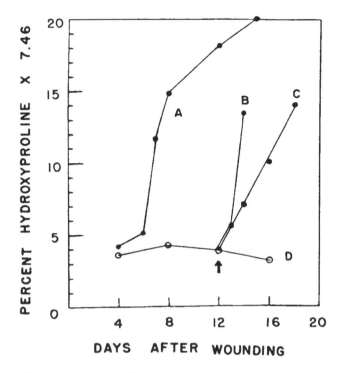

FIGURE 1.   Hydroxyproline content of hydrolysates of autoclave-extracted material from granulation tissue of guinea pig skin wounds made 7 or 14 d after withdrawal of ascorbic acid from the diet. The hydroxyproline values are multiplied by the factor 7.46 to give a measure of apparent collagen. Curve A: animals replaced on a diet supplemented by 30 mg of ascorbic acid daily. Curve B: animals wounded 7 d after withdrawal of ascorbic acid, maintained on scorbutogenic diet for 12 d, and then given 50 mg of ascorbic acid. Curve C: animals treated as in Curve B, but wounded after 14 d on a scorbutogenic diet. Curve D: animals maintained on a scorbutogenic diet throughout. The arrow indicates the point at which ascorbic acid was restored to the animals. (From Gould, B. S. and Woessner, J. F. [1957], *J. Biol. Chem.*, 226, 289. With permission.

of new collagen was provided by Robertson and Schwartz (1953) when they observed that Irish moss granulomas in scorbutic guinea pigs contained less collagen than similar granulomas from normal guinea pigs.

Using hydroxyproline formation as a measure of collagen deposition in guinea pig skin wounds, Gould and Woessner (1957) observed that animals deprived of ascorbic acid for 7 d showed an almost complete cessation of hydroxyproline formation. Normal hydroxyproline formation was resumed very rapidly after the administration of ascorbic acid (Figure 1). Indeed, the wounds of scorbutic animals, treated with ascorbic acid on the 12th postoperative day, attained in 2 d a hydroxyproline level reached by normal animals after 8 d. While normal animals show an appreciable lag in hydroxyproline formation, scorbutic animals showed an almost immediate production of hydroxyproline once ascorbic acid was administered. The scorbutic wound regenerating area seems to be in a state of cellular organization that merely requires some effect of ascorbic acid for the synthesis of hydroxyproline. It is as though some precollagenous material has been accumulated, which is rapidly fibrillated upon the dietary administration of ascorbic acid. In another experiment, these same workers studied groups of guinea pigs which were wounded 0, 2, 4, and 7 d after withdrawal of ascorbic acid from the diet. Granulation tissue was collected 12 h after wounding (Figure

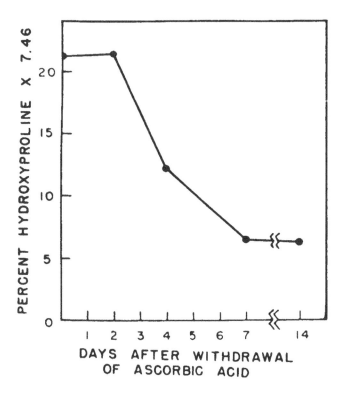

FIGURE 2. Hydroxyproline content of hydrolysates of autoclave-extracted material from granulation tissue of guinea pig skin wounds made at various times after withdrawal of ascorbic acid from the diet. Determinations were made in each case 12 d after wounding. Each point is the average of data from six or more animals. The hydroxyproline values have been multiplied by a factor of 7.46 to give a measure of apparent collagen. (From Gould, B. S. and Woessner, J. F. [1957], *J. Biol. Chem.*, 226, 289. With permission.)

2); it was found that the hydroxyproline content of the wounds was markedly reduced when the animals were wounded after 7 or more days on the ascorbic acid-deficient diet. The rapidity of the hydroxyproline accumulation after administration of ascorbic acid to scorbutic animals (Figure 1) suggested that a proline-containing precollagen accumulates in ascorbic acid deficiency and that ascorbic acid functions in the hydroxylation of this peptide-bound proline.

In a further study, Gould (1958) demonstrated a direct and specific effect of ascorbic acid on collagen synthesis. Paired polyvinyl sponges were implanted beneath the skin of ascorbic acid-depleted guinea pigs; a small amount of ascorbic acid was injected directly into one sponge and normal saline into the other. The ascorbic acid-injected sponge synthesized collagen in normal amounts, as measured by hydroxyproline synthesis, while the saline control showed little synthesis. This effect was shown to be specific for L-ascorbic acid and dehydro-L-ascorbic acid. Dihydroxymaleic acid, D-glucoascorbic acid, and D-isoascorbic acid were found to be inactive in this experiment.

Gross (1959), studying proline, hydroxyproline, and glycine in neutral salt-extractable collagen of the skin of scorbutic and pair-fed guinea pigs, confirmed that new collagen was absent from the skin of the scorbutic animals, but failed to reveal the presence of any significant amounts of a soluble collagen precursor with unusual proportions of glycine and proline to hydroxyproline.

This inability to find hydroxyproline-poor protein precursors led Mitoma and Smith (1960)

Table 1

**INCORPORATION OF [^{14}C] PROLINE INTO THE COLLAGEN PROLINE
AND HYDROXYPROLINE OF GRANULOMAS FROM NORMAL AND
SCORBUTIC GUINEA PIGS**

Specific Activity (cpm/μmol)

Normal			Scorbutic		
Proline	Hydroxyproline	Proline/ hydroxylproline	Proline	Hydroxyproline	Proline/ hydroxyproline
790	697	1.13	787	<10	>7.9
745	680	1.10	530	10	5 3

Note· Data of Stone and Meister (1962), as interpreted by Barnes (1975) Granuloma minces were
incubated in a medium containing ^{14}C-proline. Collagen was extracted with hot trichloracetic acid
and precipitated with tannic acid. Proline and hydroxyproline radioactivity were estimated in the
precipitate. The markedly increased ratio of ^{14}C-proline to ^{14}C-hydroxyproline in the collagen of
the scorbutic guinea pigs is evidence of defective hydroxylation.

From Barnes, M. J. (1975), *Ann. N.Y. Acad. Sci.*, 258, 264. With permission.

to suggest that hydroxylation of proline was not affected by ascorbic acid deficiency and
that ascorbic acid may be responsible for maturation of the fibroblasts. However, Gerber et
al. (1960), studying the distribution of proline and hydroxyproline in rats following the
administration of labeled proline, reported the existence of several different kinds of collagen
with widely differing turnover times, ranging from 25 d for citrate-soluble skin collagen to
over 150 d for insoluble skin collagen and more than 110 d for tendon. This demonstrated
the importance of distinguishing between rapid- and slow-turnover collagen in any study of
scurvy.

Studies by Martin et al. (1961) demonstrated decreased urinary excretion of hydroxyproline
in guinea pigs with scurvy. Robertson (1961) therefore suggested that ascorbic acid might
be responsible for the conversion of proline to hydroxyproline before the synthesis of the
peptide chain.

Gould (1961) suggested that more than one mechanism for the synthesis of collagen may
exist: one that is involved in normal growth and body collagen formation and another that
predominated in tissue repair. The former could be visualized as a slow and stable mechanism
relatively independent of ascorbic acid, while the other would be rapid, perhaps less stable,
and ascorbic acid dependent. It was suggested that both growth and repair collagen may be
involved in repair. This would explain why collagen synthesis was found to be markedly
impaired, but never completely inhibited, in scurvy; it was found to be about 15 to 20% of
normal in regenerating skin wounds.

Subsequent studies by Robertson and Hewitt (1961) and by Stone and Meister (1962)
showed that the incorporation of ^{14}C-proline into collagen proline and hydroxyproline in the
minced tissues from granulomas of guinea pigs was markedly inhibited in scurvy (Table 1).
The addition of ascorbic acid to the granuloma preparations *in vitro* restored the proline
incorporation to normal. These findings showed that the hydroxylation of proline was affected
by ascorbic acid deficiency, but it was not evident whether this was the basic cause or simply
a consequence of the defective collagen synthesis of scurvy.

Gould (1963) reviewed the literature on this subject and noted that the precise mechanism
of action of ascorbic acid in collagen formation and fibrogenesis was uncertain. Green and
Goldberg (1964) studied the incorporation of ^{14}C-labeled L-proline into protein-bound hydroxyproline by cultured human fibroblasts. They observed that the addition of ascorbic

acid caused a considerable increase in collagen synthesis in and on the cells, and in the aqueous phase of the medium, but did not affect the synthesis of other proteins, as measured by proline incorporation. Ascorbic acid did not seem to affect the precipitation of collagen fibres in culture, as the fraction of total collagen escaping from the cell layer to the medium was not reduced by ascorbic acid.

Smiley and Ziff (1964) observed that growing children excrete higher levels of urinary hydroxyproline peptides than adults. They concluded that hydroxyproline peptides are released into the bloodstream and urine during growth and remodeling processes and also during the repair of wounds or burns.

Peterkofsky and Udenfriend (1963) described a cell-free system derived from chick embryo which incorporated ^{14}C-proline into protein and into collagen hydroxyproline. It was found that incorporation of ^{14}C-proline into protein-bound proline began immediately and reached a maximum at about 1 h. However, radioactivity did not appear in collagen hydroxyproline until after 30 min of incubation, when about 70% of ^{14}C-proline had already been incorporated into protein. This initial 30-min period was termed the lag phase, and further experiments led to the conclusions that hydroxylation of proline did not begin until after the lag phase and that proline already incorporated into a microsomal-bound polypeptide was the substrate for hydroxylation. These findings suggested that microsomes labeled with radioactive proline during the lag phase and then isolated could serve as substrate for proline hydroxylation. Further studies by the same workers (Peterkofsky and Udenfriend, 1965) confirmed this supposition and demonstrated that the conversion of unhydroxylated microsomal-bound polypeptide to collagen requires a hydrogen donor, another nonenzymatic factor in the soluble protein fraction, and the microsomes. Ascorbate and also 2 amino-4 hydroxytetrahydrodimethylpteridine (a cofactor for phenylalanine and tyrosine hydroxylases) were found capable of serving as hydrogen donors. It was thought that ascorbic acid might be necessary to maintain this folate compound in the reduced (folinic acid) state, but it seems that folinic acid is not as active as ascorbic acid in this reaction. Collagen prolyl hydroxylase was found in the microsomal fraction.

Jeffrey and Martin (1966) reported experiments leading to the conclusion that ascorbic acid exerts a direct action on collagen synthesis by stimulating the hydroxylation of peptide-bound proline.

Studies by Hutton et al. (1966) and by Kivirikko and Prockop (1967) showed that oxygen, ferrous iron, α-ketoglutarate, and ascorbate are essential for the full activity of both proline hydroxylase and lysine hydroxylase.

It seemed at first that the same enzyme and the same cofactors were involved in the hydroxylations of both the imino acid proline and the amino acid lysine; the name "protocollagen hydroxylase" was therefore suggested for this enzyme. However, purified preparations of protocollagen proline hydroxylase were later found which contained no lysine-hydroxylating activity, indicating that two separate enzymes were involved (Barnes and Kodicek, 1972).

Juva et al. (1966) demonstrated that when hydroxylating reactions are inhibited in isolated chick embryo cartilage, protocollagen continues to be synthesized, but cannot pass out of the chondroblast cells until the hydroxyl groups are added. Thus, hydroxylation seems to be essential for extrusion of collagen. Indeed, Kivirikko and Prockop (1967) have suggested that it could be the accumulation of protocollagen within the fibroblast cells that causes cessation of further protein synthesis in ascorbic acid deficiency.

The peptidyl proline hydroxylase activities of the tissues of rats and guinea pigs were studied by Mussini et al. (1967). Ascorbic acid, ferrous iron, and α-ketoglutarate were provided in optimal mounts, so that the rate of release of tritium from 3,4,-^{3}H-proline was entirely dependent on the amount of enzyme present. These studies revealed that the peptidyl proline hydroxylase activities in homogenates of normal guinea pig lung and skin were 1200 and 240 counts per minute, respectively, whereas, activities in homogenates of lung and skin from scorbutic animals were 480 and 60 counts per minute, respectively.

It seems that in guinea pigs, ascorbic acid not only functions as a cofactor for proline hydroxylation, but is also involved in the maintenance of levels of the enzyme in the tissue. These observations may explain some of the defects in the synthesis of connective tissue that are observed in scurvy. Another very interesting feature of these studies was the finding of an abrupt rise in the peptidyl proline hydroxylase activity in the skin wounds of rats between the third and fourth day after wounding; this continued till day 5, reaching specific activities that were five times the normal. After day 5, hydroxylase activity slowly decreased and reached the normal range 2 to 3 weeks after wounding. An increase in extractable hydroxyproline first appears on days 4 to 5, which is a day or two later than the increase in peptidyl proline hydroxylase and signals the onset of the second phase of wound healing. It seems that the dividing and migrating fibroblasts of the first 3 d have a low hydroxylase level. The major enzymatic change that distinguishes the productive phase of wound healing from the collagen phase may be the induction in fibroblasts of collagen proline hydroxylase. These authors suggest that this change may be associated with the transition of dividing, migrating cells to stationary, secreting cells.

Burkley (1968), in a M.Sc. thesis cited by Barnes and Kodicek (1972), studied hydroxyproline excretion in human scurvy. Four healthy men were fed a diet deficient in ascorbic acid for 29 weeks. The subjects were considered, on the basis of clinical signs and body pool size of ascorbic acid, to be scorbutic by week 14. Repletion commenced at this time by daily dosage with ascorbic acid. Unexpectedly the urinary hydroxyproline was found to increase as subjects became scorbutic and gradually returned to normal levels on repletion. This elevation in urinary hydroxyproline excretion appeared as early as 1 week after the introduction of the ascorbic acid-free diet. Moreover, Efron et al. (1968) studied a patient with hydroxyprolinemia due to lack of the enzyme hydroxyproline oxidase. The urinary excretion of free hydroxyproline was markedly greater than normal and increased even more when the patient was fed an ascorbic acid-deficient diet for 5 months. These unexpected results have been interpreted as indicating that ascorbic acid depletion increases the turnover of mature collagen or of partially hydroxylated collagen precursors. On administration of ascorbic acid there was a further increase followed by a return of the urinary free hydroxyproline excretion to more normal levels.

An interesting study of elderly patients was conducted by Windsor and Williams (1970). These workers observed that the hydroxyproline/creatinine ratio in the urine of elderly patients with low leukocyte ascorbic acid levels (below 15 $\mu g/10^8$ cells) was significantly increased from a mean of 34 during 3 d before treatment to a mean of 43 during 6 d of treatment with ascorbic acid, 1 g daily. No increase was observed in patients with higher leukocyte ascorbic acid levels.

Caygill and Clucas (1971) studied collagen fibril formation *in vitro* and made the interesting suggestion that the role of ascorbic acid may be to prevent collagen fibril formation within the fibroblast cell. Outside the cell there is a lower ascorbic acid concentration and fibril formation can occur.

Barnes et al. (1970) studied the incorporation of labeled proline into collagen proline and hydroxyproline in guinea pigs. Incorporation into elastin was also studied, as this extracellular connective tissue also contains some hydroxyproline, although much less than collagen. It was anticipated that ascorbic acid-deficient guinea pigs would show impaired hydroxylation and, therefore, decreased radioactivity of hydroxyproline relative to proline. Precisely this situation was found in aortic elastin. Incorporation of radioactivity into elastin hydroxyproline became negligible in scurvy, while that into elastin proline was little affected (Table 2). In similar studies of skin collagen, there was a rapid fall in the incorporation of both proline and hydroxyproline around the eighth to tenth day of ascorbic acid deprivation. The proline/

Table 2
INCORPORATION OF TRITIATED PROLINE *IN VIVO* INTO ELASTIN AND COLLAGEN IN CONTROL AND SCORBUTIC GUINEA PIGS

Day of experiment	Number of animals	Elastin specific radioactivity (cpm/µmol)			Collagen specific radioactivity (cpm/µmol)		
		Proline	Hydroxyproline	Pro/Hyp	Proline	Hydroxyproline	Pro/Hyp
Ascorbic acid-deficient group							
6	6	222	149	1.5	739	667	1.11
8	6	132	62	2.1	357	289	1.24
10	5	142	15	9.5	74	69	1.07 Mean
12	6	61	3	20.3	21	17	1.24 ⎬ 1.15
14	6	—	—	—	13	12	1.08
Control group							
6	6	129	125	1.0	614	583	1.05
8	6	111	131	0.9	785	828	0.95
10	4	80	57	1.4	553	563	0.98 Mean
12	6	35	31	1.1	224	217	1.03 ⎬ 1.0
14	6	—	—	—	415	417	1.0

Note: Each guinea pig received, on the appropriate day of the experiment, 0.1 mCi of L-[G-³H] proline in a single dose. Animals were killed 24 h later. Aortas were removed for study of elastin, and dorsal skin was used for study of collagen. Controls were individually pair-fed with animals in the ascorbic acid-deficient group after day 8. Evidence of defective hydroxylation in scurvy was much more marked, as an increase in the proline/hydroxyproline ratio in the elastin, than in the collagen, but both showed a significant increase.

From Barnes, M. J. (1975), *Ann. N.Y. Acad. Sci.*, 258, 264. With permission.

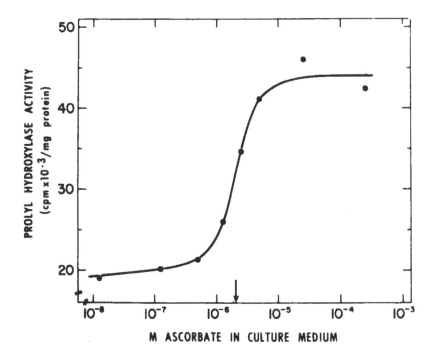

FIGURE 3.   Stimulation of prolyl hydroxylase activity in early log-phase L-929 fibroblast monolayer cultures as a function of the ascorbate concentration. The cells were harvested 1 h after the addition of ascorbate. (From Stassen, F. L. H., Cardinale, G. J., and Udenfriend, S. [1973], *Proc. Natl. Acad. Sci. U.S.A.*, 70, 1090. With permission.)

hydroxyproline specific activity ratio showed only a slight (but, nevertheless, highly significant: $p < 0.001$) rise (Table 2).

Studies of cultured fibroblasts by Comstock et al. (1970) and Comstock and Udenfriend (1970) revealed that the activity of prolyl hydroxylase could be stimulated by overcrowding of the cells or by addition of lactate to the medium. Hydroxyproline synthesis rises to a peak as the cells become more crowded and finally reaches a plateau. Further studies by Stassen et al. (1973) and Cardinale et al. (1975) have demonstrated marked ascorbic acid activation of prolyl hydroxylase in cultured fibroblasts (Figure 3).

Reviewing the extensive literature on this subject, Barnes (1975) expressed the belief that the collagen lesion, as it occurs in scurvy, is fully accountable in terms of impaired hydroxylation (of peptidyl proline and/or peptidyl lysine). He cited evidence that the hydroxyproline content of collagen is critical in regard to the stability of the triple-helical structure of the molecule and that reduction in the hydroxyproline content gives rise to a molecule in which the triple helix is unstable at body temperature. He considers that, "in the scorbutic guinea pig a slight additional reduction in hydroxylation beyond the 5 to 10 per cent detected in the tissues gives rise to a molecule that remains in the form of unassociated α chains and, lacking the protection of the triple helical structure, is rapidly subjected to the action of proteolytic enzymes."

Dell'Orco and Nash (1973) demonstrated ascorbate stimulation of collagen production by cultured human fibroblasts. Studies of cultured mouse fibroblasts by Levene and Bates (1975) demonstrated that the proline hydroxylase activity of these cells is markedly stimulated by ascorbic acid (Table 3), even though they would normally derive this substance from the liver and do not need it in the mouse diet. These authors concluded that, "Ascorbate seems to have two roles in controlling collagen proline hydroxylase activity. One, demonstrable *in vitro*, can be mimicked by high concentrations of thiol compounds and by other reducing

Table 3
EFFECT OF ASCORBATE DEFICIENCY ON PROLINE HYDROXYLATION AND LYSINE HYDROXYLATION IN COLLAGEN SYNTHESIZED BY 3T6 FIBROBLASTS IN CULTURE

	Plus ascorbate	Minus ascorbate
%Hydroxylation of proline		
Cell layer	37	15
Medium	41	7
%Hydroxylation of lysine		
Cell layer	26	24
Medium	33	23
Free hydroxylysine (%)		
Cell layer	17	12
Medium	23	12

Note: Cultures of mouse fibroblasts showed increased hydroxylation of both proline and lysine when ascorbic acid was added to the medium. *In vivo* they would presumably receive an adequate supply from the mouse liver.

From Levene, C. I. and Bates, C. J. (1975), *Ann. N.Y. Acad. Sci.*, 258, 288. With permission.

agents. The other, an activation effect that occurs only in living cells, is mimicked by such conditions as high cell density, high lactate concentration, and low oxygen tension — all of which are likely to occur in stationary phase cultures. Each of these effects can increase the level of proline hydroxylation in nascent collagen. It is not yet known which is functionally the more important, but both are probably needed to achieve complete hydroxylation. This dual action of ascorbate may help to explain its specificity as the antiscorbutic substance *in vivo*. Whereas several unrelated and naturally occurring compounds can replace it in one of its roles, none is yet known that can replace it entirely in both.'' The effects of ascorbic acid deficiency on collagen production by fibroblasts, as conceived by Levene and Bates (1975), are shown in Figure 4.

Alfano et al. (1975) demonstrated impairment of collagen biosynthesis in guinea pig mucosal epithelial explants. They expressed the belief that this accounts for weakness and increased permeability of lingual, buccal, and gingival epithelial basement membranes in scurvy and possibly might also account for the capillary fragility of scurvy.

An interesting study was conducted by Bates (1977) on 23 elderly volunteers (12 men and 11 women, aged 74 to 86 years) who were chosen on the basis of good health and a wide range of ascorbate levels when tested in an earlier survey. Blood and urine samples were drawn at 3-month intervals over one winter and also in some during the following summer. During the winter there was a significant negative correlation ($p < 0.01$) between plasma or buffy coat ascorbic acid (TAA)* and total urinary proline, when expressed per unit of total amino acids in the hydrolysates, both in men and in women, as illustrated in Figures 5 and 6. This correlation was not observed with unhydrolyzed urine, and it appeared to reside in the diffusible fraction. Later samples collected during the summer showed correlations similar to those in winter, although they did not reach as high a level of significance ($p < 0.05$). Bates observed that an increased proline/total amino acids ratio could result either from an increase in degradation of fully hydroxylated collagen or from

* TAA — total ascorbic acid, reduced and oxidized forms.

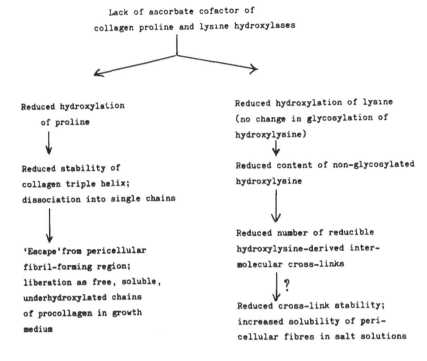

FIGURE 4.   Concept of how ascorbic acid deficiency impairs collagen synthesis, based on experimental observations with cultured fibroblasts. (From Levene, C. I and Bates, C. J. [1975], *Ann. N.Y. Acad. Sci.*, 258, 288. With permission.)

a contribution from partially hydroxylated collagen. However, fully hydroxylated collagen would yield an equivalent amount of hydroxyproline, and since the negative correlation was stronger for proline than for hydroxyproline, the involvement of underhydroxylated collagen is suggested. He concluded that his results supported the view, "that poor vitamin C status in elderly humans may be associated with a defect in collagen proline hydroxylation, reflected by increased excretion of proline-rich, collagen-derived peptides."

This complex subject was discussed in *Nutrition Reviews* (1978), in which prolyl hydroxylase is described as being a tetramer, consisting of two pairs of nonidentical subunits, one pair of M_r 64,000, the other pair 60,000. The larger subunit contains a carbohydrate moiety. Moreover, evidence that fibroblasts can concentrate ascorbic acid tenfold was amplified by the suggestion that it may be further concentrated within a subcellular compartment, possibly the endoplasmic reticulum where it functions as a cofactor for prolyl hydroxylase.

Elsas et al. (1978) observed beneficial effects following the administration of ascorbic acid to an 8-year-old boy with Ehlers-Danlos syndrome type VI (lysyl hydroxylase deficiency), a severe congenital collagen deficiency disease. This disease is characterized by extreme laxity of the skin and joints, hemorrhagic friable skin, and a prolonged bleeding time. A particularly interesting observation by these authors was a marked increase in urinary free hydroxylysine excretion following the administration of ascorbic acid (4 g daily) not only by the paitent, but also by all of five normal subjects aged 8, 9, 12, 23, and 34 years while on collagen-free diets. The greatest clinical benefit to the patient was improved wound healing: three scars produced by skin biopsies before ascorbate therapy was commenced were red, raised, and thin, in contrast to three biopsy scars similarly produced 3 and 8 months after ascorbate treatment was begun. On histological examination, preascorbate scars contained dilated vessels with extravasated old blood, whereas postascorbate scars looked normal: muscle strength, corneal growth, bleeding time, and pulmonary residual volume all

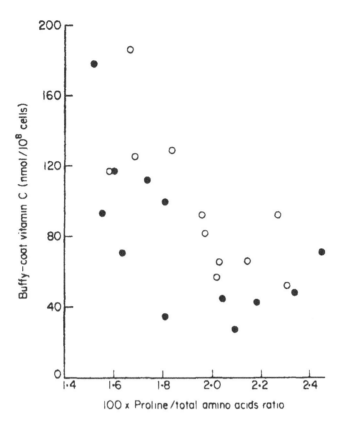

FIGURE 5. Relation between buffy coat ascorbic acid (TAA) and urinary proline/total amino acids ratio in men (●) and women (○) The urinary ratio represents the molar ratio × 100, r = −0 64, p ≈0.001. (From Bates, C. J [1977], *Clin. Sci. Mol Med*, 52, 535. © 1977 The Biochemical Society, London With permission)

improved during 2 years of treatment with ascorbic acid, but the joint laxity and the skin friability remained unchanged.

Bates et al. (1978) observed that the protein-bound hydroxyproline (PBH) content of the serum was significantly lower in guinea pigs with acute vitamin C deficiency than in pair-fed control animals ($p < 0.001$). However, these workers also observed a reduction in the total serum protein level of the scorbutic animals, so they concluded that the reduction in PBH did not represent a specific reduction in the complement Clq component, but reflected a change in the total protein concentration. Reviewing this subject in 1981, Bates summarized the principal steps in the biosynthesis of collagen as shown in Figure 7.

In addition to the need for ascorbic acid in the hydroxylation of proline to stabilize the triple-helical structure of collagen at 37°C, he pointed out that the hydroxylation of lysine is also essential for glycosylation and for the subsequent formation of hydroxylysine-derived cross-links. He descirbed recent studies using highly purified preparations of prolyl and lysyl hydroxylases, which suggest that ascorbic acid acts by maintaining enzyme-bound iron in a loosely bound ferrous form and preventing its oxidation to the more tightly bound ferric form which renders the enzyme inactive. Also, ascorbic acid seems to have a second function in maintaining essential thiol groups in the reduced form.

Studies of human infant skin fibroblasts in tissue culture by Murad et al. (1981a,b) confirmed that total collagen production was stimulated eightfold by L-ascorbate, while noncollagen protein production remained essentially unchanged. Moreover, examination of

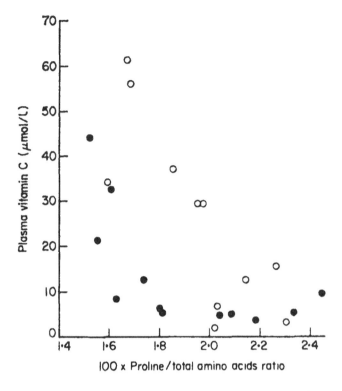

FIGURE 6.   Relation between plasma ascorbic acid (TAA) and urinary proline/total amino acids ratio in men (●) and women (○); $r = -0.59$; $p <0.01$ (From Bates, C. J. [1977], *Clin. Sci. Mol Med.*, 52, 535. © 1977 The Biochemical Society, London. With permission.)

the collagenase-digested media proteins revealed that the hydroxyproline content of the untreated cultures was lower than that of the ascorbate-treated cultures. However, prolyl and lysyl hydroxylase enzyme levels did not appear to be the limiting factors. Reviewing the literature, they cited evidence that hydroxyproline serves to stabilize the collagen triple helix and that hydroxylysine is necessary for the formation of molecular cross-links in collagen, but these workers suggested that the mechanism by which collagen synthesis is stimulated by ascorbic acid is uncertain and may be independent of hydroxylation. Bates (1981) observed that the synthesis of collagen in different tissues of guinea pigs shows wide differences in sensitivity to ascorbic acid deficiency.

Murad et al. (1983) provided evidence that ascorbic acid caused an increase of collagen polypeptide synthesis in cultured human skin fibroblasts and expressed the belief that L-ascorbic acid is a cofactor for the enzymes catalyzing the synthesis of collagen hydroxyproline and hydroxylysine. It has recently been reported by Chandrasekhar et al. (1983) that the "link protein" of cartilage which was isolated by Hascall and Sajdera (1969) binds to collagen types I and II, and the following facts have evolved:

1.   The glycoprotein link contains cystine and methionine, but no hydroxyproline; it seems to require disulfide bonds for aggregation.
2.   The binding of link protein to collagen type fibrils is higher than to monameric collagen.
3.   The binding occurs only when both link protein and collagen are native.
4.   Link protein binding to collagen fibrils is saturable and occurs in a molar ratio of collagen to link protein of 7 to 13:1.

Although link protein itself does not contain hydroxyproline, these data suggest that it binds

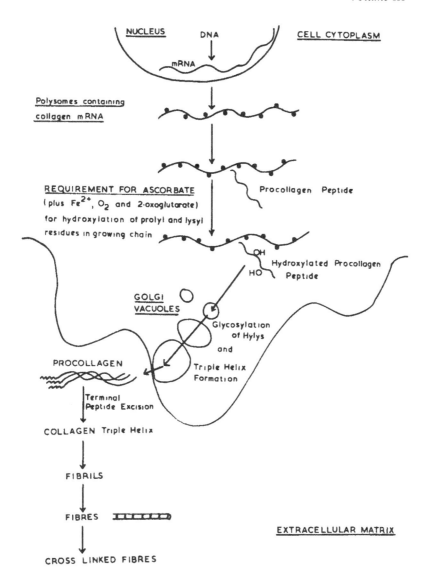

FIGURE 7   Summary of the principal steps in the biosynthesis of collagen. (From Bates, C. J. [1981], in *Vitamin C (Ascorbic Acid),* 1st ed., Counsell, J. N. and Hornig, D. H., Eds., Applied Science, London, 1 With permission.)

to collagen only when the latter is in its native triple-helical structure. Future studies may reveal whether or not this glycoprotein links at hydroxyproline sites on the collagen chains.

Chojkier et al. (1983) produced evidence confirming that the hydroxylation of proline is severely restricted in guinea pigs with scurvy, but they reached the conclusion that collagen synthesis may be controlled by other factors associated with decreased food intake. Instead of the usual pair-fed test and control groups, where dietary intakes are the same, they studied a weight-matched group, where dietary restriction was used to cause weight loss in ascorbate-fed animals as great as that in the ascorbic acid-deficient guinea pigs. They observed a temporal dissociation between the effects of ascorbic acid deficiency on proline hydroxylation and on collagen biosynthesis in guinea pig calvarial bone. As a result of their findings, Chojkier et al. proposed the hypothesis that, "vitamin C deficiency leads to decreased food intake and associated weight loss, which either directly causes a decrease in collagen synthesis because of loss of an additional nutrient or this initiates a signal to another regulatory mechanism. In either case, the mechanism would be independent of an effect on proline hydroxylation.''

Myllylä et al. (1984) have reported that ascorbate is consumed stoichiometrically in the uncoupled reactions catalyzed by prolyl-4-hydroxylase and lysyl hydroxylase *in vitro*. However, it would seem unlikely that any such ascorbate loss occurs *in vivo*.

Johnston et al. (1985) studied the effects of different ascorbic acid intakes: (1) suboptimal (0.5 mg/100 g), (2) adequate (2.0 mg/100 g), (3) ample (10 mg/100 g), and (4) tissue saturating (10 g/l of drinking water), on the PBH levels of the plasma in guinea pigs. The observed PBH levels of 2.2, 2.6, 3.5, and 4.0 μg/ml, respectively, in the four groups, demonstrated that ascorbic acid intakes 3 and 4, exceeding the optimal requirement for growth (2 mg/100 g), caused a significant increase in the PBH levels of the euglobulin fraction of the plasma ($p < 0.01$). Since the plasma complement component Clq is the only known hydroxyproline-containing protein in the euglobulin fraction of plasma, it was concluded that either the synthesis of complement Clq is increased by ascorbic acid supplements, or else Clq molecules become overhydroxylated by ascorbic acid. Subsequent work by Johnston et al. (1987) has confirmed that the increased level of PBH found in the blood plasma of ascorbic acid-supplemented guinea pigs does represent an increased level of the complement component Clq, for an increased level of plasma Clq was found both by hemolytic assay and by immunodiffusion against anti-Clq (Chapter 12, Volume II).

Plasma complement component Clq is a plasma protein containing a hydroxyproline-rich helical region similar to that of collagen, and there is immunological evidence (Chapter 12, Volume II) that Clq activity increases with dietary ascorbic acid intake, so the chemical evidence is supported by biological findings. These are important findings because plasma complement component "Clq is essential for the immune-response, in that it initiates the complement cascade, which lyses antibody-coated bacteria and viruses. Clq can also initiate complement-mediated cell lysis by binding directly to bacteria whose cell surface contains lipid A and to certain RNA viruses.'' Clq may also be important in phagocytic activity by macrophages and neutrophils. The importance of this factor is well illustrated in people with a genetic deficiency of Clq, for they suffer from repeated infections.

Buzina et al. (1986) studied the gingival tissues of 24 adult volunteers who had been found to have relatively low plasma ascorbic acid levels when they attended a dental clinic at the University of Zagreb. Biopsy specimens of periodontal tissues were obtained initially and again after 6 weeks, during which time half of them received supplementary oral ascorbic acid, 70 mg daily, and the other half acted as an untreated control group. The mean plasma ascorbic acid level of the experimental group rose from 0.53 to 1.53 mg/dl, while that of the control group remained low at 0.33 and 0.31 mg/dl. There was a statistically significant increase in both the proline and the hydroxyproline levels of the peridontal tissues of the treated group, but not until the plasma ascorbic acid rose above 0.9 mg/dl. The optimal plasma ascorbic acid level, which was associated with the highest hydroxyproline and proline tissue contents, was reported to be between 1.00 and 1.30 mg/dl, corresponding to a total daily vitamin C intake of about 100 mg daily. In this experiment, both the proline and the hydroxyproline levels of the gingival tissues rose and the mean hydroxyproline/proline ratio actually fell from 0.49 to 0.38 as a result of the vitamin supplement, which is difficult to interpret; possibly it indicates the synthesis of other proteins in addition to collagen.

In vitro studies by Tuderman et al. (1986) demonstrated that, ''the hydroxylation of peptide-bound proline or lysine by prolyl or lysyl hydroxylase, respectively, in the absence of ascorbate, results in the formation of an inactive Fe^3-S-enzyme complex.'' These authors suggest that ascorbate reactivates the hydroxylases by reducing the ferric to ferrous iron, as shown in Figure 8.

Schwarz et al. (1987), studying cultured chick embryo tendon cells in 0.2% serum,

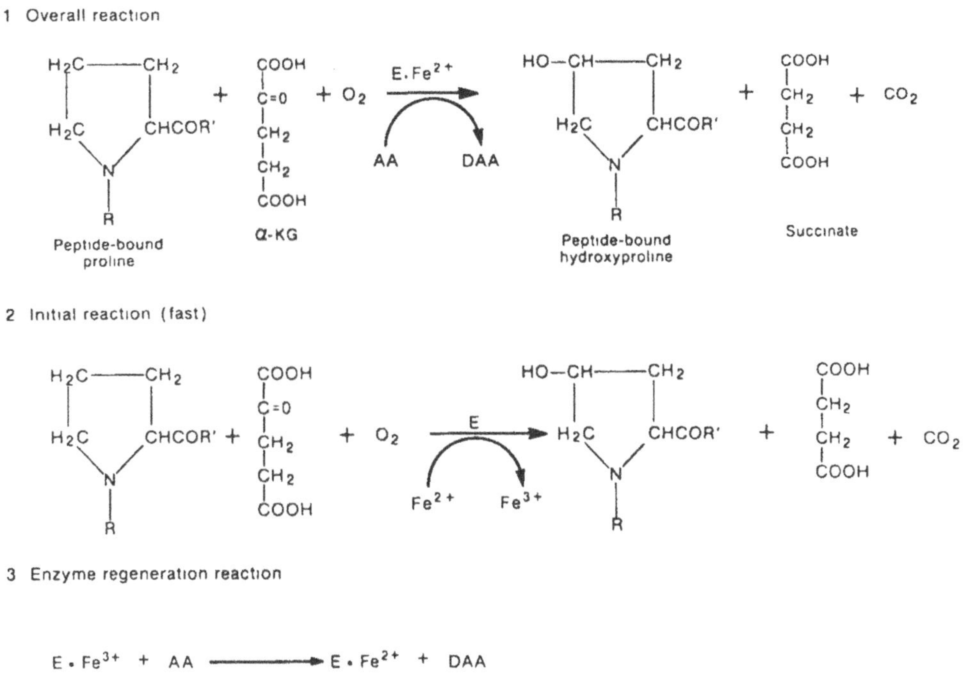

1 Overall reaction

H₂C——CH₂ ... + COOH C=O CH₂ CH₂ COOH (α-KG) + O₂ → E·Fe²⁺ (AA → DAA) → HO—CH——CH₂ ... + COOH CH₂ CH₂ COOH + CO₂

$$\text{Peptide-bound proline} + \alpha\text{-KG} + O_2 \xrightarrow[\text{AA} \to \text{DAA}]{E \cdot Fe^{2+}} \text{Peptide-bound hydroxyproline} + \text{Succinate} + CO_2$$

Peptide-bound proline   α-KG   Peptide-bound hydroxyproline   Succinate

2 Initial reaction (fast)

$$\text{Peptide-bound proline} + \begin{matrix}\text{COOH}\\ C=O \\ CH_2 \\ CH_2 \\ \text{COOH}\end{matrix} + O_2 \xrightarrow[Fe^{2+} \to Fe^{3+}]{E} \text{Peptide-bound hydroxyproline} + \begin{matrix}\text{COOH}\\ CH_2 \\ CH_2 \\ \text{COOH}\end{matrix} + CO_2$$

3 Enzyme regeneration reaction

$$E \cdot Fe^{3+} + AA \longrightarrow E \cdot Fe^{2+} + DAA$$

FIGURE 8. Mechanism of the ascorbate-dependent hydroxylation of peptide-bound proline. AA = ascorbate; DAA = dehydroascorbate. (From Tudermann, L., Myllylä, R., and Kivirikko, K. I. [1986], *Nutr Rev.*, 42, 392. With permission.)

observed that the most rapid effect of ascorbate was stimulation of the hydroxylation of proline constituents of the procollagen molecule; within 1 h the hydroxylation rose from about 10% of all proline residues to about 50%. However, these workers also observed a secondary increase in the rate of synthesis and secretion of procollagen. The induction was said to be specific, because there was a sixfold change in the absolute procollagen production in 48 h, with only minor changes in the synthesis of other proteins, but presumably procollagen is the principal product of primary tendon cells.

In summary, a very extensive literature has accumulated concerning the role of ascorbic acid in collagen synthesis; only a fraction of it is cited here. *In vitro* studies have shown evidence of the accumulation of underhydroxylated collagen peptides when ascorbic acid is absent. However, *in vivo* studies of scorbutic guinea pigs have mostly shown arrested collagen synthesis associated with only a slight degree of underhydroxylation. It would seem that ascorbic acid deficiency impairs collagen synthesis in two ways. We can envisage the primary ascorbate deficiency as impairing hydroxylation of proline and lysine and perhaps a secondary folinic acid deficiency (Chapter 4 of this volume) or some other factor as being responsible for arrest of protein synthesis *in vivo*. Ascorbic acid also seems to play an important role in the synthesis of the plasma complement component Clq, which is essential for the immune response.

REFERENCES

Alfano, M. C., Miller, S. A., and Drummond, J. F. (1975), Effect of ascorbic acid deficiency on the permeability and collagen biosynthesis of oral mucosal epithelium, *Ann. N.Y. Acad. Sci.*, 258, 253.

Barnes, M. J. (1975), Function of ascorbic acid in collagen metabolism, *Ann. N.Y. Acad. Sci.*, 258, 264.

Barnes, M. J., Constable, B. J., Morton, L. F., and Kodicek, E. (1970), Evidence for the formation and degradation of a partially hydroxylated collagen, *Biochem J*, 119, 575.

Barnes, M. J. and Kodicek, E. (1972), Biological hydroxylations and ascorbic acid with special regard to collagen metabolism, *Vitam Horm*, 30, 1.

Bates, C. J. (1981), The function and metabolism of vitamin C in man, *Vitamin C (Ascorbic Acid)*, 1st ed., Counsell, J. N. and Hornig, D H, Eds, Applied Science, London, 1.

Bates, C. J. (1977), Proline and hydroxyproline excretion and vitamin C status in elderly human subjects, *Clin Sci. Mol. Med*, 52, 535.

Bates, C. J., Levene, C. I., Oldroyd, R. G., and Lachmann, P. J. (1978), Complement component Clq is insensitive to acute vitamin C deficiency in guinea pigs, *Biochim Biophys. Acta*, 540, 423.

Burkley, K. (1968), M.Sc. thesis, University of Iowa, Iowa City.

Buzina, R., Aurer-Koželj, J., Srdak-Jorgić, K., Bühler, E., and Gey, K. F. (1986), Increase of gingival hydroxyproline and proline by improvement of ascorbic acid status in man, *Int J Vitam. Nutr. Res.*, 56, 367.

Cardinale, G. J., Stassen, F. L. H., Ramadassan, K., and Udenfriend, S. (1975), Activation of prolyl hydroxylase in fibroblasts by ascorbic acid, *Ann. N.Y. Acad Sci*, 258, 278

Caygill, J. C. and Clucas, I. J. (1971), The role of ascorbic acid in collagen biosynthesis, *Z. Klin. Chem. Klin Biochem.*, 1, 63.

Chandrasekhar, S., Kleinman, H. K., and Hassell, J. R. (1983), Interaction of link protein with collagen, *J. Biol Chem.*, 258, 6226

Chojkier, M., Spanheimer, R., and Peterkofsky, B. (1983), Specifically decreased collagen biosynthesis in scurvy dissociated from an effect on proline hydroxylation and correlated with body weight loss: in vitro studies in guinea pig calvarial bones, *J. Clin. Invest.*, 72, 826.

Comstock, J. P., Gribble, T. J., and Udenfriend, S. (1970), Further study on the activation of collagen proline hydroxylase in cultures of L-929 fibroblasts, *Arch. Biochem. Biophys.*, 137, 115.

Comstock, J. P and Udenfriend, S. (1970), Effect of lactate on collagen proline hydroxylase activity in cultured L-929 fibroblasts, *Proc. Natl. Acad. Sci. U.S.A.*, 66, 552.

Dell'Orco, R. T. and Nash, J. H. (1973), Effects of ascorbic acid on collagen synthesis in nonmitotic human diploid fibroblasts, *Pro. Soc. Exp. Biol. Med.*, 144, 621.

Efron, M. L., Bixby, E. M., Hockaday, T. D. R., Smith, L. H., and Meshorer, E. (1968), Hydroxyprolinemia. III. The origin of free hydroxyproline in hydroxyprolinemia. Collagen turnover. Evidence for a biosynthetic pathway in man, *Biochem. Biophys. Acta*, 165, 238.

Elsas, L. J., Miller, R. L., and Pinnell, S. R. (1978), Inherited human collagen lysyl hydroxylase deficiency: ascorbic acid response, *J. Pediatr.*, 92, 378.

Gerber, G., Gerber, G., and Altman, K. I. (1960), Studies on the metabolism of tissue proteins. I. Turnover of collagen labelled with proline-U-C^{14} in young rats, *J. Biol. Chem.*, 235, 2653.

Gould, B. S. (1958), Biosynthesis of collagen. III. The direct action of ascorbic acid on hydroxyproline and collagen formation in subcutaneous polyvinyl sponge implants in guinea pigs, *J. Biol. Chem.*, 232, 637.

Gould, B. S. (1961), Ascorbic acid-independent and ascorbic acid-dependent collagen-forming mechanisms, *Ann. N.Y. Acad. Sci.*, 92, 168.

Gould, B. S. (1963), Collagen formation and fibrogenesis with special reference to the role of ascorbic acid, in *International Review of Cytology*, Vol. 15, Bourne, G. H. and Danielli, J. F., Eds., Academic Press, New York, 301.

Gould, B. S. and Woessner, J. F. (1957), Biosynthesis of collagen. The influence of ascorbic acid on the proline, hydroxyproline, glycine, and collagen content of regenerating guinea pig skin, *J. Biol. Chem.*, 226, 289.

Green, H. and Goldberg, B. (1964), Collagen synthesis by human fibroblast strains, *Proc. Soc. Exp. Biol. Med.*, 117, 258.

Gross, J. (1959), Studies on the formation of collagen. IV. Effect of vitamin C deficiency on the neutral salt-extractible collagen of skin, *J. Exp. Med.*, 109, 557.

Hascall, V. C. and Sajdera, S. W. (1969), Protein polysaccharide complex from bovine nasal cartilage, *J Biol. Chem.*, 244, 2384.

Hutton, J. J., Tappel, A. L., and Udenfriend, S. (1966), Requirements for α-ketoglutarate, ferrous iron and ascorbate by collagen proline hydroxylase, *Biochem. Biophys Res. Commun.*, 24, 179.

Jeffrey, J. J. and Martin, G. R. (1966), The role of ascorbic acid in the biosynthesis of collagen. II. Site and nature of ascorbic acid participation, *Biochim. Biophys. Acta*, 121, 281.

Johnston, C. S., Cartee, G. D., and Haskell, B. E. (1985), Effect of ascorbic acid nutriture on protein-bound hydroxyproline in guinea pig plasma, *J. Nutr.*, 115, 1089.

Johnston, C. S., Kolb, W. P., and Haskell, B. E. (1987), The effect of vitamin C nutriture on complement component Clq concentrations in guinea pig plasma, *J. Nutr.*, 117, 764.

Juva, K., Prockop, D. J., Cooper, G. W., and Lash, J. W. (1966), Hydroxylation of proline and the intracellular accumulation of a polypeptide precursor of collagen, *Science*, 152, 92.

Kivirikko, K. I. and Prockop, D. J. (1967), Enzymatic hydroxylation of proline and lysine in protocollagen, *Proc. Natl. Acad. Sci. U.S.A.*, 57, 782.

Levene, C. I. and Bates, C. J. (1975), Ascorbic acid and collagen synthesis in cultured fibroblasts, *Ann N Y Acad Sci*, 258, 288

Martin, G. R., Mergenhagen, S. E., and Prockop, D. J. (1961), Influence of scurvy and lathyrism (odoratism) on hydroxyproline excretion, *Nature (London)*, 191, 1008

Mitoma, C. and Smith, T. E. (1960), Studies on the role of ascorbic acid in collagen synthesis, *J Biol Chem.*, 235, 426.

Murad, S., Sivarajah, A., and Pinnell, S. R. (1981a), Regulation of prolyl and lysyl hydroxylase activities in cultured human skin fibroblasts by ascorbic acid, *Biochem Biophys Res Commun.*, 101, 868.

Murad, S., Grove, D., Lindberg, K. A., Reynolds, G., Sivarajah, A., and Pinnell, S. R. (1981b), Regulation of collagen synthesis by ascorbic acid, *Proc. Natl Acad Sci. U.S.A.*, 78, 2879

Murad, S., Tajima, S., Johnson, G. R., Sivarajah, A., and Pinnell, S. R. (1983), Collagen synthesis in cultured human skin fibroblasts: effect of ascorbic acid and its analogs, *J Invest. Dermatol*, 81, 158.

Mussini, E., Hutton, J. J., and Udenfriend, S. (1967), Collagen proline hydroxylase in wound healing, granuloma formation, scurvy and growth, *Science*, 157, 927

Myllylä, R., Majamaa, K., Günzler, V., Hanauske-Abel, H. M., and Kivirikko, K. I. (1984), Ascorbate is consumed stoichiometrically in the uncoupled reactions catalysed by prolyl-4-hydroxylase and lysyl hydroxylase, *J. Biol Chem*, 259, 5403.

Nutrition Reviews (1978), The function of ascorbic acid in collagen formation, *Nutr. Rev*, 36, 118.

Peterkofsky, B. and Udenfriend, S. (1963), Conversion of proline to collagen hydroxyproline in a cell-free system from chick embryo, *J. Biol Chem*, 238, 3966

Peterkofsky, B. and Udenfriend, S. (1965), Enzymatic hydroxylation of proline in microsomal polypeptide leading to formation of collagen, *Proc Natl Acad Sci. U.S.A.*, 53, 335

Prockop, D. J. and Kivirikko, K. I. (1967), Relationship of hydroxyproline excretion in urine to collagen metabolism, *Ann. Intern. Med*, 66, 1243.

Robertson, W. van B. (1961), The biochemical role of ascorbic acid in connective tissue, *Ann. N Y Acad Sci*, 92, 159.

Robertson, W. van B. (1952), The effect of ascorbic acid deficiency on the collagen concentration of newly induced fibrous tissue, *J. Biol. Chem.*, 196, 403.

Robertson, W. van B. and Hewitt, J. (1961), Augmentation of collagen synthesis by ascorbic acid in vitro, *Biochim Biophys Acta*, 49, 404.

Robertson, W. van B. and Schwartz, B. (1953), Ascorbic acid and the formation of collagen, *J. Biol. Chem.*, 201, 689.

Schwarz, R. I., Kleinman, P., and Owens, N. (1987), Ascorbate can act as an inducer of the collagen pathway because most steps are tightly coupled, *Ann N.Y Acad Sci.*, 498, 172.

Sinex, F. M., Van Slyke, D. D., and Christman, D. R. (1959), The source and state of hydroxylysine of collagen. II. Failure of free hydroxylysine to serve as a source of the hydroxylysine or lysine of collagen, *J Biol. Chem.*, 234, 918.

Smiley, J. D. and Ziff, M. (1964), Urinary hydroxyproline excretion and growth, *Physiol. Rev.*, 44, 30

Stassen, F. L. H., Cardinale, G. J., and Udenfriend, S. (1973), Activation of prolyl hydroxylase in L-929 fibroblasts by ascorbic acid, *Proc. Natl. Acad Sci. U.S A.*, 70, 1090.

Stetten, M. R. (1949), Some aspects of the metabolism of hydroxyproline, studied with the aid of isotopic nitrogen, *J. Biol. Chem.*, 181, 31.

Stetten, M. R. and Schoenheimer, R. (1944), Metabolism of *l* (−)-proline studied with the aid of deuterium and isotopic nitrogen, *J. Biol. Chem.*, 153, 113.

Stone, N. and Meister, A. (1962), Function of ascorbic acid in the conversion of proline to collagen hydroxyproline, *Nature (London)*, 194, 555.

Tudermann, L., Myllylä, R., and Kivirikko, K. I. (1986), Elucidation of the biochemical role of ascorbic acid, *Nutr. Rev.*, 42, 392.

Windsor, A. C. W. and Williams, C. B. (1970), Urinary hydroxyproline in the elderly with low leucocyte ascorbic acid levels, *Br. Med. J.*, March 21, 732.

Wolbach, S. B. (1933), Controlled formation of collagen and reticulum. A study of the source of intercellular substance in recovery from experimental scorbutus, *Am. J. Pathol.*, 9, 689.

Wolbach, S. B. and Howe, P. R. (1926), Intercellular substances in experimental scorbutus, *Arch. Pathol.*, 1,

Chapter 3

CARBOHYDRATE METABOLISM

Randoin and Michaux (1925) observed that the liver glycogen stores of guinea pigs were somewhat less abundant in scurvy, but they observed no abnormality in the blood sugar or glycoprotein levels in this condition. However, Palladin and Utewski (1928) and Tomita (1928) reported hyperglycemia in guinea pigs with scurvy.

Bonsignore and Pinotti (1935), at the University of Genoa, observed a progressive increase in the dehydroascorbic acid (DHAA) levels and a fall in the ascorbic acid levels of the tissues of guinea pigs on a vitamin C-deficient diet. They reported that the ratio of oxidized to reduced ascorbic acid in the liver rose from 7.5:1 to 20:1 after 21 d on the scorbutogenic diet. These observations were not related to carbohydrate metabolism at the time, but later became pertinent to this subject when the diabetogenic effect of DHAA was described.

While studying the effects of diphtheria toxin on guinea pigs maintained on a low-ascorbic acid diet (0.25 mg daily) for several weeks, Menten and King (1935) observed diffuse hyperplastic arteriosclerosis and degenerative lesions of the brain and the pancreas. Hydropic degeneration of the β-cells of the islets of Langerhans was very notable; moreover, the glucose tolerance of the animals was found to be markedly reduced.

The relationship of vitamin C to carbohydrate metabolism was therefore investigated by Sigal and King (1936). Studying guinea pigs, they found that glucose tolerance was reduced in both the prescorbutic and the scorbutic stages of vitamin C deficiency. Successive stages of vitamin C depletion (10, 15, and 20 d) induced a corresponding rise in fasting blood sugar and a distinctly lowered glucose tolerance. The typical peak in the blood sugar curve moved characteristically upward and to the right with successive stages of vitamin C deficiency. Replenishment with ascorbic acid, 10 mg/d, after 20 d depletion, induced a return to normal within 15 d. Hirsch (1936) administered vitamin C to guinea pigs and observed an increase in the liver glycogen content. Hermann (1938) and Bartelheimer (1938) also observed that glycogen storage is increased after the administration of vitamin C. Nair (1941) observed a marked delay in the absorption of sugar after a glucose meal and also decreased glucose tolerance in scorbutic guinea pigs (Figure 1). He also noted a marked decrease in the glycogen content of the liver in scurvy. Hamne (1941), studying scorbutic and pair-fed guinea pigs, observed that in chronic scurvy the glycogen contents of both the liver and the muscle were reduced. The liver glycogen was decreased in the early stages of scurvy and the muscle glycogen at a later stage, while the glycogen of the heart was not affected. Giroud and Ratsimamanga (1941) observed that the phosphocreatine levels of the muscle and both the hepatic and the muscle glycogen levels of guinea pigs began to fall and the muscle lactic acid level began to rise as soon as ascorbic acid was withdrawn from the diet of guinea pigs. These authors concluded that a ''normal level'' of ascorbic acid is essential for normal carbohydrate metabolism and that it is not enough to provide just the amount of ascorbic acid necessary for survival.

Banerjee (1943a,b) reported that both scorbutic and partially pancreatectomized, normally fed guinea pigs showed the diabetic type of glucose tolerance curve, had diminished glycogen in the liver, and excreted sugar in the urine, which indicated that the carbohydrate metabolism was similarly disturbed in both of these conditions. Banerjee (1943c, 1944) demonstrated a reduction in the insulin content of the pancreas, associated with hyperplasia of the α-cells, and degranulation of the β-cells of the islets of Langerhans in scorbutic guinea pigs. Murray and Morgan (1946) confirmed the observation that scorbutic guinea pigs have decreased glucose tolerance and decreased liver and carcass glycogen stores. They believed that this might be due to adrenal medullary overactivity. But Banerjee and Ghosh (1946) could not

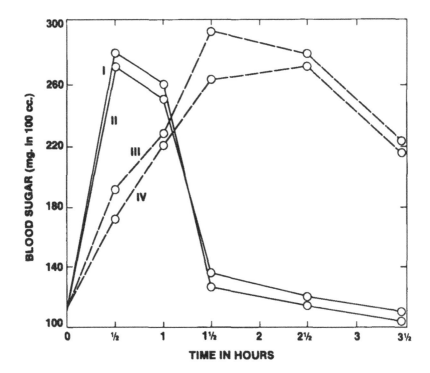

FIGURE 1.   Glucose tolerance tests on guinea pigs after (I) 10 d, (II) 18 d, (III) 24 d, and
(IV) 26 d, on a scorbutogenic diet. The animals were fasted for 4 h preceding the test. A
weighed amount of glucose (0.3 g/100 g body weight) was given orally. (From Nair, K. R.
[1941], *Ann. Biochem. Exp. Med.*, 1, 179. With permission.)

accept this hypothesis, as they found decreased glucose tolerance in scorbutic guinea pigs
even after demedullation of the adrenal glands. Banerjee and Ghosh (1947) saw the decreased
glucose tolerance, and decreased liver glycogen storage of scurvy, as being due to hypoinsulinism, for they found the insulin content of the pancreas of guinea pigs to be about one
fourth normal in scurvy. However, Murray (1948) reported that the injection of insulin failed
to correct the defect of glycogen storage in ascorbic acid-deficient guinea pigs within 6 h.

Patterson (1949, 1950) and Patterson and Mastin (1951) demonstrated that intravenous
injection of DHAA caused hypertension in rats when given in a dose of 5 to 10 mg/kg body
weight. Moreover, repeated large doses caused temporary or permanent diabetes mellitus,
due to damage to, or atrophy of, the β-cells of the islets of Langerhans. Patterson pointed
out the similarity between the chemical structures of DHAA and alloxan, both of which
cause diabetes mellitus by damaging the β-cells of the islets. Both substances have three
adjacent carbonyl groups, as shown in Figure 2, and both are unstable compounds with very
short half-lives under biological conditions. Of course, DHAA already exists in the animal
body as the oxidized form of ascorbic acid, so any factor which increases the oxidation-
reduction equilibrium or redox potential of the blood or tissues in favor of DHAA may tend
to cause the development of diabetes mellitus.

Likewise, substances that reduce the oxidation-reduction equilibrium, reducing DHAA to
ascorbic acid, tend to protect against the development of diabetes mellitus. Indeed, Patterson
and Lazarow (1950) demonstrated that cysteine, reduced glutathione, and other sulfhydryl
compounds, given intravenously 2 min before the intravenous injection of DHAA, completely
protected against the development of diabetes in rats. However, these compounds had no
protective effect if given 10 min after the DHAA. These results are similar to earlier findings
concerning the protective effect of sulfhydryl compounds against the development of alloxan
diabetes.

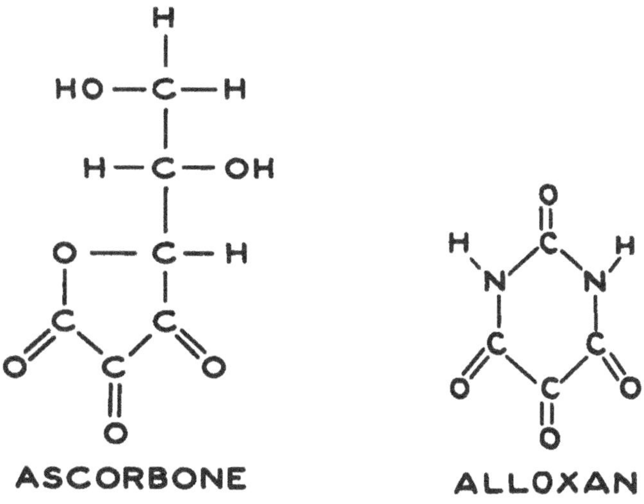

FIGURE 2. Showing the similarity between the chemical structures of dehydroascorbic acid and alloxan. Both substances have three adjacent carbonyl groups and both are diabetogenic when injected intravenously into rats or guinea pigs.

Patterson (1951) induced diabetes in rats by repeated injections of DHAA and then followed their progress over the next 7 to 11 months. It appeared that the hyperglycemia persisted indefinitely, so the damage to the β-cells of the islets seems to have been irreversible. Furthermore, the contiued steady hyperglycemia indicated that no new insulin-producing cells were being formed. In this respect it was thought that the rat may differ from the guinea pig, which is capable of producing new β-cells, according to Johnson (1950). The disease in the rats was characterized by hyperglycemia, glycosuria, polyuria, polydipsia, polyphagia, and failure to gain weight. All of the diabetic rats had blood sugars in excess of 200 mg/100 ml. Those with the highest blood sugar levels developed mature cataracts in 8 to 10 weeks; both eyes were involved at approximately the same time. It seemed that the higher the blood sugar, the sooner they developed cataracts. The minimal blood sugar for cataract formation was approximately 225 mg/100 cc. This was of particular interest as it is above the renal threshold and indicates at least partial saturation of the renal tubular reabsorption mechanism for glucose. These findings are similar to the clinical observations in human beings, that cataracts develop earliest in patients with poor diabetic control.

Princiotto (1951) studied the effects of DHAA in rabbits. Intravenous injections of DHAA produced diabetes mellitus, and intravenous injections of cysteine or glutathione, given 1 min prior to the injection of DHAA, protected the rabbits from developing diabetes. Banerjee et al. (1952) reported that DHAA was not detectable in the tissues of normal guinea pigs, but was present in considerable quantities in the tissues of scorbutic guinea pigs. These workers also reported that the glutathione contents of the blood, adrenals, pancreas, and spleen were significantly diminished in scorbutic guinea pigs. The lowering of the glutathione level was greatest in the pancreas, where it fell from 237 to 114 mg/100 ml.

Stewart et al. (1952) conducted insulin tolerance tests on rhesus monkeys in the early and late stages of scurvy and also on control animals receiving the same diet supplemented with ascorbic acid. It was found that the dose of insulin which produced a pronounced depression of blood glucose in control monkeys had little effect in scorbutic monkeys. A series of experiments was conducted to analyze the effects of epinephrine and cortisone in scurvy. The data obtained were interpreted as indicating that adrenocortical hyperactivity might account for the relative lack of effect of insulin on blood glucose in scurvy.

Bacchus and Heiffer (1954) conducted insulin tolerance and glucose-insulin tolerance tests on normal and scorbutic guinea pigs and found that the ascorbic acid-deficient animals were remarkably resistant to insulin. While the blood glucose of the normal animals remained relatively unchanged following glucose and insulin, the blood sugar of the scorbutic animals rose as though no insulin had been given. Similar results were obtained in adrenalectomized guinea pigs. The fact that the scorbutic animals were resistant to both small and large doses of insulin suggested to these authors that the diminished glucose tolerance in ascorbic acid deficiency could not be due to a decreased insulin level. Moreover, the absence of any effect of adrenalectomy seemed to rule out adrenal hyperactivity as the cause of the impaired glucose tolerance in scurvy. However, the pancreatic insulin content has been found to be decreased in scurvy; so the most likely explanation of the insulin insensitivity observed in scorbutic monkeys and guinea pigs is that the liver cells cannot resume glycogenesis immediately, after so long an insulin and ascorbate deficiency. Nadal and Mulay (1954) confirmed that there is a decrease in the rate of formation of liver glycogen from precursors in guinea pigs with scurvy.

The enzyme hexokinase catalyzes the transfer of the terminal phosphate group of adenosine triphosphate to glucose or fructose, yielding the corresponding hexose-6-phosphate and adenosine diphosphate. Banerjee and Ghosh (1955) observed a significant decrease in the hexokinase activity of the liver and skeletal muscle in scorbutic guinea pigs. Lahiri and Banerjee (1956) observed an increase in the glucose-6-phosphatase activity of both liver and kidney in scorbutic guinea pigs. These changes could have been due to the insulin deficiency of scurvy, for insulin has been reported to stimulate hexokinase and to reduce liver glucose-6-phosphatase activities. In keeping with these findings, Lahiri and Banerjee observed decreased levels of glucose-6-phosphate, fructose-6-phosphate, and fructose 1,6-diphosphate in the livers of scorbutic guinea pigs.

Sarkar and Banerjee (1957) carried out glucose tolerance tests on rhesus monkeys on a normal diet and again after 2 and 3 months on a scorbutogenic diet. When the monkeys had been depleted of ascorbic acid for 2 months, they showed higher levels of blood sugar 90 and 120 min after they were fed glucose. At the end of 3 months, when monkeys developed severe scurvy, blood sugar levels were higher after the glucose administration in all the blood samples collected. No change was observed in the fasting blood sugar levels when the monkeys developed scurvy. They concluded that vitamin C is essential for the normal utilization of glucose in monkeys.

Banerjee et al. (1958) observed that prolonged insulin treatment of scorbutic guinea pigs corrected their decreased glucose tolerance and caused a striking improvement in their impaired glycogen storage. The tissue contents of citric, malic, and lactic acids were found to be significantly increased in scorbutic guinea pigs and to be corrected by prolonged treatment with insulin. These workers therefore suggested that scurvy is associated with a metabolic block in the tricarboxylic acid cycle below the level of citrate. Banerjee and Ghosh (1960) noted that scorbutic guinea pigs display an increased total body cholesterologenesis, apparently due to hypoinsulinism, for upon administration of insulin this alteration is corrected. Merlini (1961) proposed the hypothesis that the hyperglycemia which follows the administration of DHAA results from a complex pharmacological action of this substance on the autonomic nervous system.

Ganguli and Banerjee (1961) reported that the activities of hexokinase, phosphoglucomutase, and phosphohexoisomerase were reduced in scorbutic guinea pigs (Figure 3); Maximal depression was observed with phosphoglucomutase. Contrary to the earlier findings, liver glucose 6-phosphatase was found to be unaffected by scurvy; the dehydrogenases of glucose 6-phosphate and 6-phosphogluconate showed increased activity in scurvy. The extent of oxidation of glucose-1-^{14}C to $^{14}CO_2$ was found to be similar by both scorbutic and pair-fed guinea pig tissues, whereas the oxidation of glucose-6-^{14}C to $^{14}CO_2$ was appreciably

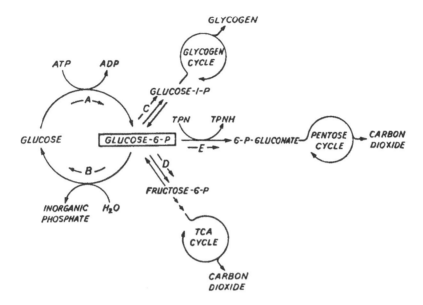

FIGURE 3. Pathways of glucose-6-phosphate metabolism. The letters A, B, C, etc. refer to different pathways by which glucose-6-P can be metabolized. TCA = tricarboxylic cycle. The observation that in scurvy, specific activities of hexokinase, phosphohexoisomerase, and particularly phosphoglucomutase were diminished reflects decreased metabolism of glucose via pathways A, C, and D in that condition. (From Ganguli, N. C. and Banerjee, A. B. [1961], *J. Biol. Chem.*, 236, 979. With permission.)

depressed in scorbutic guinea pigs. Ganguli and Banerjee concluded that the defect in glycogenesis and glycolysis may perhaps be due to hypoinsulinism associated with scurvy. Banerjee and Ganguli (1962) observed that the scorbutic liver is less capable of converting glucose-^{14}C into glycogen than is the normal liver. This defect was observed *in vivo* as well as *in vitro*. Kawishwar et al. (1963) confirmed that when insulin is administered to guinea pigs during ascorbate depletion, the abnormalities of carbohydrate and lipid metabolism are prevented; needless to say, this treatment did not prevent the development of scurvy.

Merlini and Caramia (1965), at the University of Perugia, studied the appearance of the islets of Langerhans of rats by electron-microscopy, before and after injection of DHAA (700 mg/kg). They observed marked degranulation of the β-cells after two injections, which became more evident after three injections when almost all of the β-cells were found to be severely damaged and the animals were hyperglycemic. However, these workers concluded that these changes were entirely different from those seen following alloxan treatment. They therefore suggested different mechanisms must be involved in the development of DHAA diabetes and alloxan diabetes.

While several authors have reported that the administration of ascorbic acid to scorbutic guinea pigs and human beings increased the glucose tolerance and cured the diabetes of scurvy, there is also evidence that high doses of ascorbic acid can under certain circumstances cause an increase in the blood DHAA levels and decrease glucose tolerance. Glatzel and Rüberg-Schweer (1966) observed glycosuria following high doses of ascorbic acid in human subjects. Messina et al. (1968), at the University of Catania, studied the effects of intravenous injection of DHAA on the plasma glucose and insulin levels of rats. An initial dose of 100 mg was followed by 60 mg after 24 h and a further 60 mg at 48 h, for a total dose of 260 mg of DHAA in rats with a mean weight of 230 g. The plasma glucose level rose progressively from an initial level of 75.9 ± 3.0 to 165.6 ± 24.1 on day 3, to 254.3 ± 28.1 on day 7, and was still markedly elevated at 292.2 ± 11.7 mg/100 ml on day 30. The plasma insulin level fell from 15.5 μU/ml initially to 9.6 on day 3, to 5.6 on day 7, and to 0 on day 30. The insulin content of the pancreas was also measured; it fell from 454 μU/g initially to 112 on day 3, to 17 on day 7, and to 10μU/g on day 30.

Nandi et al. (1973) observed that large doses of ascorbic acid (AA)* were toxic to guinea pigs fed high-cereal diets. At a daily dose of 50 mg of AA/100 g of body weight/day, all of 30 guinea pigs died within 25 d, and at a dose of 100 mg AA, all of 30 guinea pigs died within 16 d. The same doses of ascorbic acid were not toxic when fed to guinea pigs receiving the same diet fortified with 15% casein. Studying the cause of toxicity of large doses of AA, Chatterjee et al. (1975) found that under that dietary condition, DHAA increased markedly in blood, urine, and liver of the guinea pigs, and there was also a concomitant increase in blood sugar levels of the animals (see Figure 6 of Chapter 12, Volume I). Clearly the increase in the blood sugar level was related to the increase in the DHAA level. When a high dose of ascorbic acid (100 mg of AA per 100 g body weight per day) was administered to guinea pigs fed a casein-fortified wheat diet, the blood AA level was found to be similar to that obtained with high cereal diets, but the blood DHAA level did not increase and there was no increase in the blood sugar, nor was there any sign of toxicity. The beneficial effects of dietary protein on ascorbic acid metabolism have been discussed in Chapter 12, Volume 1.

Chatterjee et al. (1975) also reported that feeding large doses of ascorbic acid to human volunteers on high-cereal diets led to high blood DHAA levels and increased blood sugar levels in some of them (see Table 5 of Chapter 12, Volume 1). After administration of ascorbic acid (AA), 4 g per man per day, for 15 d, all of ten subjects showed a marked blood DHAA increase. The blood DHAA level also increased when the dosage of AA was 2 g per man per day for 20 d. A few of the volunteers showed higher 2-h postprandial blood sugar levels; 10 d after discontinuation of AA, both the blood DHAA and sugar levels returned to normal. These same workers observed increased DHAA levels in patients with diabetes mellitus, as noted in Chapter 2, Volume II. Clearly, it is very important to maintain ascorbic acid in the reduced form.

Pence and Mennear (1979), at Perdue University, prepared DHAA in redistilled water immediately before use and observed that giving it by intravenous injection (200 mg/kg) produced hyperglycemia, decreased insulin levels, and decreased glucose tolerance in mice. However, daily injection of doses as high as 300 mg/kg produced only transient hyperglycemia, and none of the animals developed permanent diabetes mellitus. The effect was vividly apparent when insulinogenic indices (ratio of immunoreactive insulin [IRI] in μU/ml to glucose in mg/dl) were compared. The insulinogenic indices of control mice ranged from 0.16 to 0.23 during the experiment, while those of the DHAA-treated animals ranged from only 0.07 to 0.13. The inhibition of pancreatic insulin secretion by DHAA was even more pronounced when mice were challenged with a glucose load 45 min after the DHAA injection. Tissue culture studies of isolated mouse pancreatic islets demonstrated that DHAA inhibited glucose-stimulated insulin secretion, apparently through a direct action on the pancreatic cell. In contrast to the opinion expressed by Merlini and Caramia, it was the view of Pence and Mannear, as a result of these and other experiments, that DHAA and alloxan seem to act in the same way. They found that both *in vivo* and *in vitro* the pancreotoxic effects of alloxan and DHAA can be antagonized by prior exposure to glucose, suggesting that the pancreotoxic effect of DHAA is mediated through an action at the level of the glucose receptor.

The reason why diabetics are unable to keep ascorbic acid in the reduced form in their blood and tissues was investigated by Som et al. (1981). They found that the *in vitro* rates of dehydroascorbate reduction in the hemolysate, the erythrocyte reduced-glutathione levels,

---

* AA — ascorbic acid, reduced form.

and the glucose-6-phosphate dehydrogenase activities, which regulate dehydroascorbate reduction, were similar in normal and diabetic subjects. The turnover rate of ascorbic acid was higher in the diabetics than in the normal volunteers. Experiments with diabetic rats indicated that the increased turnover of ascorbic acid was probably due to increased oxidation of ascorbate to dehydroascorbate in tissue mitochondria.

Kopylova and Shelagina (1981), studying the skin of guinea pigs with chronic vitamin C deficiency, have reported that the phosphoglucomutase activity is decreased, while the lactic dehydrogenase and glucose-6-phosphate dehydrogenase activities are increased. Rikans (1982), studying liver microsomes from ascorbic acid-deficient guinea pigs reported findings suggesting that this vitamin selectively affects the transfer of electrons from NADH to cytochrome P-450. Bandyopadyhay and Banerjee (1982) have confirmed that guinea pig liver glycogen is decreased in scurvy. They found liver phosphorylase and α-glucosidase to be increased, but the liver glycogen synthetase activity was unchanged in scurvy. In scorbutic guinea pigs given insulin, the phosphorylase activity returned to normal and the α-glucosidase activity decreased, but did not reach normal. In guinea pigs which recovered from scurvy, the phosphorylase and α-glucosidase activities decreased.

Domke and Weis (1983) observed no hyperglycemia and no decreased glucose tolerance in rats receiving intravenous injections of pure DHAA in sodium phosphate buffer at pH 6.2. These workers suggested that the diabetogenic effect of DHAA reported by others may have been due to impurities. Other possible explanations are that the rats used by Domke and Weis may have been protected by a high sulphydryl content in the liver, or that the unstable compound DHAA may have been lost by hydrolysis in the phosphate buffer, even before it was injected. This latter question could be settled by analysis of DHAA in phosphate buffer by the Hughes (1956) homocysteine method at set times after preparation, as in Figures 3C and D, Chapter 11, Volume I.

The toxic and diabetogenic effects of high-ascorbic acid/low-protein diets serve to indicate the importance of combining ascorbic acid with a direct or indirect antioxidant substance. Ascorbate oxidation in the gastrointestinal tract can be markedly reduced either by a high-protein diet or by combination of ascorbic acid with certain chelating bioflavonoids or catechins as discussed in Chapter 11, Volume I; the writer's choice is D-catechin.

The development of diabetes (bronze diabetes) by patients with hemochromatosis and by patients with thalassemia major, as discussed in Chapter 15, Volume I, are clear indications of the prooxidant effects of iron compounds accumulated in the tissues on the ratio of DHAA to AA. They indicate the need for insoluble chelating fiber in our food to draw excess heavy metals from the blood stream into the lumen of the bowel. Further research is needed to find out whether D-catechin acts in this way. There is already evidence that tea reduces the absorption of iron. Looking to the future, we may see the development of noninvasive methods, perhaps nuclear magnetic resonance, for determination of heavy metal stores within the body, and these may prove useful in developing a better understanding of the prooxidant causes of diabetes mellitus.

REFERENCES

Bacchus, H. and Heiffer, M. H. (**1954**), Carbohydrate metabolism in ascorbic acid deficiency, *Am. J. Physiol.*, 176, 262.

Bandyopadhyay, S. and Banerjee, S. (**1982**), Effects of ascorbate deficiency on liver glycogen metabolism in guinea pigs, *Indian J. Exp. Biol.*, 20, 44.

Banerjee, A. B. and Ganguli, N. C. (**1962**), Metabolic studies on scorbutic guinea pigs. II. Hepatic glycogen synthesis *in vitro* and *in vivo*, *J. Biol. Chem.*, 237, 14.

Banerjee, S. (**1943a**), Vitamin C and carbohydrate metabolism. I. The effect of vitamin C on the glucose tolerance test in guinea-pigs, *Ann. Biochem. Exp. Med.*, 3, 157.

Banerjee, S. (1943b), Vitamın C and carbohydrate metabolısm. II. The effect of vitamin C on the glycogen value of the lıver of guınea pıgs, *Ann. Biochem. Exp. Med.,* 3, 165.

Banerjee, S. (1943c), Vitamın C and carbohydrate metabolism, *Nature (London),* 152, 329.

Banerjee, S. (1944), The relation of scurvy to histologıcal changes in the pancreas, *Nature (London),* 153, 344.

Banerjee, S., Biswas, D. K., and Singh, H. D. (1958), Studıes on carbohydrate metabolısm ın scorbutıc guinea pıgs, *J. Biol Chem.,* 230, 261.

Banerjee, S., Deb, C., and Belavady, B. (1952), Effect of scurvy on glutathıone and dehydroascorbıc acıd ın guınea pıg tissues, *J. Bıol. Chem.,* 195, 271

Banerjee, S. and Ghosh, N. C. (1946), Adrenalın in scurvy, *J. Biol. Chem.,* 166, 25.

Banerjee, S. and Ghosh, N. C. (1947), Relation of scurvy to glucose tolerance test, liver glycogen, and insulin content of pancreas in guinea pıgs, *J. Biol. Chem.,* 168, 207

Banerjee, S. and Ghosh, P. K. (1955), Effect of scurvy on hexokinase activity of tissues of guinea pigs, *Proc. Exp. Biol Med.,* 88, 415

Banerjee, S. and Ghosh, P. K. (1960), Metabolism of acetate in scorbutic guinea pigs, *Am. J. Physiol.,* 199, 1064.

Bartelheimer, H. (1938), C-Vitamin und Diabetes, *Dtsch. Arch. Klin. Med.,* 182, 546.

Bonsignore, A. and Pinotti, F. (1935), Acido ascorbico e deıdrascorbico nello scorbuto sperimentale. L'indice scorbutıco, *Quad. Nutr.* 2, 333.

Chatterjee, I. B., Majumder, A. K., Nandi, B. K., and Subramanian, N. (1975), Synthesis and some major functions of vıtamin C in anımals, *Ann. N.Y. Acad. Sci.,* 258, 24.

Domke, I. and Weis, W. (1983), Reinvestigation of the diabetogenic effect of dehydroascorbic acid, *Int. J. Vit. Nutr. Res.,* 53, 51.

Ganguli, N. C. and Banerjee, A. B. (1961), Metabolıc studies on scorbutic guınea pıgs. I. Hepatic glucose 6-phosphate metabolism, *J. Biol. Chem.,* 236, 979.

Giroud, A. et Ratsimamanga, A. R. (1941), Variations chimiques, en particulier des glucides, en fonction de l'acide ascorbıque. Nécessité de la réalisation du taux normal, *Bull Soc. Chim. Bıol.,* 23, 102.

Glatzel, von H. and Rüberg-Schweer, M. (1966), Zur Fráge der Glukosurie nach hohen Askorbınsäuredosen, *Med. Klin.,* 61, 1249.

Hamne, B. (1941), Studien über die Biologie des C-Vitamins mit Besonderer Berücksichtigung des Kohlehydratstoffwechsels, *Acta Paediatr ,* 28 (Suppl. IV), 1.

Hermann, V. S. (1938), Uber die Wirkung des Follikelhormons, des Vitamin C und deren gleichzeıtiger Verabreichung auf den Glykogengehalt der Leber bei Meerschweinchen, *Hoppe-Seyler's Z. Physiol. Chem.,* 251, 78.

Hirsch, L. (1936), Über den Einfluss der Ascorbinsäure auf den Glykogengehalt der Leber hyperthyreotisierter Meerschweinchen, *Biochem. Z.,* 287, 126.

Hughes, R. E. (1956), The use of homocysteine in the estimation of dehydroascorbic acid, *Biochem. J.,* 64, 203. 208.

Johnson, D. D. (1950), Alloxan administration in the guinea pig. a study of the histological changes in the islands of Langerhans, the blood sugar fluctuations, and changes in the glucose tolerance, *Endocrinology,* 46, 135

Kawishwar, W. K., Chakrapani, B., and Banerjee, S. (1963), Carbohydrate and lipid metabolısm in scurvy: effect of vıtamın C supplement, *Indian J. Med. Res.,* 51, 488.

Kopylova, Z. A. and Shelagina, N. A. (1981), Energy metabolism of carbohydrates in skin in experimental chronic vitamin C deficiency, *Tr. Leningr. Sanit. Gig. Med. Inst.,* 141, 39.

Lahiri, S. and Banerjee, S. (1956), Carbohydrate metabolism and phosphate turnover rate in scorbutıc guinea pigs, *Proc. Soc. Exp. Bıol. Med.,* 93, 557.

Menten, M. L. and King, C. G. (1935), The influence of vitamin C level upon resistance to diphtheria toxin, *J. Nutr.,* 10, 141.

Merlini, D. (1961), Sul diabete da acido deidroascorbico, *Rass. Clin. Sci. Inst. Biochim. Ital.,* 37, 109.

Merlini, D. and Caramia, F. (1965), Effect of dehydroascorbic acid on the islets of Langerhans of the rat pancreas, *J. Cell Biol.,* 26, 245.

Messinà, A., Bruchieri, A., and Gasso, G. (1968), Diabete sperimentale da acido deidroascorbico, *Boll Soc. Ital. Biol. Sper.,* 44, 1138.

Murray, H. C. (1948), Effect of insulin, adrenal cortical hormones, salt and *dl*-alanine on carbohydrate metabolism in scurvy, *Proc. Soc. Exp. Biol. Med.,* 69, 351.

Murray, H. C. and Morgan, A. F. (1946), Carbohydrate metabolism in the ascorbic acid-deficient guinea pig under normal and anoxic conditions, *J. Biol. Chem.,* 163, 401.

Nadal, E. M. and Mulay, A. S. (1954), Compound E and glycogen deposition in the livers of normal and scorbutic guinea pigs, *Proc. Fed. Am. Soc. Exp. Biol.,* 13, 440.

Nair, K. R. (1941), Biochemical changes in experimental scurvy, *Ann. Biochem. Exp. Med.,* 1, 179.

Nandi, B. K., Majumder, A. K., Subramanian, N., and Chatterjee, I. B. (1973), Effects of large dose of vitamin C in guinea pigs and rats, *J. Nutr.,* 103, 1688.

Palladin, A. and Utewski, A. (1928), Beiträge zur Biochemie der Avitaminosen Nr. 9. Über den Einfluss des Charakters der Nahrung auf die Blutzuckerkurve bei experimentellem Skorbut und auf die Empfindlichkeit der Meerschweinchen gegen Insulin, *Biochem. Z.*, 199, 377.

Patterson, J. W. (1951), Course of diabetes and development of cataracts after injecting dehydroascorbic acid and related substances, *Am J. Physiol.*, 165, 61.

Patterson, J. W. (1949), The diabetogenic effect of dehydroascorbic acid, *Endocrinology*, 45, 344.

Patterson, J. W. (1950), The diabetogenic effect of dehydroascorbic acid and dehydroisoascorbic acids, *J. Biol. Chem*, 183, 81.

Patterson, J. W. and Lazarow, A. (1950), Sulfhydril protection against dehydroascorbic acid diabetes, *J. Biol Chem.*, 186, 141.

Patterson, J. W. and Mastin, D. W. (1951), Some effects of dehydroascorbic acid on the central nervous system, *Am. J. Physiol.*, 167, 119.

Pence, L. A. and Mennear, J. H. (1979), The inhibitory effect of dehydroascorbic acid on insulin secretion from mouse pancreatic islets, *Toxicol. Appl. Pharmacol.*, 50, 57.

Princiotto, J. V. (1951), Experimental diabetes produced by dehydroascorbic acid, *J. Clin. Endocrinol*, 11, 775.

Randoin, L. and Michaux, A. (1925), Réserves glycogéniques et glycemie artérielle (effective et proteique) au cours du scorbut expérimental, *C. R. Soc. Biol.*, 181, 1179.

Rikans, L. E. (1982), NADPH-dependent reduction of cytochrome P-450 in liver microsomes from vitamin C-deficient guinea pigs: effect of benzphetamine, *J. Nutr.*, 112, 1796.

Sarkar, A. K. and Banerjee, S. (1957), Studies on the glucose tolerance test in scorbutic monkeys, *Indian J. Physiol. Pharmacol.*, 1, 27.

Sigal, A. and King, C. G. (1936), The relationship of vitamin C to glucose tolerance in the guinea pig, *J. Biol. Chem.*, 116, 489.

Som, S., Basu, S., Mukherjee, D., Deb, S., Choudhury, P. R., Mukherjee, S., Chatterjee, S. N., and Chatterjee, I. B. (1981), Ascorbic acid metabolism in diabetes mellitus, *Metabolism*, 30, 572.

Stewart, C. T., Salmon, R. J., and May, C. D. (1952), Factors determining effect of insulin on metabolism of glucose in ascorbic acid deficiency and scurvy in the monkey, *AMA Am. J. Dis. Child.*, 84, 677.

Tomita, K. (1928), Experimental studies on vitamin C: on carbohydrate and nitrogen metabolism of experimental scurvy in guinea-pigs fed with exclusive oat diet, *Sei-I-Kai Med. J.*, 47, 6.

Chapter 4

FOLIC ACID METABOLISM

Folic acid (pteroylglutamic acid, PGA) is a member of the B complex group of vitamins, so a detailed discussion of its discovery and chemistry would not be appropriate here. Suffice it to say that much work in this field arose from the fact that folic acid (and folinic acid) stimulate the growth rates of *Lactobacillus casei* and *Streptococcus faecalis*. The most significant advances arose from the discovery that a biologically much more potent compound (folinic acid) was active in promoting the rapid growth of *Leuconostoc citrovorum* 8081. Thus, folinic acid was variously known as the *Leuconostoc citrovorum* factor, the citrovorum factor, LCF, CF, leucovorin, *Pediococcus cerevesiae* factor, PCF, and by its chemical name, the N^5-formyl derivative of 5,6,7,8-tetrahydrofolic acid. The whole business was greatly simplified by use of the terms folic acid and folinic acid for these substances which play an important role in nucleic acid synthesis, but in fact there are several nutritionally related pteroyl glutamates, differing in the number of glutamic acid residues, which may all be considered as folic acids. There are also several compounds including the 10 formyl and the 5-10 formyl, as well as the 5-formyl tetrahydrofolic acids, which are all folinic acids. Moreover, the 5-formyl can be converted to the 10-formyl derivative by the action of an enzyme, formyl, tetrahydrofolic acid isomerase.

It was biological studies by Woodruff and Darby (1948, 1949) on guinea pigs, by May et al. (1949) on monkeys, and by May et al. (1950a) on human infants with megaloblastic anemia which led to the suggestion that ascorbic acid might be involved in the metabolism of folic acid. Working at the University of Minnesota Medical School, May (1950) concluded that a chronic deficiency of ascorbic acid leads to a deficiency of pteroylglutamic acid or some difficulty in the metabolism of pteroyl glutamic acid or related compounds, which results in a megaloblastic pattern in the bone marrow of human infants and monkeys.

In vitro studies by Nichol and Welch (1950) and Nichol (1952) soon showed that the conversion of folic acid to folinic acid by rat liver slices is significantly augmented by ascorbic acid. Moreover, Nichol (1953), working with homogenates of chick liver, confirmed these findings and also observed that ascorbic acid suppresses the enzymatic destruction of natural or synthetic folinic acid. Hill and Scott (1952), also working with chick liver, reached the conclusion that a folinic acid-liberating enzyme in the liver is activated by ascorbic acid.

May (1950b) reported that a dose of 7.5 to 30 µg of folinic acid was as effective as 15 mg of folic acid in causing the disappearance of megaloblasts from the bone marrows of scorbutic megaloblastic monkeys. Moreover, May et al. (1951) demonstrated that megaloblastic anemia could be produced regularly in monkeys by feeding milk diets deficient in ascorbic acid, and that vitamin B_{12} would neither cure it nor prevent it. It could, however, be cured with folic acid, or with very small doses of folinic acid, or by the use of ascorbic acid alone, without any folic or folinic acid. This experimental megaloblastosis is due to a disturbance in the metabolism of folic acid caused by a deficiency of ascorbic acid. Subsequent experiments by the same group of workers, Proehl and May (1952), May et al. (1952a,b 1953) confirmed that the conversion of folic acid to folinic acid in monkeys may be less efficient in scurvy, but does occur to some extent. They concluded that the severe deficiency of folic acid compounds which occurs regularly as a complication of scurvy, in monkeys fed milk diets deficient in ascorbic acid, is probably due to nonspecific factors operating in scurvy. The net effect of scurvy is to increase requirements for folic acid.

Studies of folic acid metabolism in men and women by Welch et al. (1951) demonstrated that the administration of ascorbic acid (750 mg/d) promptly increased the conversion of folic acid to folinic acid in seven normal subjects (Figure 1). The folinic response by two

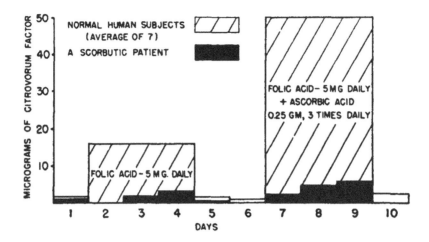

FIGURE 1.   The effects of folic acid and of a combination of folic acid and ascorbic acid on the urinary excretion of folinic acid (citrovorum factor) in seven normal subjects and a severely scorbutic man The poor folinic acid excretory response by the man with scurvy may have been partly due to liver damage, as he was known to be an alcoholic. accustomed to drinking 12 to 14 bottles of beer a day (From Welch, A. D., Nichol, C A., Anker, R. M., and Boehne, J. W. [1951], *J. Pharmacol. Exp. Ther.*, 103, 403 © American Society for Pharmacology and Experimental Therapeutics. With permission.)

Table 1
EFFECT OF ADMINISTRATION OF PTEROYLGLUTAMIC ACID (PGA) AND ASCORBIC ACID ON URINARY EXCRETION OF CITROVORUM FACTOR (CF)

Subject	CF content of normal urine (μg/ml)	Total CF excretion in 6 h following oral administration of 50 mg PGA (μg)	Total CF excreted in 6 h following oral administration of 50 mg PGA + 1 g ascorbic acid (μg)
1	0.0006	55	103
2	0.0004	36	116
3	0.0004	32	108
4	0.0006	52	121
5	0 0008	40	
6	0.0006	36	124
Average	0.0006	42	114

Note: The folinic acid (CF) content of the urine of six normal men first on an ordinary diet, then following the administration of folic acid (PGA) and again later following the same dose of folic acid 50 mg plus 1 g of ascorbic acid.

From Broquist, H. P., Stokstad, E. L. R., and Jukes, T. H (1951), *J. Lab. Clin Med.*, 38, 95. With permission.

scorbutic patients was markedly delayed, but one of them was an alcoholic with an increased clotting time, suggesting cirrhosis of the liver; the other scorbutic also had a prolonged clotting time, suggesting liver damage. Comparable studies of six normal men by Broquist et al. (1951) led to a similar conclusion; within the first 6 h after giving 50 mg of PGA (folic acid) by mouth, they found that 0.1% of that dose appeared in the urine as CF (folinic acid). A threefold higher level of CF was found in the urine if 1 g of ascorbic acid was given simultaneously with the 50 mg of PGA (Table 1). In contrast, when leucovorin (folinic

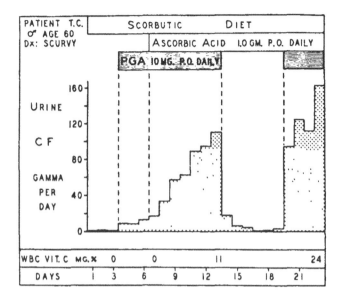

FIGURE 2. Urinary excretion of folinic acid (CF) following oral folic acid (PGA) in patient with scurvy before and after ascorbic acid therapy (From Gabuzda, G. J. Phillips, G. B., Schilling, R. F., and Davidson, C. F. [1952], *J. Clin. Invest.*, 31, 756. © American Society for Clinical Investigation. With permission.)

acid) was fed by mouth, ascorbic acid administration did not affect the recovery of CF (folinic acid) in the urine. When liver homogenate from PGA deficient rats was incubated in the presence of PGA, a marked increase in the amount of CF formed was noted on addition of ascorbic acid. Studies of CF (folinic acid) excretion by two men with scurvy were conducted by Gabuzda et al. (1952). A very small amount of CF was excreted by these patients with scurvy. It was increased slightly when folic acid was administered, but maximal CF excretion, comparable to that in a nonscorbutic individual, occurred in the patients with scurvy only after treatment with ascorbic acid (Figure 2). The authors concluded that one role of ascorbic acid in man is to provide for the conversion of PGA to CF (folic to folinic acid). This effect of ascorbic acid deficiency on folate metabolism also seems to be responsible for the abnormalities of tyrosine and phenylalanine metabolism, which are seen in scorbutic infants on high-protein diets (Chapter 6 of this volume). Darby et al. (1953) reviewed the work carried out in their own laboratory and also the work of others. Folic acid reduces the urinary excretion of *p*-hydroxyphenyl pyruvic and *p*-hydroxyphenyllactic acids by scorbutic infants, without in any way relieving the scurvy. It seems that neither folic nor folinic acid is as effective as ascorbic acid in reducing the hydroxyphenyluria of scurvy, but apparently folic acid does enhance the oxidation of tyrosine by liver tissue *in vitro* and large doses may have a similar effect *in vivo*.

Jandl and Gabuzda (1953), treating two men with megaloblastic anemia associated with scurvy, observed a good reticulocyte response and a rise in the hemoglobin level as a result of folic acid administration. A potentiation of this response was noted on subsequent administration of ascorbic acid, 1 g daily, in addition to the folic acid. Zalusky and Herbert (1961) reported macrocytic anemia in a 60-year-old bachelor with hemosiderosis and scurvy who had a normal vitamin B_{12} level. The anemia failed to respond to large intravenous doses of ascorbic acid (1 g daily) while he was receiving a synthetic diet which was totally devoid of folic acid, but subsequently responded well to small doses of folic acid (50 μg intramuscularly daily) after correction of the ascorbate deficiency (Figure 3). In contrast, Asquith et al. (1967) observed a 70-year-old man with scurvy who showed a complete response of

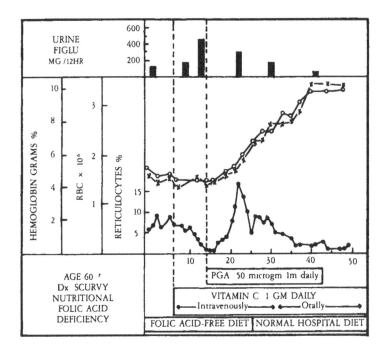

FIGURE 3. This patient with scurvy and megaloblastic anemia was placed on a synthetic folate-free diet and was treated at first with ascorbic acid, 1 g daily by intravenous injection. The anemia did not improve with ascorbic acid alone, but responded well to a combination of ascorbic acid and folic acid. (From Zalusky, R. and Herbert, V. [1961], *N. Engl. J. Med.*, 265, 1033. With permission.)

his megaloblastic erythropoeisis following treatment with ascorbic acid alone (Figure 3, Chapter 3, Volume II). Clearly, both ascorbic acid and folic acid are essential; ascorbic acid can only correct the metabolism of folic acid when some folic acid is present in the diet and is being absorbed.

Stokes et al. (1975) studied the metabolism of folinic acid, administered before and during treatment of scorbutic megaloblastic anemia in a 64-year-old bachelor. They found an increase in the oxidized folate derivatives in scurvy and concluded that an important role of ascorbic acid in human metabolism is to prevent irreversible oxidative losses of folinic acid. Bates et al. (1980) observed no significant changes in the plasma or red cell folate levels of 21 healthy elderly men during a period of supplementation with ascorbic acid, even in subjects who had low initial levels of circulating vitamin C. Lewis et al. (1982) have reported that experimental ascorbic acid deficiency, either alone, or in combination with folate restriction did not affect tissue folate levels, nor did ascorbic acid deficiency significantly exacerbate the anemia and leukopenia caused by folate deficiency in guinea pigs. However, there are several unsatisfactory features of that study: (1) none of the guinea pigs had any signs of scurvy, (2) there was only a modest difference between the plasma ascorbic acid levels of the "control" (0.23 mg/100 ml) and the "AA deficient" (0.18 mg/100 ml) animals, possibly because the specimens of plasma were frozen and stored at minus 70°C for up to 2 weeks before analysis. Unlike other substances, ascorbic acid has been reported by Grant and Alburn (1965) to be more stable just above freezing than it is below 0°C, (3) ascorbic acid was added to all specimens to be stored for folate analysis, so some folic acid may have been reduced to folinic acid before analysis for folinic acid, (4) the folate-deficient guinea pigs and especially the folate- and ascorbate-deficient animals developed leukopenia and a slight anemia, but none of them showed any increase in the mean corpuscular volume of the red cells, and (5) the results of the red cell folinic acid (*Pediococcus cerevesiae*) assays were recorded as being almost identical in the control, AA*-deficient, folate-deficient, and folate/AA-deficient animals — 29, 27, 27, and 32 μg/ml, respectively; but, of course, all were apparently analyzed in the presence of ascorbic acid, which was intended as an antioxidant, but may have reduced folic to tetrahydrofolic or folinic acid. It is probably true that ascorbic acid deficiency does not cause folinic acid deficiency and megaloblastic anemia unless the ascorbic acid deficiency is severe and the folic acid supply is also low. However, it just so happens that ascorbic acid-deficient diets, like boiled milk, are usually also low in folic acid. In such circumstances, ascorbic acid deficiency can so affect the folate metabolism of monkeys and man that it causes a folinic acid deficiency and megaloblastic anemia which can be cured with folinic acid or with ascorbic acid alone.

Whether ascorbic acid functions by aiding in the conversion of folic to folinic acid, or by preventing oxidative losses of folinic acid within the body, still seems to be a matter of academic dispute. It probably has both effects, as originally suggested by Nichol and Welch.

In view of the fact that X-ray irradiation of guinea pig and rat tissues has been shown to cause oxidation of ascorbic acid and release of ascorbate free radical (Figures 1, 2, and 3, Chapter 21, Volume I), one can understand the findings of Schreurs et al. (1985) that women undergoing radiotherapy for cancer of the cervix or the endometrium showed a highly significant fall in their blood ascorbic acid level from an original mean of 53.9 before treatment to 38.5 μmol/l after 4 weeks of X-ray therapy ($p < 0.01$). It is worthy of note that these women also developed significant reductions in serum vitamin E from 24.6 to 21.1 μmol/l, vitamin B_{12} from 329 to 262 pmol/l, and in serum folic acid from 8.06 to 6.50 nmol/l. We may conjecture that the folic acid losses may have been directly due to the oxidative effect of irradiation, involving a rise in the oxidation-reduction potential of the tissues, or indirectly due to a fall in the ascorbate/dehydroascorbate ratio.

REFERENCES

Asquith, P., Oelbaum, M. H., and Dawson, D. W. (1967), Scorbutic megaloblastic anaemia responding to ascorbic acid alone, *Br. Med. J.*, November 18, p. 402.

Bates, C. J., Fleming, M., Paul, A. A., Black, A. E., and Mandal, A. R. (1980), Folate status and its relation to vitamin C in healthy elderly men and women, *Age Ageing*, 9, 241.

Broquist, H. P., Stokstad, E. L. R., and Jukes, T. H. (1951), Biochemical studies with the "citrovorum factor", *J. Lab. Clin. Med.*, 38, 95.

Darby, W. J., McGanity, W. J., Stockell, A., and Woodruff, C. A. (1953), Ascorbic acid, pteroylglutamates, and other factors in scorbutic hydroxyphenyluria, *Proc. Nutr. Soc.*, 12, 329.

Gabuzda, G. J., Phillips, G. B., Schilling, R. F., and Davidson, C. S. (1952), Metabolism of pteroylglutamic acid and the citrovorum factor in patients with scurvy, *J. Clin. Invest.*, 31, 756.

Grant, N. H. and Alburn, H. E. (1965), Fast reactions of ascorbic acid and hydrogen peroxide in ice, a presumptive early environment, *Science*, 150, 1589.

Hill, C. H. and Scott, M. L. (1952), The effect of ascorbic acid on the citrovorum factor-liberating enzyme of the chick liver, *J. Biol. Chem.*, 196, 195.

Jandl, J. H. and Gabuzda, G. J. (1953), Potentiation of pteroylglutamic acid by ascorbic acid in anemia of scurvy, *Proc. Soc. Exp. Biol. Med.*, 84, 452.

Lewis, C. M., McGown, E. L., Rusnak, M. G., and Sauberlich, H. E. (1982), Interactions between folate and ascorbic acid in the guinea pig, *J. Nutr.*, 112, 673.

May, C. D., Hamilton, A., and Stewart, C. T. (1952a), Nature of the relation of ascorbic acid deficiency to the metabolism of folic acid compounds, *J. Lab. Clin. Med.*, 40, 924.

May, C. D., Hamilton, A., and Stewart, C. T. (1952b), Experimental megaloblastic anaemia and scurvy in the monkey. IV. Vitamin B_{12} and folic acid compounds in the diet, liver, urine and feces and effects of therapy, *Blood*, 7, 978.

---

* AA — ascorbic acid, reduced form.

May, C. D., Hamilton, A., and Stewart, C. T. (1953), Experimental megaloblastic anemia and scurvy in the monkey. V. Nature of the relation of ascorbic acid deficiency to the metabolism of folic acid compounds, *J Nutr.*, 49, 121.

May, C. D., Nelson, E. N., and Salmon, R. J. (1949), Experimental production of megaloblastic anemia; an interrelationship between ascorbic acid and pteroylglutamic acid, *J. Lab. Clin. Med.*, 34, 1724.

May, C. D., Nelson, E. N., Salmon, R. J., Lowe, C. U., Lienke, R. I., and Sundberg, R. D. (1950a), Experimental production of megaloblastic anemia in relation to megaloblastic anemia in infants, *Bull. Univ. Minn. Hosp.*, 21, 208.

May, C. D., Sundberg, R. D., and Schaar, F. (1950b),Comparison of effects of folic and folinic acid in experimental megaloblastic anemia, *J. Lab. Clin. Med.*,36, 963.

May, C. D., Sundberg, R. D., Shaar, F., Lowe, C. U., and Salmon, R. J. (1951), Experimental nutritional megaloblastic anemia and pteroylglutamic acid. I. Nutritional data and manifestations of animals, *Am. J Dis. Child.*, 82, 282.

Nichol, C. A. (1952), Enzymatic conversion of folic acid to citrovorum factor, *Fed. Proc. Fed. Am. Soc. Exp. Biol.*, 11, 452.

Nichol, C. A. (1953), The effect of ascorbic acid on the enzymatic formation of the citrovorum factor, *J. Biol. Chem.*, 204, 469.

Nichol, C. A. and Welch, A. D. (1950), Synthesis of citrovorum factor from folic acid by liver slices; augmentation by ascorbic acid, *Proc. Soc. Exp. Biol. Med.*, 74, 52.

Proehl, E. C. and May, C. D. (1952), Experimental nutritional megaloblastic anemia and scurvy in the monkey. III. Protoporphyrin, coproporphyrin, urobilinogen and iron in blood and excreta, *Blood*, 7, 671.

Schreurs, W. H. P., Odink, J., Wedel, M., and Bruning, P. F. (1985), The influence of radiotherapy and chemotherapy on the vitamin status of cancer patients, *Int. J. Vitam. Nutr. Res.*, 55, 425.

Stokes, P. L., Melikian, V., Leeming, R. L., Porter-Graham, H., Blair, J. A., and Cooke, W. T. (1975), Folate metabolism in scurvy, *Am. J. Clin. Nutr.*, 28, 126.

Welch, A. D., Nichol, C. A., Anker, R. M., and Boehne, J. W. (1951), The effect of ascorbic acid on the urinary excretion of citrovorum factor derived from folic acid, *J. Pharmacol. Exp. Ther.*, 103, 403.

Woodruff, C. W. and Darby, W. J. (1948), An in vivo effect of pteroylglutamic acid upon tyrosine metabolism in the scorbutic guinea pig, *J. Biol. Chem.*, 172, 851.

Woodruff, C. W. and Darby, W. J. (1949), Influence of pteroylglutamic acid on tyrosine metabolism in the scorbutic guinea pig, *Am. J. Dis. Child.* 77, 128.

Zalusky, R. and Herbert, V. (1961), Megaloblastic anemia in scurvy with response to 50 microgm of folic acid daily, *N. Engl. J. Med.*, 265, 1033.

Chapter 5

CHOLESTEROL METABOLISM

I. GUINEA PIG STUDIES

The cause of cholesterol accumulation in vitamin C deficiency has been the subject of much research in the last 30 years. Initial studies of guinea pigs which were cachectic and terminally ill, due to complete deprivation of ascorbic acid, led to the suggestion that there was increased synthesis of cholesterol from acetate in scurvy. This was attributed to decreased utilization of acetate in the Krebs cycle by terminally ill animals. However, subsequent studies of chronic hypovitaminosis C in animals which were not losing weight revealed a decreased conversion of cholesterol to bile acids as the principal abnormality of cholesterol metabolism due to ascorbic acid deficiency.

Using isotopically labeled acetate, Bloch and Rittenberg (1942, 1945), Bloch (1944), and Little and Bloch (1950) demonstrated that acetic acid or acetate is the most important source of carbon and hydrogen for cholesterol synthesis in rats and mice. Banerjee and Deb (1951) and Oesterling and Long (1951) reported that both the ascorbic acid and the cholesterol contents of the adrenals were diminished in guinea pigs with scurvy. However, Becker et al. (1953), working with King in his laboratory in New York City, reported a threefold increase in the rate of conversion of acetate-1-^{14}C into cholesterol by scorbutic guinea pig adrenals, compared with pair-fed controls. They also observed a more modest (61%) increase in the rate of cholesterol formation by the liver in scorbutic animals. These observations were of particular interest because Willis (1953), working in Montreal, at the same time observed histological evidence of markedly accelerated subendothelial lipid deposition in the aortas of scorbutic guinea pigs (Chapter 8, Volume II). Belavady and Banerjee (1954) showed that during the development of scurvy, guinea pigs excreted in the urine lesser amounts of *p*-aminobenzoic acid in the acetylated form than did normal controls. So the decrease in adrenal cholesterol in scurvy was at first attributed to diminished acetylation in the body. However, while the cholesterol contents of the adrenals, spleen, and lungs decreased, those of the testes and the small intestine increased significantly in scurvy; there was no change in the cholesterol content of the liver or kidneys. The blood cholesterol increased during the early part of the experiment; it became normal in the later stages when the animals developed scurvy. Bolker et al. (1956) reported that the serum cholesterol levels of scorbutic guinea pigs were significantly elevated compared with controls, but these workers observed no significant difference between the rates of incorporation of acetate-1-^{14}C into cholesterol and fatty acids by isolated portions of liver, aorta, and adrenal from scorbutic and from normal guinea pigs. Banerjee and Singh (1958) suspected that the abnormal cholesterol metabolism in scurvy might be related to the decreased insulin content of the pancreas which had already been observed; so they studied normal, scorbutic and insulin treated scorbutic guinea pigs. They found that the intestinal wall and total body cholesterol content was increased in scorbutic guinea pigs in comparison with normal controls; but prolonged treatment of the scorbutic animals with insulin lowered the total cholesterol content to the normal level (Table 1). They suggested that the "increased cholesterologenesis" in scurvy might be due to the utilization of an increased acetate pool which is not burned through the tricarboxylic acid cycle.

In a subsequent study, Banerjee and Ghosh (1960) studied guinea pigs fed a scorbutogenic diet, with or without a daily ascorbic acid supplement (5 mg daily), by a paired-feeding technique; 10 d later, while still on the diet, half of the animals of each group were given 20 mg of sodium acetate per 100 g or body weight daily by mouth. After feeding sodium

Table 1

TOTAL BODY CHOLESTEROL CONTENT OF NORMAL, SCORBUTIC, AND INSULIN-TREATED GUINEA PIGS[a]

Total cholesterol content[b] (mg/100 g of fresh tissue)			t-values		
Normal	Scorbutic	Insulin-treated scorbutic	Between normal and scorbutic	Between normal and insulin-treated scorbutic	Between scorbutic and insulin-treated scorbutic
189 ± 14	277 ± 27	173 ± 15	2.9	0.7	3.0

[a] There were six animals in each group.
[b] Mean ± standard error.

From Banerjee, S. and Singh, H. D. (1958). *J. Biol. Chem.*, 233, 336. With permission

Table 2
CATABOLISM OF (4-^{14}C) CHOLESTEROL
IN THE LIVER AND GALL BLADDER OF SCORBUTIC
GUINEA PIGS *IN VIVO*[a]

	10^{-3} × Radioactivity			
	In bile acids		In nonsaponifiable fraction	
Condition of animals	Range	Average	Range	Average
Liver				
Scorbutic	1.12—1.18	1.16	80.24—84.25	83.56
Pair fed	2.81—3.23	2.90	20.28—22.56	20.58
Gall bladder				
Scorbutic	10.12—11.02	10.50	33.44—41.12	39.93
Pair fed	40.65—44.87	43.18	13.33—16.98	15.33

Note: The formation of bile acids from cholesterol in the liver is depressed in the scorbutic guinea pigs compared with pair-fed controls.

[a] Results are from six individual pairs of animals and are expressed as counts per minute per gram of liver and gall bladder bile.

From Guchhait, R., Guha, B. C., and Ganguli, N. C. (1963), *Biochem. J.*, 86, 193. © 1963 The Biochemical Society, London. With permission.

acetate for 14 d glucose tolerance tests were conducted, the animals were killed and their tissues were analyzed. In scorbutic guinea pigs of this experiment, the cholesterol content of the whole body and of blood, small intestine, and testes increased; that of the adrenals diminished; there was no change in liver cholesterol. The cholesterol contents of the whole body, blood, small intestine and testes were further increased when the scorbutic animals were fed acetate. The glucose tolerance was lowered in both the scorbutic and the acetate-fed scorbutic guinea pigs. Liver glycogen was considerably increased when the normal guinea pigs were fed acetate, and diminished when the animals developed scurvy. Not only were the glucose tolerance and the "increased cholesterologenesis" of scurvy rectified by insulin administration, as previously demonstrated, but in this paper these authors reported that increased urinary excretions of ketone bodies, citric acid, and malic acid by scorbutic guinea pigs fed sodium acetate were also returned to normal after insulin treatment. This was taken as further evidence that the "increased cholesterologenesis" of scurvy results from the availability of extra acetate due to decreased utilization of acetate via the Krebs cycle, as a result of hypoinsulinism.

Guchhait et al. (1963) reported having found an increased rate of hepatic synthesis of cholesterol from [1-^{14}C] acetate and from [2-^{14}C] acetate in scorbutic guinea pigs. These authors, noting that the liver cholesterol level was not increased in spite of increased synthesis of this substance in scurvy, therefore decided to study cholesterol catabolism in scurvy. Contrary to expectation, they found that the rate of conversion of [4-^{14}C] cholesterol into bile acids was depressed in scurvy, both *in vitro* and *in vivo*. Both in the liver and in the gall bladder bile from scorbutic guinea pigs, the bile acid fraction had less radioactivity in it than that from pair-fed animals after the injection of [4-^{14}C] cholesterol (Table 2). "Thus the hypercholesterolaemia in scurvy seemed to be due not only to an increased cholesterol synthesis, but also to a depressed catabolism of cholesterol."

In their studies of guinea pigs, Kawishwar et al. (1963) confirmed that the total cholesterol content of the small intestine and of the whole body is increased in scurvy, even though there was no change in the cholesterol content of the liver, kidneys, and testes. There was

Table 3
**CHOLESTEROL CONCENTRATION (mg/100 ml) IN
THE BLOOD SERUM AND THE LIVER (mg/100 g)
OF MALE GUINEA PIGS WITH
HYPOVITAMINOSIS C AND CONTROL ANIMALS
RECEIVING THE SAME DIET WITH 10 mg
ASCORBIC ACID DAILY**[a]

Series	Control	Hypovitaminosis	Significance
1. 49 d			
Liver	319 ± 21	458 ± 95	—
Blood serum	95 ± 6	158 ± 25	$p < 0.05$
2 70 d			
Liver	422 ± 21	493 ± 28	$p < 0.05$
Blood serum	162 ± 14	184 ± 22	—
3 104d			
Liver	368 ± 29	659 ± 40	$p < 0.001$
Blood serum	167 ± 14	215 ± 17	$p < 0.05$

[a] Average values ± mean error. The duration of hypovitaminosis C is
given for each series.

From Ginter, E , Bobek, P., Zopec, Z., Ovečka, M., and Čerey, K. (1967),
Z Versuchstierkd., 9, 228. With permission.

a diminution in the total lipid and phospholipid contents of the body. They confirmed the
findings of Banerjee and Ghosh that the defects in carbohydrate and lipid metabolism
observed in scorbutic guinea pigs could be reversed either by insulin treatment or by providing
ascorbic acid. Needless to say, insulin did not cure the scurvy.

Ginter et al. (1965) did much to clarify this subject and to explain some of the contradictory
observations in the literature. Working at the Institute of Human Nutrition, Bratislava,
Czechoslovakia, they observed that, "acute scurvy and chronic hypovitaminosis C are two
metabolically different conditions." In acute scurvy the serum β-lipoprotein fraction showed
a gradual increase by 18 d, but had fallen even below normal by 25 d when the animals
were refusing food and were losing weight as they were terminally ill with scurvy. In chronic
hypovitaminosis C there was an increase in the β-lipoprotein fraction and in the cholesterol
level of the serum. So Ginter et al. (1967) made a detailed study of chronic borderline
ascorbic acid deficiency in guinea pigs. Giving the test animals no ascorbic acid for 2 weeks
and then 0.5 mg daily by oral pipette, they were able to keep them in a state of hypovitaminosis C without any clinical or gross pathological signs of scurvy for as long as 1 year,
without even any significant loss of weight; control guinea pigs received 5 mg of ascorbic
acid daily. In this way they were able more closely to simulate the common condition of
people living during the long winters in northern regions of the world. They demonstrated
that chronic hypovitaminosis C results in an accumulation of cholesterol in the serum and
the liver of guinea pigs (Table 3). They also observed markedly decreased liver glycogen
storage and decreased pancreatic protease and lipase activities in guinea pigs with chronic
borderline ascorbic acid deficiency.

Further investigations of chronic hypovitaminosis C in guinea pigs by Ginter et al. (1969c)
confirmed the accumulation of cholesterol in the serum and the livers of male guinea pigs
(Table 4). They also demonstrated an increased content of saturated fatty acids (up to chain
length C_{16}) and a decreased content of mono- and polyunsaturated fatty acids in the cholesterol
esters of the liver in guinea pigs with C hypovitaminosis. Similar changes were observed
after the administration of the same diet supplemented with 0.25% cholesterol to guinea

Table 4
TOTAL CHOLESTEROL CONCENTRATION IN BLOOD SERUM AND LIVER, AND CHOLESTEROL CONTENT IN WHOLE LIVER OF CONTROL AND VITAMIN C-DEFICIENT GUINEA PIGS

Sample	Control	Hypovitaminosis C	Statistical significance
No. of animals	12	12	
Blood serum (mg/100 ml)	118 ± 14	171 ± 18	$p < 0.05$
Liver (mg/100 g)	456 ± 56	627 ± 65	border $p = 0.05$
Liver (mg/organ)	109 ± 12	181 ± 24	$p < 0.02$

Note After 16 to 20 weeks of chronic hypovitaminosis C (no ascorbic acid for 2 weeks and then 0.5 mg daily), there was an increase in the concentration of cholesterol in the blood serum and in the content and concentration of cholesterol in the livers of guinea pigs. The control guinea pigs received 10 mg of ascorbic acid daily. No such changes were observed in female guinea pigs fed the same hypovitaminosis C diet for 12 weeks. It is not evident whether this represents a sex difference, as the duration of the experiment was so much shorter in the females.

ᵃ Mean ± SEM.

From Ginter, E., Ondreička, R., Bobek, P., and Šimko, V. (1969), *J. Nutr.*, 99, 261. © American Institute of Nutrition. With permission.

pigs saturated with vitamin C. It was therefore suggested that the primary cause of the observed changes was the abnormality of cholesterol metabolism produced by chronic vitamin C deficiency. However, studies by Ginter and Nemec (1969) showed that the endogenous synthesis of cholesterol from acetate-1-^{14}C is not increased during chronic hypovitaminosis C: these findings stand in contrast to the earlier observations of Becker et al. (1953) in acute scurvy, but are consistent with the observations of Bolker et al. (1956).

In a study of acute scurvy in guinea pigs receiving 0.3% cholesterol in their diet, Ginter et al. (1969b) observed that the presence or absence of ascorbic acid in the diet did not affect tissue cholesterol levels in 10 or 20 d. However, a study of chronic hypovitaminosis C in guinea pigs receiving 0.3% cholesterol in their diet showed significantly increased accumulation of cholesterol in the brain, small intestine, and aorta of guinea pigs after 12 weeks on the diet. The mean serum, liver, and adrenal cholesterol levels were not significantly increased by ascorbate deficiency (Table 5) in this experiment. Under the influence of chronic vitamin C deficiency, cholesterol deposition in the wall of the aorta was significantly more marked than in the group receiving high doses of vitamin C.

In another study, Ginter et al. (1969a) compared the effects of 20 weeks of exposure to high, medium, and low (50, 5.0, and 0.5 mg daily) ascorbic acid intakes on guinea pigs fed a diet containing 0.3% cholesterol. The animals of the low-ascorbic group were found to have significantly more cholesterol accumulation in the liver, adrenals, and small intestine than those of the high-ascorbic acid group, but ascorbic acid intake did not influence the accumulation of cholesterol in the serum or the brain. The most advanced atheromatous changes were found in the aorta and coronary arteries of the guinea pigs with chronic hypovitaminosis C. It would seem that chronic hypovitaminosis C does increase the serum cholesterol level of guinea pigs to 215 vs. 167 mg/100 ml (Table 3) and to 171 vs. 118 mg/100 ml (Table 4), but in guinea pigs on an atherogenic diet (containing 0.3% cholesterol) the serum cholesterol is already markedly elevated after 12 weeks to 355 mg/100 ml in the

Table 5

THE EFFECT OF DIFFERENT DOSES OF VITAMIN C ON CHOLESTEROL STORAGE IN TISSUES OF GUINEA PIGS FED A DIET CONTAINING 0.3% CHOLESTEROL

Tissue	Atherogenic diet (doses of vitamin C/24 h)				Statistical significance (50 mg vs. 0.5 mg)	Correlation (cholesterol-vitamin C)	
	Control	50 mg	5 mg	0.5 mg		Correlation coefficient	p
Blood serum	103 ± 9	355 ± 11	377 ± 58	437 ± 134	—	−0.331	—
Aorta	271 ± 21	409 ± 29	545 ± 96	548 ± 48	<0.01	−0.313	0.1
Brain	1098 ± 42	1356 ± 85	1219 ± 69	2235 ± 186	<0.001	−0.572	<0.002
Liver	486 ± 80	3404 ± 42	3652 ± 310	4017 ± 485	0.1	−0.320	0.1
Adrenals	6612 ± 386	8651 ± 527	10647 ± 1047	10774 ± 1621	0.1	−0.247	<0.05
Small intestine	281 ± 25	272 ± 32	345 ± 35	387 ± 19	<0.002	−0.437	<0.02

Note: Cholesterol concentrations are expressed in mg/100 g of wet tissue. The data presented in the seventh column show the closeness of the negative correlation between cholesterol content in the respective tissue and vitamin C concentration in the spleen.

From Ginter, E., Bobek, P., Babala, J., and Barbierikova, E. (1969), *Cor Vasa*, 11, 65. With permission.

high-ascorbate group and shows no significantly greater increase at 437 mg/100 ml in the chronic vitamin C deficiency group (Table 5).

Ginter et al. (1971) observed that guinea pigs show a significant reduction in the rate of transformation of cholesterol to its principal catabolic product, the bile acids, in chronic hypovitaminosis C, just as Guchhait et al. had observed in scurvy. Ginter et al. observed that the basic biological role of ascorbic acid seems to involve its participation in hydroxylation reactions, as in the hydroxylation of proline, the catabolism of aromatic amino acids, the hydroxylation of dopamine, tryptophan, acetanilid, tyramine, and other substances. They therefore suggested that ascorbic acid plays an improtant role in the hydroxylation reactions involved in the transformation of cholesterol to bile acids.

Studies by Björkhem et al. (1967) suggested that the principal pathway in the conversion of cholesterol to chenodeoxycholic acid in guinea pigs might be

$$Cholesterol \rightarrow cholest\text{-}5\text{-}ene\text{-}3\beta,7\alpha\text{-diol}$$
$$\rightarrow 7\alpha\text{-hydroxycholest-4-en-3-one}$$
$$\rightarrow 7\alpha\text{-hydroxy-5}\beta\text{-cholestan-3-one}$$
$$\rightarrow 5\beta\text{-cholestane-3}\alpha,7\alpha\text{-diol}$$
$$\rightarrow 5\beta\text{-cholestane-3}\alpha,7\alpha,26\text{-triol}$$
$$\rightarrow chenodeoxycholic\ acid.$$

During the first reaction, the cholesterol nucleus is hydroxylated in position 7; during the second to last, there is hydroxylation of the side chain at position 26. Ginter et al. (1971) therefore suggested as a working hypothesis that ascorbic acid is needed for one or both of these hydroxylation reactions. The direct participation of ascorbic acid in this process is supported by the data on the stimulatory effect of ascorbic acid on the synthesis of bile acids *in vitro* in liver mitochondria from scorbutic guinea pigs, as reported by Guchhait et al. (1963).

Fujinami et al. (1971) studied guinea pigs after 2 weeks on a scorbutogenic diet supplemented with coconut oil, ascorbic acid, or both. None of the animals had reached the stage of frank scurvy, but elevations of plasma cholesterol, cholesterol esters, phospholipids, and nonesterified fatty acids were observed in all those receiving no ascorbic acid. The ascorbate-deprived coconut oil-fed group had the highest lipid levels. Addition of ascorbic acid restored the lipid levels to normal even in those receiving a coconut supplement. The plasma ascorbic acid levels of the guinea pigs receiving coconut oil and ascorbic acid remained low. Studies by Kritchevsky et al. (1973) showed a twofold increase in the rate of 7α-hydroxylation of cholesterol by guinea pig liver microsomes when ascorbic acid was added *in vitro*, but this was not a statistically significant change. A more definite change might have been noted if the ascorbic acid had been given to the guinea pigs while they were alive, so as to promote the synthesis of new cytochrome P-450.

Further work on latent hypovitaminosis C in guinea pigs by Ginter (1973) demonstrated a significant negative correlation between the concentration of vitamin C in the liver and the concentrations of cholesterol in the serum and liver ($p < 0.001$). That is, the higher the vitamin C concentration, the lower the cholesterol concentration in the serum and the liver. It was also confirmed that latent hypovitaminosis C significantly reduced the rate of transformation of cholesterol to bile acids (Table 6). Indeed, a significant linear correlation was observed between the concentration of vitamin C in the liver and the rate of transformation of cholesterol to bile acids ($p < 0.001$). Furthermore, Ginter (1974) observed early atheromatous lesions in the intima of the thoracic aorta of guinea pigs maintained in a state of hypovitaminosis C for 28 weeks, without the addition of cholesterol to the diet. Fujinami et al. (1975) observed a significant increase in cholesterol and triglyceride concentration and in lipase and esterase activity in the arch of the aorta of vitamin C-deficient guinea pigs

Table 6
THE EFFECT OF HYPOVITAMINOSIS C ON THE CONCENTRATIONS OF ASCORBIC ACID AND CHOLESTEROL, AND ON THE RATE OF TRANSFORMATION OF CHOLESTEROL TO BILE ACIDS IN GUINEA PIGS[a]

Animal	Vitamin C (mg/100 g) in		Cholesterol in		Cholesterol → bile acids (mg/24 h/500 g of body weight)
	Liver	Spleen	Serum (mg/100 ml)	Liver (mg/100 g)	
Control	8.2 ± 0.4	21 6 ± 0 8	126 ± 9	359 ± 15[b]	11 8 ± 0 6 (23)
Deficient in vitamin C	1.6 ± 0.1	4 7 ± 0 2	218 ± 17	443 ± 19[b]	8 3 ± 0 4 (21)

Note· The number of animals observed for each determination was 26, except where otherwise indicated in parentheses. The statistical significance of results between control and vitamin C-deficient animals was *p* <0 001, unless otherwise indicated

[a] Means ± standard errors.
[b] *p* <0.002.

From Ginter, E. (1973), *Science,* 179, 702. © 1973 AAAS With permission

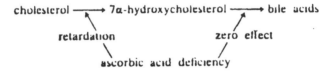

FIGURE 1.   The first step in the transformation of cholesterol to bile acids is the production of 7α-hydroxycholesterol; this reaction is rate limiting and ascorbate dependent. (From Ginter, E. [1975], *Ann. N.Y. Acad. Sci.,* 258, 410. With permission.)

fed on a diet with no added cholesterol. Additional work by Ginter (1975) demonstrated that ascorbic acid almost certainly intervenes in the biosynthesis of bile acids at the stage of 7α-hydroxylation of the cholesterol nucleus (Figure 1). Cholesterol 7α-hydroxylase is located in the microsomal fraction of the liver cell and requires NADPH and oxygen for minimal activity; it belongs to the group of enzymes classified as mixed-function oxidases. Ginter believes that cytochrome P-450 has a role in 7α-hydroxylation of cholesterol. He points out that the concentration of cytochrome P-450 in liver microsomes of guinea pigs rapidly decreases when the supply of ascorbic acid is discontinued.

Guinea pigs on a marginal ascorbic acid-deficient diet were also studied by Sulkin and Sulkin (1975). These workers reported an increase in the cholesterol content of the liver (*p* <0.05), but no significant change in the serum cholesterol level in the small number of samples studied. They did, however, observe marked endothelial proliferation and arterio-sclerotic plaque formation in the aortas of many of the animals after 100 to 110 d on the low-ascorbic acid diet. Studies of guinea pigs by Nambisan and Kurup (1975) showed that a high dose of ascorbic acid decreased the cholesterol levels of the liver and aorta, but not the serum when the cholesterol content of the diet was low. However, when the animals were fed on a high-cholesterol diet, the high-dose ascorbic acid caused a decrease in the cholesterol level of the serum and liver, but not in the aorta. Serum triglycerides were not affected by the dose of ascorbic acid in the animals on a normal diet, but in animals receiving an atherogenic diet, the high dose of ascorbic acid caused serum triglyceride levels to decrease. Hepatic and aortic triglycerides in groups on normal and atherogenic diets were decreased by high-dose ascorbic acid.

Björkhem and Kallner (1976) demonstrated that when guinea pigs were fed a scorbutogenic diet, the activity of the 7α-hydroxylating system of the liver microsomes was decreased. The effect of ascorbate on the 7α-hydroxylation of cholesterol was not a direct effect on the enzyme activity per se, since addition of ascorbate to the incubation mixture had no effect on the rate of 7α-hydroxylation, regardless of whether microsomal fraction from deficient guinea pigs or microsomal fraction from guinea pigs treated with ascorbate was used. The decreased cholesterol catabolism in ascorbate-depleted animals appears to be due to a more or less selective reduction of the specific type of cytochrome P-450 involved in the hydroxylation. Whether this was due to decreased synthesis or an increased rate of degradation of cytochrome P-450 was not determined. Hughes (1976) reviewed the evidence suggesting that chronic ascorbic acid deficiency leads to cholesterol accumulation in guinea pigs and in human beings. Iwamoto et al. (1976) reported that ascorbic acid lowers the absorption of bile salts from the intestine in guinea pigs, in particular, the ileal absorption of sodium taurochenodeoxycholate. Hornig and Weiser (1976) showed in guinea pigs that total bile acids were significantly reduced when the animals were deficient in vitamin C. In animals receiving an adequate ascorbic acid intake, conversion of cholesterol to bile acids could not be increased by supplying more ascorbic acid. Pavel et al. (1969) observed the frequent occurrence of bile pigment gallstones in liver-fed guinea pigs with hypovitaminosis C. This association between ascorbic acid deficiency and gallstones was later confirmed by Di Filippo and Blumenthal (1972) and by Bellmann et al. (1974). Jenkins (1977, 1978) observed that guinea pigs on a high-cholesterol diet receiving a marginal ascorbic acid intake developed cholesterol gallstones, while guinea pigs on the same diet, but receiving an adequate ascorbic acid intake did not. It seems that the formation of cholesterol gallstones results from the secretion of a cholesterol-rich and bile acid-poor bile by the liver. Ginter (1978) has observed that the incidence of gallstones in both hamsters and guinea pigs on a fat-free, high-glucose diet is reduced by ascorbic acid supplementation. He also wrote an extensive review of his own work and that of others concerning the effects of ascorbic acid on cholesterol metabolism, concluding that a defect in 7α-hydroxylation of cholesterol is the primary defect in hypovitaminosis C and that 25- and 26-hydroxylation are not affected (Figure 2). Harris et al. (1979) observed a 54% decrease in fecal bile acid excretion and a 55% decrease in the bile acid pool size of guinea pigs fed an ascorbic acid-free diet for 10 to 14 d. These changes occurred before the animals showed any loss of weight or signs of scurvy. Hanck and Weiser (1979) observed that increasing the ascorbic acid content of the diet of guinea pigs from 100 ppm to 1250 ppm caused a significant reduction in plasma total cholesterol, triglycerides, and phosphatides and an increase in lecithin cholesterol acyltransferase (LCAT) activity of the plasma. They also found a 30% decrease in the cholesterol content of the aortas of the animals receiving the higher ascorbate intake.

Further work by Jenkins (1980) demonstrated that guinea pigs on a low-ascorbic acid diet (0.2 mg/100 g body weight per day) developed a significant increase in serum cholesterol ($p < 0.01$), in hepatic bile cholesterol ($p < 0.05$), and in gall bladder bile cholesterol ($p < 0.05$) in late pregnancy. These changes were not seen in nonpregnant animals on the same low-ascorbic acid diet for the same length of time (60 d), nor in pregnant or nonpregnant guinea pigs receiving a full diet (ascorbic acid 2.0 mg/100 g/d). The bile acid concentrations in the hepatic bile and in the gall bladder bile were not altered by pregnancy while on the low-ascorbic acid diet, but were significantly increased in late pregnancy in the animals receiving the higher dose of ascorbic acid ($p < 0.05$). Thus, pregnancy in the group of guinea pigs receiving the higher dose of ascorbic acid was associated with a significantly higher bile acid concentration, while the hepatic bile of those pregnant animals receiving the lower dose of ascorbic acid contained significantly more cholesterol. Thus, pregnancy in the group

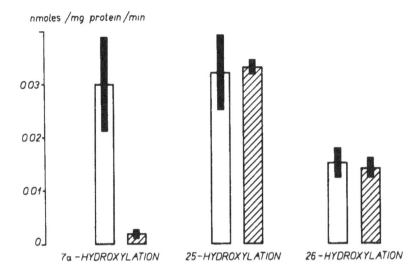

FIGURE 2   7α-Hydroxylation of endogenous microsomal cholesterol and 25- and 26-hydroxylation of 5β-cholestane-3α,7α-diol by microsomal fraction of liver homogenate from control (unshaded columns) and vitamin C-deficient (shaded columns) guinea pigs. The figure was constructed from data provided by Björkhem and Kallner (1976). (From Ginter, E. [1978], *Adv. Lipid Res.*, 16, 167 With permission.)

of guinea pigs receiving the higher dose of ascorbic acid was associated with a significant increase in the bile acid to cholesterol ratio, favoring solubilization of cholesterol. Conversely, pregnancy in the guinea pigs with the lower ascorbic acid intake caused a decrease in the bile acid to cholesterol ratio, a condition favoring cholelithiasis or gall stone formation. These findings are particularly interesting because they occurred in guinea pigs on diets which were adequate for the maintenance of maternal and fetal weight and survival.

Holloway and Rivers (1981) studied the influence of chronic ascorbic acid deficiency and excessive ascorbate consumption on bile acid metabolism, liver, and plasma cholesterol levels, hepatic microsomal cytochromes and biliary lipid composition in guinea pigs fed a cereal-based diet. They found that dietary extremes of ascorbic acid intake caused similar alterations. Relative to the control group, the deficient and excess groups exhibited reduced cytochrome P-450 concentration, lower cholesterol 7α-hydroxylase activity, lower bile acid turnover rate, prolonged bile acid half-life, and increased plasma and liver cholesterol concentrations. Similarly, Holloway et al. (1982) observed that increasing the daily ascorbic acid intake of guinea pigs on a cereal-based diet from 2 to 20 mg/100 g resulted in a 44% increase in bile acid synthesis and a 28% decrease in plasma cholesterol. A further increase of ascorbic acid intake to 65 mg/100 g resulted in a decrease of bile acid synthesis. This adverse effect of very high doses of ascorbic acid in guinea pigs on a cereal-based diet is reminiscent of the toxicity and diabetogenic effect of a high-ascorbate, high-cereal diet observed by Chatterjee et al. (1975), which was found to be associated with elevated blood dehydroascorbic acid (DHAA) levels. These findings provide further evidence that modest doses of ascorbic acid, given with a reducing agent such as cysteine, an antioxidant such as α-tocopherol, or indirect chelating antioxidants such as proteins, amino acids, and certain bioflavonoids or catechins will be preferable to the administration of high doses of ascorbic acid alone. In this respect it is pertinent that Ginter et al. (1982) have observed a synergism between vitamin C and vitamin E on microsomal hydroxylation reactions in guinea pig liver.

Relatively short-term studies of less than 12 weeks by Weight et al. (1982) compared guinea pigs fed a low-ascorbic acid diet (0.5 mg/d), after an initial 14 d of zero ascorbate intake, with guinea pigs receiving 10 mg of ascorbic acid a day. These workers observed an increase in the terminal serum cholesterol levels of the low ascorbate group. Surprisingly the bile acid secretion of the *in situ* perfused liver was higher in the latently scorbutic animals than in controls, even though cholesterol synthesis was not increased. Ginter et al. (1982) have confirmed their earlier work on hypovitaminosis C in guinea pigs and demonstrated that the increase in total serum cholesterol is conditioned by an increase of the atherogenic low-density lipoprotein (LDL) cholesterol. The concentration of high-density lipoprotein (HDL) cholesterol, on the other hand, moderately declines in deficient animals. As a result, the ratio of total cholesterol to HDL cholesterol which, according to many authors, expresses the risk of atherogenesis, increases in vitamin C-deficient animals to twice the control values ($p < 0.001$). Studies by Odumosu (1982) have shown that vitamin C controls the cholesterol-lowering and weight-reducing actions of the drugs clofibrate and diosgenin, which deplete tissue ascorbic acid levels. Dietary ascorbic acid enhances the action of these drugs in guinea pigs.

II. OTHER ANIMALS

Myasnikov (1958), Zaitsev et al. (1964), and Sokoloff et al. (1966) studying cholesterol-fed rabbits and rats, demonstrated that ascorbic acid supplements provided marked protection against hypercholesterolemia. Sokoloff et al. observed serum cholesterol levels of 88.5 mg/100 ml in a control group of rabbits, 1234 mg/100 ml in a cholesterol-fed group, and 308 mg/100 ml in a group receiving the same dose of cholesterol and ascorbic acid (150 mg of ascorbic acid per kilogram of body weight). Bellmann et al. (1974), studying rabbits on a 1% dihydrocholesterol diet, observed that an ascorbic acid supplement prevented the threefold increase in serum cholesterol that occurred in the control animals. Similar observations have been reported by Sadava et al. (1982) who observed that ascorbic acid supplements protected rabbits receiving injections of cholesterol and ergocalciferol in ethylene glycol against hypercholesterolemia. These results are truly remarkable when we consider that these animals can and do synthesize ascorbic acid from simple sugars. Even so, they are afforded protection against hypercholesterolemia by additional ascorbic acid. It certainly illustrates the need for more ascorbic acid than just enough to prevent scurvy.

Nambisan and Kurup (1974) reported a decrease in the cholesterol level of the serum, liver, and aorta of weanling rats when their normal diet was supplemented with ascorbic acid, and more so with ascorbic acid plus methionine. The decrease of the cholesterol level was greater in both groups in the aorta and in the liver than in the serum. Klevay (1976) observed a somewhat higher plasma cholesterol level (129 vs. 109 mg/100 ml) in rats when an ascorbate supplement was added to an artificial diet, but Holloway et al. (1984) reported that ascorbate supplementation had no effect upon the plasma and liver cholesterol levels or cholesterol 7α-hydroxylase activity in the rat. Kotzé et al. (1974) observed that ascorbic acid feeding decreased the cholesterol production rate in young baboons. However, in a subsequent study, Kotzé et al. (1975), studying young baboons after 3 months on a diet low in vitamin C, noted that ascorbic acid administration caused a temporary increase in their serum cholesterol levels. They concluded that the ascorbic acid mobilized cholesterol from body depots into the blood stream for subsequent degradation. Machlin et al. (1976) observed no significant difference in serum cholesterol between ascorbic acid-deficient and ascorbic acid-supplemented rhesus monkeys, but they did report that ascorbic acid caused a significant reduction of the serum triglyceride levels of deficient animals. Ginter and Mikus (1977) have reported that the formation of gallstones in hamsters can be decreased by high doses of ascorbate. John et al. (1979) have reported a significant increase in the plasma cholesterol and triglycerides and a decrease in plasma level of free fatty acids in rainbow trout when these fish are fed an ascorbic acid-deficient diet. Moreover, it is now known that many trout die of coronary artery disease. Horio et al. (1987) have reported a most interesting study of mutant Wistar rats which have lost the ability to synthesize ascorbic acid. They require dietary ascorbic acid 300 mg/kg/d to prevent signs of vitamin C deficiency and to achieve maximal growth. Ascorbic acid deficiency was found to be associated with a slight elevation of the serum cholesterol level in one experiment and no change in another. However, ascorbic acid deficiency induced the accumulation of cholesterol in the liver and reduced the ratio of HDL cholesterol to total cholesterol in the serum of these mutant rats in both experiments.

III. HUMAN STUDIES

Myasnikova (1947) seems to have been the first to show that ascorbic acid has the ability to affect the serum cholesterol levels of human beings. She found that the intravenous injection of high doses of ascorbic acid in patients with high cholesterol levels resulted in a definite decrease, while in people with low levels it caused an increase in the serum cholesterol level. Sedov (1956) also studied the effects of ascorbic acid supplements on the serum cholesterol levels of patients. He observed that a 0.5-g dose of ascorbic acid given to patients with hypercholesterolemia caused an abrupt decrease in the serum cholesterol level. Daily intravenous administration of 0.5 to 1.0 g of ascorbic acid for 10 to 30 d resulted in a significant decrease (up to 30%) in blood cholesterol in 92 of 106 atherosclerotic patients. Sokoloff et al. (1956) cited similar findings by Fedorova in 1960 and by Gandzha et al. in 1961. The latter group of workers observed that ascorbic acid, 0.5 g three times a day orally, brought the cholesterol level down by 35 to 40%. Several days after the ascorbic acid treatment was discontinued, the serum cholesterol level rose again. Myasnikov (1958) also reported a reduction of the cholesterol levels when high doses of ascorbic acid were administered to hypercholesterolemic patients. Anderson et al. (1958) found that an ascorbic acid supplement of 1 g daily for 3 weeks had no effect on the serum cholesterol levels of 24 "healthy" schizophrenic men aged 41 to 46 years. Mašek (1960), at the Institute of Nutrition in Prague, reported simultaneous blood analyses for cholesterol and vitamin C in volunteers chosen at random. It was found that low serum vitamin C levels were equally often associated with low as with high cholesterol levels. However, high vitamin C levels were more commonly associated with low cholesterol levels.

Monthly serum cholesterol determinations were carried out on 16 healthy male prisoners at the Maryland State Penitentiary and were reported by Thomas et al. (1961). They found a very marked decrease in the mean serum cholesterol level from 260 and 265 mg/100 ml, in December and January, to 215 and 217 in May and June (Figure 3). When data from all of 24 volunteers were included, this seasonal change was found to be highly significant ($p < 0.001$). The authors discussed the possibilities of seasonal changes in stress, physical activity, and fat intake, but did not mention the other possible cause of these findings, which is variation of the vitamin C content of the food which has been so widely reported, as discussed in Chapter 19, Volume I, entitled "Season".

The findings of Bronte-Stewart et al. (1963) in human scurvy stand in contrast to the many other observations, but are similar to the findings of Kotzé et al. (1975) in baboons. They found the mean serum cholesterol levels in 14 adult Bantu men with frank scurvy (99.0 mg/100 ml) to be significantly lower than in a comparable group of healthy men (140.4 mg/100 ml; $p < 0.001$). Treatment with ascorbic acid alone led to a rise in the serum cholesterol level. It was noted that feeding a high-fat diet did not cause the expected increase in the serum cholesterol level of these patients until ascorbic acid was given. Ascorbic acid administered orally caused a more marked and more rapid increase in serum cholesterol than did intramuscularly administered ascorbic acid. It would seem that ascorbic acid supplements may reduce the serum cholesterol levels when they are high and increase cholesterol levels when they are low, as originally suggested by Myasnikova in 1947. This is entirely possible

SEASONAL VARIATIONS

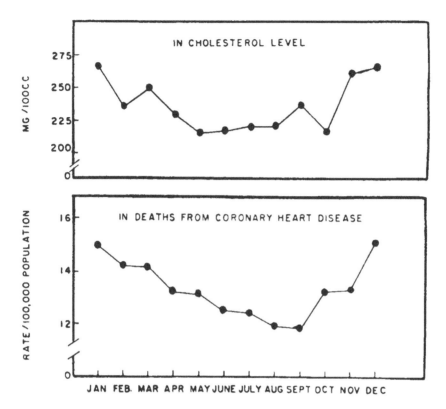

FIGURE 3. Comparison of the monthly cholesterol levels of 16 prisoners with monthly death rates from coronary heart disease in the U.S. in 1958. (From Thomas, C. B., Holljes, H. W. D., and Eisenberg, F. F. [1961], *Ann. Intern. Med.*, 54, 413. With permission.)

as, "dietary inadequacy in vitamin C is associated indirectly with a lowering of cholesterol absorption, this effect resulting from a reduction in the availability of bile acids, monoglycerides and fatty acids," as pointed out by Turley et al. (1976) in their review of this subject. We must also consider the likelihood that these Bantu men had hemosiderosis, in which case all ascorbic acid given would have entered the bloodstream as DHAA, which seems to have a deleterious effect on both carbohydrate and cholesterol metabolism.

Samuel and Shalchi (1964) administered high doses of ascorbic acid to 14 patients with elevated serum cholesterol levels, but observed a significant lowering of the cholesterol in only one. Sokoloff et al. (1966) reported that during a 5-year period, 122 patients were under observation from 4 to 30 months. They were given ascorbic acid in a dosage of 1.5 to 3.0 g daily. Blood tests (fasting state) were taken every 3 to 4 weeks. The patients were divided into two groups: "Group A included 62 patients showing normal blood fat metabolism or only moderate deviations in lipoprotein lipase (LPL) or total cholesterol. In 40 cases the serum total cholesterol triglycerides and LPL activity were relatively normal. Administration of ascorbic acid (1.5 gm daily for four to five months) had no significant effects on the fat metabolism factors in these subjects. In 22 patients aged 55 to 72, there was a moderate decrease in LPL, and the total cholesterol level was at or below 300 mg/100 ml; ascorbic acid caused a trend toward normal values in 14 of these.

"Group B included 60 cases of pronounced hypercholesterolaemia (>300 mg/100 ml) and/or cardiac disease. In some cases the triglyceride levels were 300 mg per 100 ml or

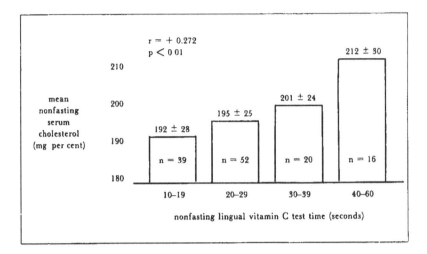

FIGURE 4   Showing the positive correlation between nonfasting serum cholesterol level and nonfasting lingual vitamin C test (decolorization times) in healthy dental students. (From Cheraskin, E. and Ringsdorf, W. M [1968], *Int. J. Vitam. Nutr. Res*, 38, 415. With permission.)

higher, with low LPL activity. In ten cases, the beneficial effect of ascorbic acid was insignificant. In the remaining 50, improvement ranged from moderate to impressive.''

Cheraskin and Ringsdorf (1968) obtained blood samples from 127 dental students at 10 a.m., approximately 3 h after breakfast. They also carried out a ''lingual vitamin C test'' using a drop of dichloroindophenol on the tongue of each subject, recording the time taken for the dye to become decolorized. In this study they observed a statistically significant positive correlation between the nonfasting serum cholesterol level and the dye decolorization time ($p < 0.01$) (Figure 4), but no significant correlation between the nonfasting plasma ascorbic acid level and the serum cholesterol. They believe that the lingual vitamin C test is more representative of tissue ascorbate levels than is plasma ascorbate. Hodges et al. (1969), in their study of experimental scurvy in man, observed that the serum cholesterol level tended to rise in some subjects, but this increase was not considered statistically significant.

Ginter et al. (1970) reviewed the literature concerning the effects of vitamin C administration on human serum cholesterol levels. Although several papers reported a reduction of the serum cholesterol levels of hypercholesterolemic subjects, they observed that these results had mostly been obtained under the simultaneous effects of other factors such as therapeutic diets or drugs. Moreover, the vitamin C status of the subjects had not been reported. Ginter et al. therefore chose to study 18 selected agricultural families in the Voderady region of southwest Slovakia, where there was known to be a highly significant negative correlation between some lipid parameters, especially the serum cholesterol level and the blood ascorbic acid content. Earlier studies had established that the average daily dietary vitamin C intake of people in the area was 54, 49, 115, and 71 mg in the four trimesters of the year, and that the blood ascorbic acid level did not change appreciably between December and March. Fifty persons were selected with relatively low blood ascorbic acid levels (about 0.6 mg/100 ml) and serum cholesterol levels at the upper limit of normal or beyond (232 to 312 mg/100 ml). They were divided into a test and a control group; the test group took 100 mg of ascorbic acid three times a day for 7 weeks from January 23 to March 12, while the control group were unaware that they were in a continuing experiment until they were approached and blood samples were drawn again at the end of the 7 weeks.

The results (Table 3, Chapter 19, Volume I) show that the mean serum cholesterol level

Table 7

MEAN AND SD OF SERUM CHOLESTEROL LEVELS IN THREE GROUPS OF VOLUNTEERS AND IN PATIENTS WITH ATHEROSCLEROSIS BEFORE AND AFTER VITAMIN C SUPPLEMENTS[a]

| Group | No. | Mean SD serum cholesterol (mg/100 ml) | | Mean change |
		Weeks 1—6 (before vitamin C)	Weeks 7—12 (on vitamin C)	
Healthy, under 25	20	194 (25.9)	177 (18.3)	− 17[b] (− 8%)
Healthy, 25—45	19	207 (32.8)	215 (28 3)	+ 8[c] (+ 4%)
Healthy, over 45	19	236 (32.2)	236 (28 1)	0
Atherosclerosis	25	242 (39 2)	261 (46.5)	+ 19[d] (+ 8%)

[a]   Effect of an ascorbic acid supplement of 1 g daily for 6 weeks. See text.
[b] $0.05 > p > 0.02$.
[c] $p > 0.1$
[d] $p < 0.001$.

From Spittle, C R. (1971), *Lancet*, 2, 1280 With permission

of the controls in December and March was 251 and 263 mg/100 ml, unchanged, while the levels of those who had received ascorbic acid had fallen significantly from 255 to 238 mg/100 ml. The hypocholesterolemic effect of vitamin C was most pronounced in persons with serum cholesterol levels above 240 mg/100 ml. In this group, the fall in the serum cholesterol was most significant, averaging 34% and being especially great in some people. They concluded that the seasonal fluctuations of serum cholesterol observed by several authors are most likely due to the different intakes of vitamin C in winter and summer.

Elwood et al. (1970) discussed the relationships between cigarette smoking, low plasma ascorbate levels, and an increased incidence of atherosclerosis. Studying 254 women in a Welsh mining valley, they did find a negative correlation between cigarette smoking and plasma ascorbic acid levels, but no significant correlation between plasma ascorbic acid and serum cholesterol levels. Hodges et al. (1971) reported a fall in the serum cholesterol levels of five prisoners during experimental ascorbic acid depletion. This could have been due to the use of a formula diet which contained no cholesterol and large amounts of polyunsaturated fats. However, there was a significant rise of the cholesterol levels during repletion with ascorbic acid, while still on the same diet.

Spittle (1971) observed that she could vary her own serum cholesterol between 140 and 230 mg/100 ml simply by varying her ascorbic acid intake. This led her to conduct a study of cholesterol and ascorbic acid levels in others. Serum cholesterol levels were estimated every week for 6 weeks in 58 healthy hospital workers and their relatives. The volunteers were then given ascorbic acid, 1 g daily, and cholesterol levels were estimated for a further 6 weeks. No dietary restrictions were imposed. Similar studies were also conducted on 25 patients with atherosclerotic heart disease. In the control group under the age of 25, cholesterol levels tended to fall after vitamin C (Table 7). There was remarkably little change in the 25-to-45 age group. Among the volunteers over the age of 45 there was no consistent pattern, but some of them showed a rise in serum cholesterol after vitamin C. In patients with atherosclerotic heart disease, there was a significant upward trend in cholesterol level. Spittle suggested that the rise in serum cholesterol noted after vitamin C in patients with atherosclerosis may be due to mobilization of arterial cholesterol. Another possibility is that these are patients who are unable to keep the majority of the ascorbic acid in the reduced form, and who might benefit from the combination of ascorbic acid with a sulfydryl amino acid or with a chelating antioxidant such as D-catechin. Morin (1972), discussing Dr. Spittle's paper, suggested caution in the interpretation of her results. Wilson and Kevany (1972), using the lingual vitamin C test as an indication of tissue ascorbate status, reported a significant positive correlation between serum cholesterol and the tongue-test time in the oldest age group. Anderson et al. (1972) observed a somewhat higher mean serum cholesterol level (193 mg/100 ml) in 18 young adults who had received ascorbic acid, 1 g daily for 14 weeks, than in 23 who had received placebo tablets for the same length of time (mean 185 mg/100 ml). Bradley et al. (1973) studied the total serum cholesterol and the whole blood ascorbic acid levels of 284 Mexican-American children aged 3 to 6 years, who were not receiving any ascorbic acid supplements. No correlation was found between the cholesterol and the ascorbate levels of these children who were not ascorbate deficient. Their mean whole blood ascorbic acid level was 1.559 ± .395 mg/100 ml, with a range of 0.3 to 3.020 mg/100 ml.

Reviewing his work on chronic ascorbic acid deficiency and atherogenesis, Ginter (1974) suggested that long-term latent vitamin C deficiency in people should be considered as a factor enhancing the risk of atherogenesis. Moreover, Krumdieck and Butterworth (1974) pointed out that, "Vitamin C seems to occupy a position of unique importance by virtue of its involvement in two systems: the maintenance of vascular integrity and the metabolism of cholesterol to bile acids." However, a small pilot study of pastoral peoples in Kenya by Davies and Newson (1974) ran counter to these opinions. These workers found a positive correlation between the levels of serum cholesterol and both plasma and leukocyte ascorbate levels. Peterson (1975) reported no significant changes in the biliary composition or serum cholesterol levels of ten healthy subjects receiving ascorbic acid supplements (1 g daily for 2 weeks). Likewise, Petersen et al. (1975), studying nine hypercholesterolemic patients with type IIA or type IIB lipoprotein electrophoretic patterns, observed no significant change in their plasma triglyceride or cholesterol levels as a result of ascorbic acid administration (4 g daily for 8 weeks). These patients had unusually high plasma ascorbic acid (TAA)* levels, mean 1.49 mg/100 ml, before receiving the supplementary ascorbic acid. Clearly, their hypercholesterolemia was due to something other than ascorbic acid deficiency.

In further studies of human subjects, Ginter (1975) reported that the administration of ascorbic acid, 1 g daily for a period of 3 months, to subjects 50 to 75 years of age with a starting concentration of plasma cholesterol below 200 mg/100 ml, had no effect on plasma cholesterol levels. On the other hand, in a group of subjects of similar age with a starting concentration of cholesterol above 200 mg/100 ml, the same dose given for 6 months brought about a very significant decline in the plasma cholesterol levels (Table 8). Kevany et al. (1975) studied 41 middle-aged men in Dublin and confirmed that the leukocyte ascorbic acid levels of men who smoked cigarettes were significantly lower than nonsmokers. They also found a significant negative correlation between leukocyte ascorbic acid levels and serum cholesterol in smokers, which was not found in nonsmokers. Coyne et al. (1976) reported diminished activity of the cytochrome P 450-dependent 7α-hydroxylase system in gallstone forming humans. This is surely very pertinent in view of the gallstones found by Jenkins and others in guinea pigs with latent hypovitaminosis C, as mentioned earlier. Biliary cholesterol is the major component of most gallstones and a high-cholesterol, low-bile acid content in the bile predisposes to gallstone formation. Indeed, Vlahcevic et al. (1970) have shown that persons with cholesterol gallstones have a markedly decreased bile acid pool size. Clearly, chronic hypovitaminosis C could be an important factor predisposing to gallstones in humans, as it is in guinea pigs. The increase in biliary cholesterol secretion and the decrease in bile acid secretion in people with suboptimal ascorbic acid levels decreases the proportion of bile acids to cholesterol, which results in the precipitation of cholesterol

* TAA — total ascorbic acid, reduced and oxidized forms.

Table 8

INFLUENCE OF ASCORBIC ACID ON PLASMA CHOLESTEROL
CONCENTRATION IN HUMANS[a]

Person examined, age, and sex	Before treatment	Intake of 1000 mg ascorbic acid daily	
		After 3 months Total cholesterol (mg/100 ml, blood plasma)	After 6 months Total cholesterol (mg/100 ml blood plasma)
N.J., 75, F	228	214	178
J M., 74, F	241	222	183
N.T., 72, F	227	200	207
K.I , 72, F	246	208	215
P.J , 71, M	206	192	159
H.J., 69, M	207	184	193
R.A., 69, F	248	192	175
S.F., 68, F	252	216	205
P.A., 67, F	327	253	303
K.M., 65, F	258	240	228
P.Z., 64, F	276	232	189
K.R., 63, F	285	258	243
S.M , 63, F	274	264	261
J.J., 60, F	248	227	179
B.R., 58, F	244	235	217
C.P., 58, F	276	213	235
M.V., 57, F	240	243	175
M.J., 57, M	213	202	157
G.M., 56, F	289	267	210
B.A., 55, F	232	210	179
M.A., 52, F	255	273	228
N.M., 52, F	213	218	186
M M., 52, M	264	227	239
S.R., 51, F	311	338	302
Mean ± SEM	253 ± 6	230 ± 7	210 ± 8
Statistical significance (comparison with starting level)		$p < 0.02$	$p < 0.001$

[a] Effect of an ascorbic acid supplement of 1 g daily for 6 months.

From Ginter, E (1975), *Ann. N.Y Acad. Sci.*, 258, 410. With permission.

and the aggregation of cholesterol crystals into gallstones. Pederson (1975) reported that large doses of vitamin C did not significantly alter the composition of the bile in healthy subjects, but this was only a 1- to 2-week trial and the subjects were not vitamin C deficient.

Kallner (1977) reported a study in which blood samples from 14 young adult male volunteers were studied before, during, and for 9 days after high-dose ascorbic acid administration (5 g daily for 4 weeks). There was a significant increase in the serum chenodeoxycholic acid level on interruption of the ascorbic acid supplementation, but the overall results were not conclusive. Similar studies of people with a seasonal deficit of vitamin C would be more informative.

Bates et al. (1977) studied plasma and leukocyte ascorbic acid levels and plasma cholesterol fractions in 23 relatively healthy elderly people (aged 72 to 86) in the north of England, every 3 months for 18 months. The levels of the HDL cholesterol showed a strong positive correlation with the plasma ascorbic acid levels in men but not in women (Figure 5). They cited similar findings by other workers in guinea pigs and monkeys. This is of special interest as HDL cholesterol levels appear to be potent risk predictors for cardiovascular disease, low levels being associated with high risk. The serum cholesterol levels of 10 healthy adults

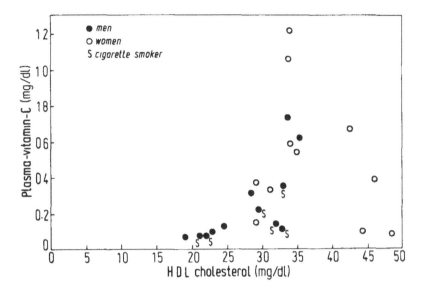

FIGURE 5.   Relation between plasma vitamin C and HDL cholesterol in elderly people.
(From Bates, C. J., Mandal, A. R., and Cole, T. J. [1977], *Lancet*, 2, 611. With permission.)

were measured by Van Steirteghem et al. (1978) for 1 month before and during the administration of ascorbic acid (3 g daily) for 18 d. Some individuals showed a definite decrease, while others showed a definite increase in their serum cholesterol levels, but the different cholesterol fractions were not studied. Subsequently, Bates et al. (1979), studying blood samples from 337 elderly men and women living at home in South Wales, found that, "both HDL cholesterol and low and very low density lipoprotein (LDL + VLDL) cholesterol levels tended to increase with increasing plasma vitamin C but this reached significance only for the LDL + VLDL fraction."

In a study of people in the Jura region of Switzerland, Hanck and Weiser (1979) found a significant negative correlation of ascorbic acid with free fatty acids, triglycerides, cholesterol and β-lipoproteins. These same authors also found a negative correlation between aortic ascorbic acid and cholesterol in post mortem studies. Severely diseased aortas had an ascorbic acid concentration half that of the less diseased ones, but a four times higher concentration of cholesterol (Figure 7). They reached the conclusion that the existing recommended daily allowances of ascorbic acid are adequate only for the prevention of scurvy and that, "far higher daily intakes of ascorbic acid are advisable to avoid undesirably high blood lipid levels and their negative long-term health implications." Moreover, Hanck (1979), studying healthy human subjects aged 25 to 40 years, observed that the administration of ascorbic acid, 4 g daily for 3 weeks, brought about a significant decrease in plasma cholesterol levels. Heine and Norden (1979) also observed that patients aged 35 to 73 years with abnormalities of lipid metabolism showed a significant decrease in plasma cholesterol levels (from 330 to 293 mg/100 ml) after administration of vitamin C, 1 g daily, for an average of 16.3 months (3 to 53 months). They reported that initial cholesterol levels less than 300 mg/100 ml returned to normal, but this could not be achieved in patients whose cholesterol levels exceeded 400 mg/100 ml.

Ginter (1979) reasserted his belief that the cholesterol-lowering effect of ascorbic acid depends on the initial serum-cholesterol level. The lower the serum-cholesterol is to start with, the lower is the decrease brought about by vitamin C. Indeed, his graphic analysis of the results of 13 human studies (Figure 6) suggests that cholesterol levels below 190 mg/100 ml will not be affected by ascorbic acid; they are already normal. So it was not surprising that Johnson and Obenshain (1981) found no change in the total cholesterol, or HDL

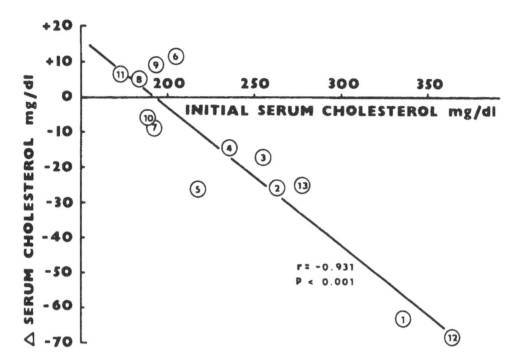

FIGURE 6. Mean change in serum cholesterol after ascorbic acid treatment plotted against initial serum cholesterol levels. Regression line obtained from mean values from 13 studies reported in the literature. For details, see original article which provides the references to all of the studies. (From Ginter, E. [1979], *Lancet*, November 3, 958. With permission.)

cholesterol, or the triglycerides of healthy young men as a result of ascorbic acid supplementation; the subjects all had normal ascorbate and lipid values before the study. Likewise, Elliott (1982) found no significant differences in the grouped means for cholesterol, HDL cholesterol, triglycerides, or bile acids as a result of ascorbate supplementation (3 g daily for 12 weeks) in subjects who were already replete with ascorbic acid (mean baseline serum ascorbic acid 1.2 mg/100 ml). However, Fidanza et al. (1982) reported that vitamin C treatment, 3 g daily for 3 weeks, reduced plasma cholesterol levels in their normocholesterolemic, as well as in their hypercholesterolemic subjects. In the first group, plasma cholesterol values decreased from 225 mg/100 ml to 194 mg/100 ml (p <0.01). Plasma triglycerides also showed a statistically significant (12.6% decrease (p <0.01). Burr et al. (1982a) found a significant positive correlation between HDL cholesterol and plasma ascorbic acid levels in 97 men. Burr et al. (1982b) made similar observations in a study of 121 healthy women ranging in age from 18 to 75 years. Greco and La Rocca (1982) observed a negative correlation between plasma ascorbic acid and cholesterol levels in elderly people, aged 70 to 84, living at the "Casa di Riposa Piccole Suore di Carita" of Naples. This correlation was significant for men (p <0.05) and for men and women (p <0.05), but not for the women alone, who had significantly higher plasma ascorbic acid levels than the men (p <0.01).

Ginter, Bobek et al. (1982), studying blood from 600 healthy blood donors, found a significant negative correlation between their leukocyte ascorbic acid levels and their serum lipids. When the donors were divided into three groups with low, medium, and high vitamin C status, respectively, the serum cholesterol and triglycerides declined as the ascorbate level rose (Figure 8). Thus, the incidence of hyperlipemia was much commoner in people with low ascorbic acid levels. In clinical trials these workers reported that the most striking cholesterol-lowering effect was achieved in elderly persons and in hypercholesterolemic

Diseased aortas	Chol (mg/g aorta)	AA (μg/g aorta)
A (more)	5.08 ± 0.39	10.51 ± 1.55
B (less)	1.26 ± 0.07	19.90 ± 1.60

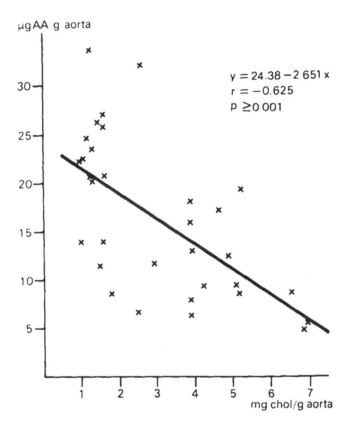

FIGURE 7.   Correlation of ascorbic acid (AA) and cholesterol (chol) concentrations in human aortas. From the Institute of Pathology, University of Basle, 1978. (From Hanck, A. and Weiser, H [1979], *Int. J. Vitam. Nutr. Res. Suppl.*, 19, 83. With permission.)

diabetics treated with ascorbic acid in doses of 500 to 1000 mg/d for a year (Figure 9). In more resistant forms of hypercholesterolemia, such as the genetic type IIA, they are now using a combination of pectin with vitamin C to bind bile acids in the lumen of the bowel, thereby reducing their reabsorption and preventing their feedback inhibition of 7α-hydroxylation of cholesterol in the liver. Cholestyramine may also be used to bind bile acids and prevent their reabsorption. However, the literature on ascorbic acid and cholesterol metabolism is still somewhat contradictory; for Hooper et al. (1983), studying healthy elderly men and women in Albuquerque, NM, found no correlation between ascorbic acid intake and lipid or lipoprotein levels.

Jacques et al. (1987) found no correlation between plasma total ascorbic acid (TAA) and total cholesterol levels, but they did find a positive correlation between plasma TAA and high-density lipoprotein cholesterol (HDL-C) in a study of 680 men and women living in the greater Boston area. The correlation coefficients were 0.13 ($p < 0.01$) for women and

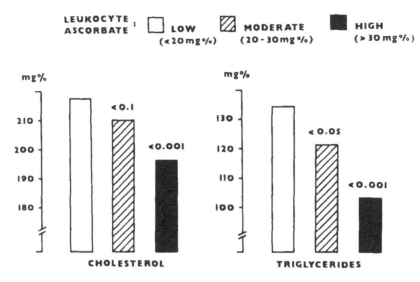

FIGURE 8   Indirect correlation between leukocyte ascorbic acid concentration and serum total cholesterol and triglycerides found in the blood of 600 blood donors. (From Ginter, E., Bobek, P , Kubec, F., Vozar, J., and Urbanová, D. [1982], *Int. J Vitam. Nutr. Res. Suppl* , 23, 137 With permission.)

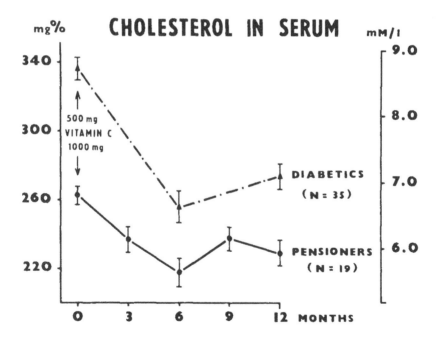

FIGURE 9.   Total serum cholesterol concentration in the serum of two hypercholesterolemic groups treated with ascorbic acid. (From Ginter, E., Bobek, P., Kubec, F., Vozar, J., and Urbanová, D. [1982], *Int. J. Vitam. Nutr. Res. Suppl.*, 23, 137. With permission.)

0.12 ($p = 0.07$) for men. Age was a strong modifier of the relationship. Partial correlation coefficients showed a statistically significant ($p < 0.05$) correlation for those aged 60 to 69 years (r = 0.14), but not for those aged 70 to 79 years, nor for those aged 80 plus. Nevertheless, these workers believe that this modest relationship could have major public health implications. They calculate that a 4- to 5-mg/dl increase in HDL-C levels could translate into a 10 to 20% decrease in the risk of coronary heart disease.

IV. SUMMARY

Ginter and Bobek (1981) summarized their own work and that of others on this subject as follows: "The activity of the cholesterol 7α-hydroxylating system containing cytochrome P-450 is depressed in the liver microsomes of guinea pigs with marginal vitamin C deficiency. Slowing-down of this rate-limiting reaction of cholesterol transformation to bile acids causes cholesterol accumulation in the liver, plasma and arteries, an increase of the index of total to HDL-cholesterol, the prolongation of plasma-cholesterol half-life, atherosclerotic changes in arteries and cholesterol-gallstone formation."

V. CONCLUSIONS

There are several different causes of cholesterol accumulation in man, but chronic borderline ascorbic acid deficiency seems to be the commonest. Indeed, the reason that hypercholesterolemia and atherosclerosis are so common in human beings may be the fact that we are all defective mammals, lacking the enzyme L-gulonolactone-oxidase which is needed for the synthesis of ascorbic acid. Other genetic defects affecting lipid metabolism are rare by comparison.

It would be nice to think that this vitamin deficiency could be avoided simply by ensuring a proper supply of fresh fruits and vegetables and by avoidance of over cooking, but it may not always be as easy as that. The underlying causes of any abnormality of ascorbic acid metabolism must also be corrected. This may involve abstinence from smoking, discontinuance of high-dose estrogenic birth control pills, and avoidance of copper pipes in the drinking and cooking water supply. Antibiotics and possibly surgery may be necessary for removal of any source of chronic infection. Also, antioxidant (vitamin E) therapy or chelation therapy may be necessary for certain individuals.

Modest doses of ascorbic acid such as 200 mg three times a day, given as catechin-coated tablets, are likely to be preferable to high doses of ascorbic acid alone for most people, both on theoretical grounds (Chapter 11, Volume I) and also because of the report by Holloway and Rivers (1981) in guinea pigs that dietary ascorbic acid deficiency and dietary ascorbic acid excess causes similar abnormalities of cholesterol metabolism.

REFERENCES

Anderson, J. T., Grande, F., and Keys, A. (1958), Dietary ascorbic acid and serum cholesterol, *Fed. Proc. Fed. Am. Soc. Exp Biol.*, (Abstr.), 17, 568.

Anderson, T. W., Reid, D. B. W., and Beaton, G. H. (1972), Vitamin C and serum cholesterol, *Lancet*, 2, 876.

Banerjee, S. and Deb, C. (1951), Effect of scurvy on cholesterol and ascorbic acid in guinea pig adrenals, *J. Biol. Chem.*, 190, 177.

Banerjee, S. and Ghosh, P. K. (1960), Metabolism of acetate in scorbutic guinea pigs, *Am. J. Physiol.*, 199, 1064.

Banerjee, S. and Singh, H. D. (1958), Cholesterol metabolism in scorbutic guinea pigs, *J. Biol. Chem.*, 233, 336.

Bates, C. J., Burr, M. K., and St. Leger, A. S. (1979), Vitamin C, high density lipoproteins and heart disease in elderly subjects, *Age Ageing*, 8, 177.

Bates, C. J., Mandal, A. R., and Cole, T. J. (1977), H. D. L. cholesterol and vitamin-C status, *Lancet*, 2, 611.

Becker, R. R., Burch, H. B., Salomon, L. L., Venkitasubramanian, T. A., and King, C. G. (1953), Ascorbic acid deficiency and cholesterol synthesis, *J. Am. Chem. Soc.*, 75, 2020.

Belavady, B. and Banerjee, S. (1954), Metabolism of cholesterol in scorbutic guinea pigs, *J. Biol. Chem.*, 209, 641.

Bellmann, H., Rauchfuss, E., Wohlgemuth, B., Schubert, S., Fuchs, K. F., Geissler, F., Haupt, R., Conradi, G., Schönlebe, W., Daniel, E., and Günther, O. (1974), Zur Pathogenese, Prophylaxe und Regression der Cholelithiasis, *Zentralbl. Inner. Med ,* 29, 997

Björkhem, I., Danielsson, H., and Einarsson, K. (1967), On the conversion of cholesterol to 5β-cholestane-3α,7α-diol in guinea pig liver homogenates, *Eur. J. Biochem.,* 2, 294.

Björkhem, I. and Kallner, A. (1976), Hepatic 7α-hydroxylation of cholesterol in ascorbate deficient and ascorbate supplemented guinea pigs, *J. Lipid Res.,* 17, 360

Bloch, K. (1944), Some aspects of the metabolism of leucine and valine, *J. Biol. Chem.,* 155, 255.

Bloch, K. and Rittenberg, D. (1942), On the utilization of acetic acid for cholesterol formation, *J. Biol. Chem.,* 145, 625.

Bloch, K. and Rittenberg, D. (1945), An estimation of acetic formation in the rat, *J. Biol. Chem.,* 159, 45

Bolker, H. I., Fishman, S., Heard, R. D. H., O'Donnell, V. J., Webb, J. L., and Willis, G. C. (1956), The incorporation of acetate-1-C[14] into cholesterol and fatty acids by surviving tissues of normal and scorbutic guinea pigs, *J. Exp. Med.,* 103, 199.

Bradley, D. W., Maynard, J. E., and Emery, G. E. (1973), Serum-cholesterol and whole-blood ascorbic acid in children, *Lancet,* 2, 201.

Bronte-Stewart, B., Roberts, B., and Wells, V. M. (1963), Serum cholesterol in vitamin C deficiency in man, *Br. J. Nutr.,* 17, 61.

Burr, M. L., Bates, C. J., Milbank, J. E., and Yarnell, J. W. G. (1982a), The relationship between plasma ascorbate and lipid concentrations in fasting men, *Hum. Nutr. Clin. Nutr.,* 36C, 135.

Burr, M. L., Bates, C. J., Sweetnam, P. M., and Barasi, M. E. (1982b), Plasma ascorbate and HDL-cholesterol in women, *Hum. Nutr. Clin. Nutr.,* 36C, 399.

Chatterjee, I. B., Majumder, A. K., Nandi, B. K., and Subramanian, N. (1975), Synthesis and some major functions of vitamin C in animals, *Ann. N.Y. Acad. Sci.,* 258, 24.

Cheraskin, E. and Ringsdorf, W. M. (1968), A lingual vitamin C test. VII. Relationship of non-fasting serum cholesterol and vitamin C state, *Int. J. Vitam. Nutr. Res.,* 38, 415.

Coyne, M. J., Bonorris, G. G., Goldstein, L. I., and Schoenfield, L. J. (1976), Effect of chenodeoxycholic acid and phenobarbital on the rate-limiting enzymes of hepatic cholesterol and bile acid synthesis in patients with gallstones, *J. Lab. Clin. Med.,* 87, 281.

Davies, J. D. and Newson, J. (1974), Ascorbic acid and cholesterol levels in pastoral peoples in Kenya, *Am. J. Clin. Nutr.,* 27, 1039.

Elliott, H. C. (1982), Effects of vitamin C loading on serum constituents in man, *Proc. Soc. Exp. Biol. Med.,* 169, 363.

Elwood, P. C., Hughes, R. E., and Hurley, R. J. (1970), Ascorbic acid and serum cholesterol, *Lancet,* 2, 1197.

Fidanza, A., Audisio, M., and Mastroiacovo, P. (1982), Vitamin C and cholesterol, *Int. J. Vitam. Nutr. Res. Suppl.,* 23, 153.

Fujinami, T., Okado, K., Senda, K., Sugimura, M., and Kishikawa, M. (1971), Experimental atherosclerosis with ascorbic acid deficiency, *Jpn. Circ. J.,* 35, 1559.

Fujinami, T., Okado, K., Senda, K., Nakano, S., Higuchi, R., Nakayama, K., Hayashi, K., and Sukuma, N. (1975), *Jpn J. Atheroscler.,* 3, 117 (cited unseen).

Ginter, E. (1973), Cholesterol: Vitamin C controls its transformation to bile acids, *Science,* 179, 702.

Ginter, E. (1974), Vitamin C in lipid metabolism and atherosclerosis, in *Vitamin C: Recent Aspects of Its Physiological and Technological Importance,* Birch, G. G. and Parker, K. J., Eds., John Wiley & Sons, New York, 179.

Ginter, E. (1975), Ascorbic acid in cholesterol and bile acid metabolism, *Ann. N.Y. Acad. Sci.,* 258, 410.

Ginter, E. (1978), Marginal vitamin C deficiency, lipid metabolism and atherogenesis, *Adv. Lipid Res.,* 16, 167.

Ginter, E. (1979), Pretreatment serum-cholesterol and response to ascorbic acid, *Lancet,* November 3, 958.

Ginter, E., Babala, J., and Červeň, J. (1969a), The effect of chronic hypovitaminosis C on the metabolism of cholesterol and atherogenesis in guinea pigs, *J. Atheroscler. Res.,* 10, 341.

Ginter, E. and Bobek, P. (1981), The influence of vitamin C on lipid metabolism, in *Vitamin C: Ascorbic Acid,* Counsell, J. N. and Hornig, D. H., Eds., Applied Science, London.

Ginter, E., Bobek, P., Babala, J., and Barbierikova, E. (1969b), The effect of ascorbic acid on the lipid metabolism of guinea pigs fed on atherogenic diet, *Cor Vasa,* 11, 65.

Ginter, E., Bobek, P., and Gerbelová, M. (1965), The influence of scorbut and prolonged low intake of vitamin C on serum lipoproteins in guinea pigs, *Nutr. Diet.,* 7, 103.

Ginter, E., Bobek, P., Kubec, F., Vozar, J., and Urbanová, D. (1982), Vitamin C in the control of hypercholesterolemia in man, *Int. J. Vitam. Nutr. Res. Suppl.,* 23, 137.

Ginter, E., Bobek, P., Zopec, Z., Ovečka, M., and Čerey, K. (1967), Metabolic disorders in guinea-pigs with chronic vitamin C hyposaturation (in English), *Z. Versuchstierkd.,* 9, 228.

Ginter, E., Červeň, J., Nemec, R., and Mikuj, L. (1971), Lowered cholesterol catabolism in guinea pigs with chronic ascorbic acid deficiency, *Am. J. Clin. Nutr.,* 24, 1238.

Ginter, E., Kajaba, I., and Nizner, O. (1970), The effect of ascorbic acid on cholesterolemia in healthy subjects with seasonal deficit of vitamin C, *Nutr. Metab.*, 12, 76.

Ginter, E., Košinova, A., Hudecová, A., and Madarić, A. (1982), Synergism between vitamins C and E: effect of microsomal hydroxylation in guinea pig liver, *Int. J. Vitam. Nutr. Res.*, 52, 55.

Ginter, E. and Mikus, L. (1977), Reduction of gallstone formation by ascorbic acid in hamsters, *Experientia*, 33, 716.

Ginter, E. and Nemec, R. (1969), Metabolism of [1-^{14}C] acetate in guinea pigs with chronic vitamin C hyposaturation, *J. Atheroscler. Res.*, 10, 273.

Ginter, E., Ondreička, R., Bobek, P., and Šimko, V. (1969c), The influence of chronic vitamin C deficiency on fatty acid composition of blood serum, liver triglycerides and cholesterol esters in guinea pigs, *J. Nutr*, 99, 261.

Greco, A. M. and La Rocca, L. (1982), Correlation between chronic hypovitaminosis in old age and plasma levels of cholesterol and triglycerides, *Int. J. Vitam. Nutr. Res. Suppl.*, 23, 129.

Guchhait, R., Guha, B. C., and Ganguli, N. C. (1963), Metabolic studies on scorbutic guinea pigs. III. Catabolism of [4-^{14}C] cholesterol *in vivo* and *in vitro*, *Biochem. J.*, 86, 193.

Hanck, A. (1979), in *Aggiornamento in Vitaminologia*, Centro Internazionale di Vitaminologia, Roma, 1979 (cited unseen).

Hanck, A. and Weiser, H. (1979), The influence of vitamin C on lipid metabolism in man and animals, *Int. J. Vitam. Nutr. Res. Suppl.*, 19, 83.

Harris, W. S., Kottke, B. A., and Subbiah, M. T. R. (1979), Bile acid metabolism in ascorbic acid deficient guinea pigs, *Am. J. Clin. Nutr.*, 32, 1837.

Heine, H. and Norden, C. (1979), Vitamin C therapy in hyperlipoproteinemia, *Int. J. Vitam. Nutr. Res Suppl.*, 2, 45.

Hodges, R. E., Baker, E. M., Hood, J., Sauberlich, H. E., and March, S. C. (1969), Experimental scurvy in man, *Am. J. Clin. Nutr.*, 22, 535.

Hodges, R. E., Hood, J., Canham, J. E., Sauberlich, H. E., and Baker, E. M. (1971), Clinical manifestations of ascorbic acid deficiency in man, *Am. J. Clin. Nutr.*, 24, 432.

Holloway, D. E., Guiry, V. C., Holloway, B. A., and Rivers, J. M. (1984), Influence of dietary ascorbic acid on cholesterol 7 alpha hydroxylase activity in the rat, *Int. J. Vitam. Nutr. Res.*, 54, 333.

Holloway, D. E., Peterson, F. J., and Rivers, J. M. (1982), Effects of dietary ascorbic acid on bile acid metabolism in guinea pigs fed Krehl or Reid-Briggs diets, *Nutr. Rep. Int.*, 25, 941.

Holloway, D. E. and Rivers, J. M. (1981), Influence of chronic ascorbic acid deficiency and excessive ascorbic acid intake on bile acid metabolism and bile composition in the guinea pig, *J. Nutr.*, 111, 412.

Hooper, P. L., Hooper, E. M., Hunt, W. C., Garry, P. J., and Goodwin, J. S (1983), Vitamins, lipids and lipoproteins in a healthy elderly population, *Int. J. Vitam. Nutr. Res.*, 53, 412

Horio, F., Ozaki, K., Oda, H., Makiro, S., Hayashi, Y., and Yoshida, A. (1987), Effect of dietary ascorbic acid, cholesterol and PCB on cholesterol concentrations in serum and liver in a rat mutant unable to synthesize ascorbic acid, *J. Nutr.*, 117, 1036.

Hornig, D. and Weiser, H. (1976), Ascorbic acid and cholesterol: effect of graded oral intakes on cholesterol conversion to bile acids in guinea-pigs, *Experientia*, 32, 687.

Hughes, R. E. (1976), Vitamin C and cholesterol metabolism, *J. Hum. Nutr.*, 30, 315.

Iwamoto, K., Ozawa, N., Ito, F., Okamoto, N., and Watanabe, J. (1976), Effect of ascorbic acid on the intestinal absorption of bile salts and metabolism of cholesterol in guinea pigs, *Chem. Pharm. Bull.*, 24, 2014.

Jacques, P. F., Hartz, S. C., McGandy, R. B., Jacob, R. A., and Russell, R. M. (1987), Vitamin C and blood lipoproteins in an elderly population, *Ann. N.Y. Acad. Sci.*, 498, 100.

Jenkins, S. A. (1977), Vitamin C and gallstone formation: a preliminary report, *Experientia*, 33, 1616.

Jenkins, S. A. (1978), Biliary lipids, bile acids and gallstone formation in hypovitaminotic C guinea-pigs, *Br. J. Nutr.*, 40, 317.

Jenkins, S. A. (1980), Vitamin C status, serum cholesterol levels and bile composition in the pregnant guinea pig, *Br J. Nutr.*, 43, 95.

John, T. M., George, J. C., Hilton, J. W., and Slinger, S. J. (1979), Influence of dietary ascorbic acid on plasma lipid levels in rainbow trout, *Int. J. Vitam. Nutr. Res.*, 49, 400.

Johnson, G. E. and Obenshain, S. S. (1981), Nonresponsiveness of serum high-density lipoprotein-cholesterol to high dose ascorbic acid administration in normal men, *Am. J. Clin. Nutr.*, 34, 2088.

Kallner, A. (1977), Serum bile acids in man during vitamin C supplementation and restriction, *Acta Med. Scand.*, 202, 283.

Kawishwar, W. K., Chakrapani, B., and Banerjee, S. (1963), Carbohydrate and lipid metabolism in scurvy: effect of vitamin C supplement, *Indian J. Med. Res.*, 51, 488.

Kevany, J., Jessop, W., and Goldsmith, A. (1975), The effect of smoking on ascorbic acid and serum cholesterol in adult males, *Ir. J. Med. Sci.*, 144, 474.

Klevay, L. M. (1976), Hypercholesterolemia due to ascorbic acid, *Proc. Soc. Exp. Biol. Med.*, 151, 579.

Kotzé, J. P. Weight, M. J., DeKlerk, W. A., Menne, I. V., and Weight, M. J. A. (1974), Effect of ascorbic acid on serum cholesterol levels and on die-away curves of ¹⁴C-4-cholesterol in baboons, *S. Afr. Med J.*, 48, 1182.

Kotzé, J. P., Menne, I. V., Spies, J. H., and De Klerk, W. A. (1975), Effect of ascorbic acid on serum lipid levels and depot cholesterol of the baboon (Papio ursinus), *S. Afr. Med J.*, 49, 906.

Kritchevsky, D., Tepper, S. A., and Story, J. A. (1973), Influence of vitamin C on the hydroxylation and side chain oxidation of cholesterol in vitro, *Lipids*, 8, 482.

Krumdieck, C. and Butterworth, C. E. (1974), Ascorbate-cholesterol-lecithin interactions: factors of potential importance in the pathogenesis of atherosclerosis, *Am. J. Clin. Nutr.*, 27, 866.

Little, H. N. and Bloch, K. (1950), Studies on the utilization of acetic acid for the biological synthesis of cholesterol, *J. Biol. Chem.*, 183, 33.

Machlin, L. J., Garcia, F., Kuenzig, W., Richter, C. B., Spiegel, H. E., and Brin, M. (1976), Lack of antiscorbutic activity of ascorbate 2-sulfate in the rhesus monkey, *Am. J. Clin. Nutr.*, 29, 825.

Mašek, J. (1960), Étude de la cholestérolemie dans différents groupes de population, *J. Nutr. Diet.*, 2, 193.

Morin, R. J. (1972), Arterial cholesterol and vitamin C, *Lancet*, 1, 594.

Mjasnikova, I. A (1947), O vlijanii vodorastvorimych vitaminov na nekotoryje storony obmena veščestv, *Tr. Vojenno-Morskoj Med. Akad (Leningrad)*, 8, 140 (cited unseen).

Myasnikov, A. L. (1958), Influence of some factors on development of experimental cholesterol atherosclerosis, *Circulation*, 17, 99.

Nambisan, B. and Kurup, P. A. (1974), Effect of massive doses of ascorbic acid and methionine on the levels of glycosaminoglycans in the aorta of weanling rats, *Atherosclerosis*, 19, 191.

Nambisan, B. and Kurup, P. A. (1975), Ascorbic acid and glycosaminoglycan and lipid metabolism in guinea pigs fed normal and atherogenic diets, *Atherosclerosis*, 22, 447.

Odumosu, A. (1982), How vitamin C, clofibrate and diosgenin control cholesterol metabolism in male guinea pigs, *Int. J. Vitam. Nutr. Res. Suppl.*, 23, 187.

Oesterling, M. J. and Long, C. N. H. (1951), Adrenal cholesterol in the scorbutic guinea pig, *Science*, 113, 241.

Pavel, I., Chisiu, N., and Strobici, D. (1969), La lithiase biliaire chez le cobaye avec dysnutrition scorbutique, *Nutr. Diet.*, 11, 60.

Pedersen, L. (1975), Biliary lipids during vitamin C feeding in healthy persons, *Scand. J. Gastroenterol.*, 10, 311.

Peterson, V. E., Crapo, P. A., Weininger, J., Ginsberg, H., and Olefsky, J. (1975), Quantification of plasma cholesterol and triglyceride levels in hypercholesterolemic subjects receiving ascorbic acid supplements, *Am. J. Clin. Nutr.*, 28, 584.

Sadava, D., Watumull, D., Sanders, K., and Downey, K. (1982), The effect of vitamin C on the rapid induction of aortic changes in rabbits, *J. Nutr. Sci. Vitaminol.*, 28, 85.

Samuel, P. and Shalchi, O. B. (1964), Effect of vitamin C on serum cholesterol in patients with hypercholesterolemia and arteriosclerosis, *Circulation*, 29, 24.

Sedov, K. R. (1956), Prevention and therapy of atherosclerosis with ascorbic acid, *Ter. Arkh.*, 28, 58.

Sokoloff, B., Hori, M., Saelhof, C. C., and Wrzolek, T. (1966), Aging, atherosclerosis and ascorbic acid metabolism, *J. Am. Geriatr. Soc.*, 14, 1239.

Spittle, C. R. (1971), Atherosclerosis and vitamin C, *Lancet*, 2, 1280.

Sulkin, N. M. and Sulkin, D. F. (1975), Tissue changes induced by marginal vitamin C deficiency, *Ann. N.Y. Acad. Sci.*, 258, 317.

Thomas, C. B., Holljes, H. W. D., and Eisenberg, F. F. (1961), Observations on the seasonal variations in the total serum cholesterol level among healthy young prisoners, *Ann. Intern. Med.*, 54, 413.

Turley, S. D., West, C. E., and Horton, B. J. (1976), The role of ascorbic acid in the regulation of cholesterol metabolism and in the pathogenesis of atherosclerosis, *Atherosclerosis*, 24, 1.

Van Steirteghem, A. C., Robertson, E. A., and Young, D. S. (1978), Influence of large doses of ascorbic acid on laboratory test results, *Clin. Chem.*, 24, 54.

Vlahcevic, Z. R., Bell, C. C., Buhac, I., Farrar, J. T., and Swell, L. (1970), Diminished bile acid pool size in patients with gallstones, *Gastroenterology*, 59, 165.

Weight, M. J., Compton-James, K., Van Schalkwyk, D. J., and Seier, J. V. (1982), Cholesterol-ascorbic acid interactions in guinea pig liver homogenates, and in the in situ perfused liver: relationship between ascorbic acid status and cholesterol and bile acid synthesis from mevalonate, *Int. J. Vitam. Nutr. Res.*, 52, 298.

Willis, G. C. (1953), An experimental study of the intimal ground substance in atherosclerosis, *Can. Med. Assoc. J.*, 69, 17

Wilson, C. M. W. and Kevany, J. P. (1972), Screening for vitamin C status, *Br. J. Prev. Soc. Med.*, 26, 53.

Zaitsev, V. F., Myasnikov, L. A., and Scheikman, M. B. (1964), Effect of ascorbic acid on the distribution of cholesterol-4-C¹⁴ in tissues of animals with experimental atherosclerosis, *Kardiologiya*, 4, 30

Chapter 6

TYROSINE AND PHENYLALANINE METABOLISM

I. INTRODUCTION

Phenylalanine is an essential amino acid, but tyrosine is not; *l*-phenylalanine is readily converted to *l*-tyrosine by phenylalanine hydroxylase in the livers of normal individuals, but this reaction is not reversible. Four principal metabolic pathways for the metabolism of tyrosine are shown in Figure 1. Three of these are anabolic, leading to the production of thyroxine, epinephrine, and melanin, while the other is catabolic, leading to acetoacetic acid. Ascorbic acid seems to be involved to some extent in each of these four pathways; they are treated separately below.

II. HYDROXYPHENYLURIA

Evidence of a relationship between tyrosine metabolites and dietary ascorbic acid intake was provided by Levine et al. (1939). They found parahydroxyphenyllactic (*p*HPL) and parahydroxyphenylpyruvic acid (*p*HPP), the deaminated residues of tyrosine, in the urine of premature infants fed on cow's milk and observed that these compounds disappeared from the urine when ascorbic acid was added to the diet. In the same year, Sealock and Silberstein (1939) reported that alcaptonuria could be produced by feeding tyrosine to vitamin C-deficient guinea pigs. Provision of ascorbic acid, 5 mg daily, as a dietary supplement prevented the appearance of all but traces of homogentisic acid in the urine. These authors also reported similar findings when feeding tyrosine to two normal human subjects on a vitamin C-deficient diet.

In further studies of guinea pigs Sealock et al. (1940) and Sealock and Silberstein (1940) observed that ingested tyrosine and phenylalanine do not undergo complete metabolism in scorbutic guinea pigs and that not only homogentisic acid, but also *p*HPL and *p*HPP appear in the urine. The specificity of *l*-ascorbic acid in preventing this abnormality of tyrosine metabolism was demonstrated by the observation that 10 mg of *d*-isoascorbic acid was ineffective in replacing a similar weight of *l*-ascorbic acid; 200 mg of *d*-isoascorbic acid was required to give the same effect as 10 mg of *l*-ascorbic acid. Studies by Sealock et al. (1940) showed that vitamin C will not suppress the urinary excretion of homogentisic acid in hereditary alcaptonuria. Nevertheless, ascorbic acid may be of use in reducing the degree of ochronosis in patients with alkaptonuria, as Lustberg et al. (1969) have shown that ascorbic acid supplementation reduces the binding of oxidized and/or polymerized homogentisic acid, both to cartilage and to collagen in *l*-tyrosine-fed rats ($p < 0.001$). Ascorbic acid does not suppress the excretion of phenylpyruvic acid in phenylpyruvic oligophrenia.

Levine et al. (1941) conducted further investigations on the defect in tyrosine and phenylalanine metabolism, which appeared as early as the sixth day of life in ascorbic acid-deficient premature infants fed on cow milk. Their plasma ascorbic acid (AA)* levels were found to range from 0.0 to 0.1 mg/100 ml; they excreted *p*HPL and *p*HPP acids in the urine. Levine et al. (1941) observed that this was completely eradicated by feeding vitamin C. Other vitamins which they administered in large doses, either singly or in combination, were without effect: thiamine, nicotinic acid, riboflavin, pyridoxine, biotin, pantothenic acid, choline chloride, yeast powder, rice polishings, α-tocopherol, and adrenal cortical extract were ineffective. Whole-liver extract did completely eradicate tyrosine-induced phenyluria

---

* AA — ascorbic acid, reduced form.

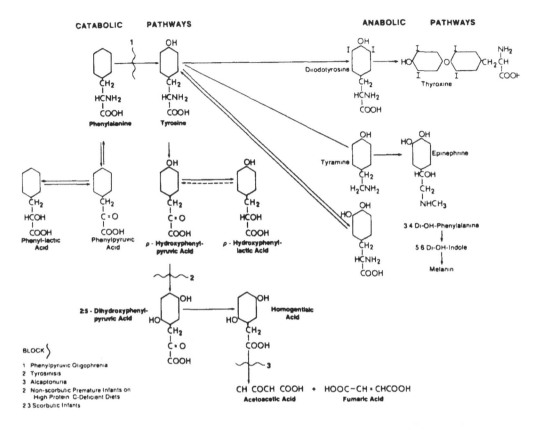

FIGURE 1   Metabolism of phenylalanine and tyrosine (modified from Peters and Van Slyke [1946]). (From Nitowsky, H. M., Govan, C. D., Jr., and Gordon, H. H. [1953], *AMA Am. J. Dis. Child.*, 85, 462. © 1953 American Medical Association. With permission.)

in one mature infant, but was ineffective in three premature infants. In one infant, 70 mg of ascorbic acid daily failed to prevent a rise in excretion following the administration of the aromatic amino acid, but subsequent administration of 80 mg of ascorbic acid daily proved effective in a later test. It is interesting to note that the administration of phenylalanine resulted in an increased output, not only of hydroxyphenyl derivatives, but also of phenylpyruvic acid. However, the administration of tyrosine increased the output of phenolic derivatives without appreciably affecting the output of phenylpyruvic acid. This observation is in accord with expectation, as the oxidation of phenylalanine to tyrosine is an irreversible process.

The ingestion by these infants of other amino acids, including glycine, methionine, and the aromatic amino acid tryptophan, in comparable or even larger dosage, failed to provoke the appearance or raise the urinary excretion of these intermediary products.

Full-term infants fed similar diets showed no spontaneous defect in their metabolism of aromatic amino acids, but the defect was precipitated in one mature infant by the ingestion of a single dose of 1.0 g/kg of phenylalanine and by a similar dose of tyrosine in another. The artificially induced defect in these subjects was readily abolished by the administration of *l*-ascorbic acid in one infant and by parenteral administration of whole-liver extract in the other. This subject was reviewed by Sealock (1942) who suggested that the point of action of the vitamin lay between the phenylalanine molecule and the keto acid of tyrosine. The defect in the metabolism of tyrosine and phenylalanine observed by Levine et al. in premature infants differed from that produced by Sealock and Silberstein in guinea pigs and in human adults in that more *p*HPL than *p*HPP acid was excreted and in that a significant amount of homogentisic acid was not found in the urine of the infants.

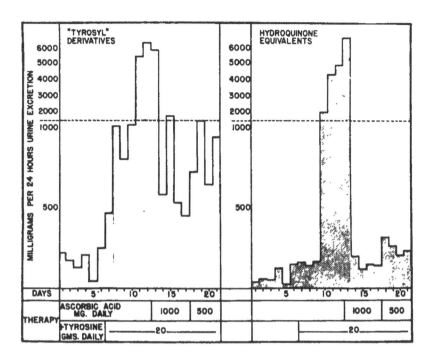

FIGURE 2.   A 62-year-old man was admitted to hospital with classical signs of scurvy. There was no ascorbic acid detectable in the leukocyte-platelet layer of his blood. The hydroxyphenyl (tyrosyl) compounds and the phosphomolybdic reducing power (hydroquinone equivalents) of his urine increased markedly when he was given tyrosine, 20 g daily orally, and were markedly reduced when he received ascorbic acid as well as tyrosine. (From Rogers, W. F. and Gardner, F. H [1949], *J. Lab. Clin. Med.*, 34, 1491. With permission.)

Studies by Lan and Sealock (1944) demonstrated the *in vitro* oxidation of *l*-tyrosine by liver slices from normal guinea pigs and the inability of the same tissue from vitamin C-deficient animals to exhibit similar metabolic activity. Painter and Zilva (1947), feeding large doses of tyrosine to guinea pigs, found excessive urinary excretion of parahydroxyphenyl compounds and parahydroxypyruvic acid soon after removal of dietary ascorbic acid, reaching a maximum in 4 to 5 d, but no homogentisic acid was found. Rodney et al. (1947) reported that liver suspensions from pteroylglutamic acid (folic acid)-deficient rats were better able to oxidize tyrosine after the addition of folic acid. These findings, and the one infant of Levine et al. whose hydroxyphenyluria responded to whole-liver extract, prompted further studies of the relationship between folic acid and hydroxyphenyluria by Woodruff and Darby (1948, 1949) and by Woodruff et al. (1949). These workers demonstrated that large doses of folic acid caused a marked reduction in both tyrosyl and keto acid excretion by tyrosine-fed scorbutic guinea pigs. Folic acid was not, however, quite as effective as ascorbic acid in this regard and did not protect the animals from scurvy. Sealock and Lepow (1948) showed that injection of liver extracts improved, but did not correct the abnormal tyrosine metabolism of scorbutic guinea pigs. Moreover, Johnson and Dana (1948) reported that the administration of ascorbic acid to folic acid-deficient rats relieved their hemorrhagic syndrome, but only delayed their death.

Studies of four patients admitted to hospital with scurvy and of three normal subjects were reported by Rogers and Gardner (1949). These workers found that the quantities of hydroxyphenyl compounds in the urines of the scorbutic patients were markedly increased by the administration of tyrosine and were promptly reduced by the administration of ascorbic acid (Figure 2). When not receiving tyrosine, the excretion of parahydroxyphenyl compounds by scorbutic patients did not differ appreciably from that of normal subjects. The metabolic abnormality was noted only when they were given exogenous tyrosine. These observations were similar to those of earlier workers studying guinea pigs and premature infants, but differed from the findings of Sealock and Silberstein in that homogentisic acid was not found.

Studying normal premature infants fed on cow's milk, Govan and Gordon (1949) observed that folic acid administration sometimes caused a diminution in the tyrosyluria induced by feeding cow's milk. Woolf and Edmunds (1950) observed hydroxyphenyluria in tyrosine- or phenylalanine-fed premature infants and confirmed that it was markedly reduced by ascorbic acid supplements. Morris et al. (1950) studied six infants, aged 5 to 10 months, with either scurvy or prescurvy, exhibiting hydroxyphenyluria following tyrosine ingestion. Ascorbic acid caused a marked decrease in the urinary excretion of tyrosyl compounds. Folic acid administration failed to influence established tyrosyluria, but administration of massive doses of folic acid (45 mg intramuscularly daily) during tyrosine administration prevented the development of *p*-hydroxyphenyluria when tested in one scorbutic infant. Woodruff (1950) studied four scorbutic infants aged 7 to 11 months and observed tyrosine-induced hydroxy-phenyluria which responded to ascorbic acid, but not to daily injections of 2 mg of folic acid per kilogram of body weight. Salmon and May (1950) studied the urinary excretion of *p*-hydroxyphenyl compounds and keto acids by rhesus and cynamologus monkeys with megaloblastic anemia due to prolonged feeding of ascorbic acid-deficient milk diets. Neither the *p*-hydroxyphenyl compounds nor the keto acid excretion of the scorbutic animals differed from normal until a supplement of l-tyrosine was added to the diet. After tyrosine loading, the excretion of *p*-hydroxyphenyl compounds and keto acids by the scorbutic animals increased. However, this increased excretion of deaminated tyrosine metabolites did not occur in ascorbic acid-supplemented monkeys. Folic acid had no effect on the amount of these abnormal metabolites excreted by scorbutic monkeys. Moreover, ascorbic acid was effective in relieving the tyrosyluria of scurvy, even when the action of folic acid had been blocked by aminopterin.

Rienits (1950) followed the oxidation of *l*-tyrosine in guinea pig liver slices by measurement of oxygen consumption and by the disappearance of hydroxyphenyl groups. Both methods showed that the oxidation of this amino acid in liver slices was ascorbic acid dependent. The oxidation of tyrosine by liver slices from scorbutic guinea pigs was markedly retarded; it could be restored to normal by the addition of *l*-ascorbic acid, but not by isoascorbic acid nor by glucoascorbic acid. However, folic acid gave results similar to *l*-ascorbic acid. When studying liver homogenates instead of slices, Rienits found no difference in the rate of metabolism of tyrosine between those from scorbutic and normal animals. This was probably because the ascorbic acid in the liver was destroyed by hemolysis during the homogenizing process. Addition of *l*-ascorbic acid stimulated tyrosine oxidation in supernatant preparations from liver homogenates. Painter and Zilva (1950) observed that the rate of disappearance of the phenolic group of *l*-tyrosine in suspensions of liver from scorbutic guinea pigs was slower than in those from normal animals, but could be accelerated by addition of ascorbic acid. Contrary to expectation these workers reported that D-glucoascorbic acid accelerated the degradation of tyrosine *in vitro* to the same extent as did *l*-ascorbic acid. Painter and Zilva took the view "that this interaction between *l*-ascorbic acid and *l*-tyrosine is a physiological response to an unusual situation and has no bearing on normal nutrition." This may be true in adults, but hydroxyphenyluria does occur spontaneously in premature infants fed on cow's milk.

In the original work of Sealock and Silberstein it was implied that the feeding of tyrosine resulted in a depletion or excess utilization of the ascorbic acid of the body. This phenomenon was also evident in the scorbutic patients of Rogers and Gardner, for the capillary fragility and other signs of scurvy seemed to worsen in their patients after the administration of tyrosine. However, there was no significant decrease in the ascorbic acid content of the whole blood when these same individuals were given ascorbic acid before and during the administration of tyrosine.

Steel et al. (1952) reported a study of ten men and women volunteers who lived on a diet containing less than 7 mg of ascorbic acid daily for 11 weeks. Their mean serum ascorbic acid level fell from 1.1 to 0.2 mg/100 ml, and their white cell ascorbate level had fallen to 11.5 mg/100 g when they were studied after 73 to 78 d on the diet, but the hydroxyphenyl compounds in their urine, while they were receiving tyrosine, 5 g twice a day orally, showed hardly any change (from 420 to 465 mg/24 h) as a result of the ascorbic acid deficiency. Clearly the abnormality of tyrosine metabolism occurs only in frank scurvy and not in the prescorbutic state.

The first step in the catabolism of tyrosine in mammalian liver seems to be a transamination involving deamination of tyrosine and requiring α-ketoglutarate as an amino group acceptor. Following this, there is a series of oxidative steps whereby *p*HPP may be broken down to acetoacetic acid. Schepartz and Nadel (1952) confirmed the need for ascorbic acid in the catabolism of tyrosine by guinea pig liver extracts *in vitro*. The work of Sealock et al. (1952) suggested that ascorbic acid, by virtue of its enediol linkage, acts as a specific coenzyme for the oxidation of tyrosine. La Du (1952) suggested that ascorbic acid stimulated a first oxidative step, the transformation of *p*HPP acid to 2,5-dihydroxyphenylpyruvic acid; he presented evidence that copper was involved in this pathway.

When Nichol and Welch (1950) demonstrated that ascorbic acid enhanced the conversion of folic acid to folinic acid, it was hoped that the role of folates in tyrosine metabolism would be clarified; for folinic acid is the active form of the vitamin as regards hematopoiesis. However, Darby et al. (1953) found folinic acid (citrovorum factor) to be no more effective than folic acid in abolishing the tyrosyluria of scorbutic guinea pigs. Moreover, the doses of folates comparable to those required to eliminate hydroxyphenyluria in infants caused signs of renal toxicity in guinea pigs. Nitowsky et al. (1953) tested the effects of various hemopoietic and other factors in abolition of hydroxyphenyluria in 35 premature infants fed high-protein, vitamin C-deficient diets. Folic acid was effective 5 times in 20 trials, folinic acid, 1 time in 7, citrovorum factor, 0 in 6, vitamin B_{12}, 2 in 8, and liver extract, 2 in 10. Dimercaprol was effective one in eight trials, methylene blue and α-tocopherol were ineffective twice each. Corticotrophin was effective in each of two, and cortisone ineffective in each of three trials. Ascorbic acid was always effective in completely abolishing the hydroxyphenyluria. These authors suggested that ascorbic acid is normally used in the metabolism of tyrosine, but that when ascorbic acid is absent from the diet of premature infants, another metabolic pathway may be called upon for the degradation of the deaminated residues of tyrosine and phenylalanine. This hypothetical alternative may represent a "chain reaction" requiring for completion a factor or series of factors, of which the presence of all or the absence of any one may be responsible, respectively, for the positive or negative results obtained when the various hemopoietic factors were tested.

Swendseid et al. (1943, 1947) and Abbott and James (1950) observed hydroxyphenyluria in patients with pernicious anemia. It failed to respond to ascorbic acid, but was promptly suppressed by the use of liver extract and vitamin B_{12}. These findings certainly indicated an action of these agents in the utilization of the deaminated residues of tyrosine which is independent of ascorbic acid.

Studying the enzymatic oxidation of tyrosine by liver extracts *in vitro*, La Du and Greenberg (1953) found D-isoascorbic acid, D-ascorbic acid, and hydroquinone, i.e., dehydroascorbic acid and quinone, to be just as effective as L-ascorbic acid in potentiating the oxidation. These authors suggested that the role of ascorbic acid in the enzymatic oxidation of tyrosine to acetoacetic acid is not as specific as was thought, and that the substance required as a cofactor by the enzyme is not ascorbic acid per se, but a compound having the appropriate oxidation-reduction potential. It was suggested that such compounds may completely replace ascorbic acid or they may protect small amounts of ascorbic acid present in the liver powder used.

Using paper chromatography Aterman et al. (1953) identified *p*-hydroxyphenylacetic acid (*p*HPA) and *p*HPL as abnormal components of the strongly acidic fraction of urine from scorbutic guinea pigs. Feeding tyrosine increased the excretion of these compounds, but they promptly disappeared from the urine following the administration of ascorbic acid. These workers also found *p*HPA in the urines of 3 out of 23 normal human subjects, and observed that it disappeared after the administration of ascorbic acid. Moreover, studying ether extracts of acid-hydrolyzed urine specimens, Boscott and Cooke (1954) demonstrated *p*HPA excretion by 6 patients with steatorrhea and megaloblastic anemia, by 5 patients with megaloblastic anemia of pregnancy, by 5 patients with thyrotoxicosis, 5 out of 6 patients with rheumatoid arthritis, and by 16 out of 18 patients with liver disease. The *p*HPA disappeared from the urine of most of these patients after the administration of relatively small amounts of ascorbic acid, such as 100 mg daily for 3 or 4 d, but the urinary *p*HPA of patients with steatorrhea was slow to respond to ascorbic acid therapy, even when the ascorbic acid was given by injection. Oral doses totaling 13.5 to 20 g were administered over periods varying from 23 to 40 d before significant amounts of the vitamin were excreted in the urine. Folic acid, vitamin B_{12}, and ferrous sulfate were also used in the treatment of the megaloblastic anemia in these patients.

Kretchmer et al. (1956) obtained samples of liver at autopsy, as soon as possible after death, from premature and mature newborn infants and from adults. They found that the tyrosine-oxidizing activity of human adult liver was 10 to 30 times greater and, of the mature infant, 3 to 5 times greater than that of the premature infant. The addition of large amounts of ascorbic acid did not stimulate the tyrosine-oxidizing activity of the liver from premature infants. These authors concluded that the premature infant in the first few days of life lacks tyrosine transaminase apoenzyme. They theorized that there may also be a lack of *p*HPP oxidase apoenzyme in premature infants, and that this deficiency, persisting after the transaminase activity has matured, may account for the hydroxyphenyluria. They suggested that extrinsic ascorbic acid is necessary to activate the increased *p*HPP oxidase apoenzyme.

Crandon et al. (1958) reported that ascorbic acid-deficient patients with the greatest inflammatory processes showed the greatest increase in tyrosyl excretion after test dosing with tyrosine. A return to normal levels of excretion was found in each case retested after ascorbic acid therapy. Guinea pig studies led Zannoni and La Du (1960) to conclude that, "the ability of ascorbic acid to maintain normal tyrosine metabolism *in vivo* appears to be mediated by its capacity to protect *p*-hydroxyphenylpyruvic acid oxidase from inhibition by its substrate."

Summarizing their own work and that of others in the preceding decade, La Du and Zannoni (1961) illustrated the pathway of tyrosine oxidation, as in Figure 3. The first step is a transamination with α-ketoglutarate as the amine acceptor. Pyridoxal phosphate is the coenzyme for the transaminase, and neither pyruvate nor oxaloacetate can replace α-keto-glutarate. The next enzyme, *p*HPP oxidase, is the one with which ascorbic acid participates. The enzyme catalyzes the oxidation of *p*HPP to homogentisic acid; this step requires two atoms of oxygen and liberates one molecule of carbon dioxide. The reaction is complicated since it involves hydroxylation of the aromatic ring, migration of the side chain, and an oxidative decarboxylation of the pyruvate to an acetate side chain. They accepted convincing evidence that 2,5-dihydroxyphenylpyruvic acid is not an intermediate in this reaction. They cited evidence that ferrous iron is required for homogentisic acid oxidase activity and that ascorbic acid plays a role, but only as a nonspecific reducing agent to keep the iron in the ferrous form *in vitro*. They concluded by reiterating the belief that ascorbic acid has the ability to protect an enzyme *p*HPP oxidase from inhibition by its own substrate when there is an exceptionally high tyrosine load. Ascorbic acid does not seem to be necessary to

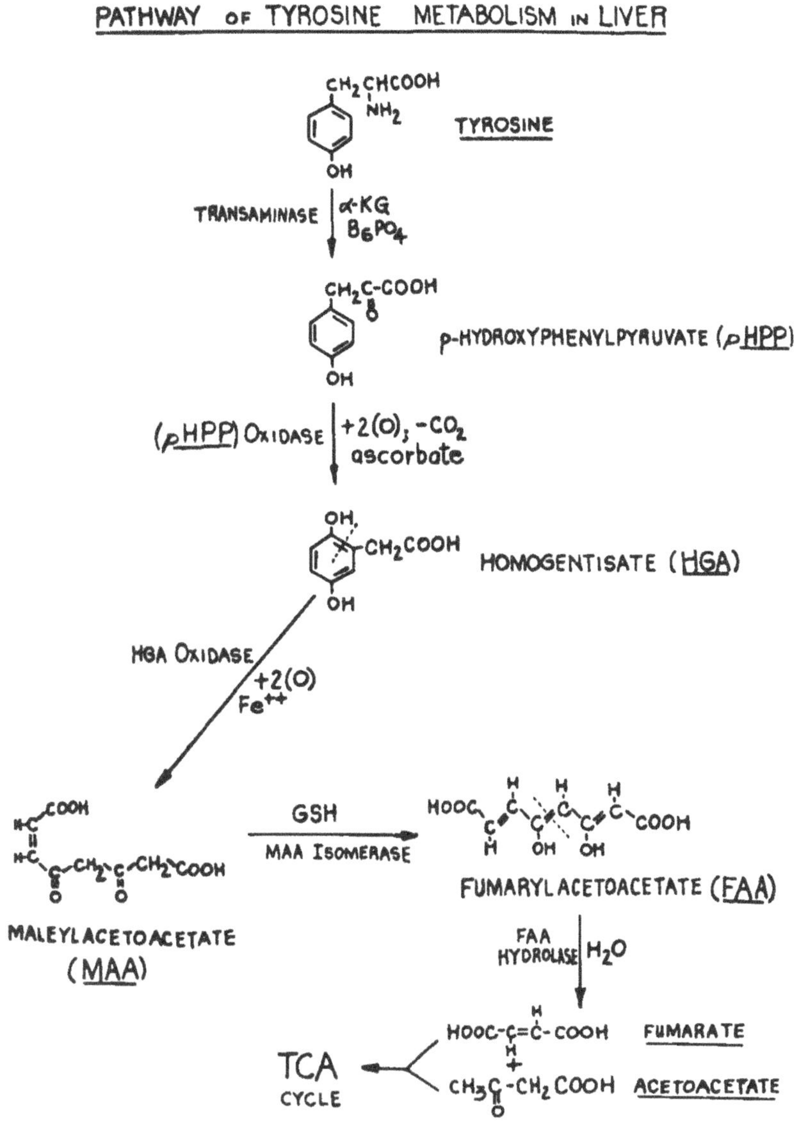

PATHWAY of TYROSINE METABOLISM in LIVER

FIGURE 3.   Pathway of tyrosine metabolism in mammalian liver. The following abbreviations are used: B_6PO_4, pyridoxal phasphate; α-KG, α-ketoglutarate; GSH, glutathione. (From La Du, B. N. and Zannoni, V. G. [1961], *Ann. N.Y. Acad. Sci.*, 92, 175. With permission.)

maintain normal tyrosine metabolism under normal dietary conditions. Folic acid was found to act only in very high concentrations and seemed to have no direct effect on tyrosine metabolism.

Knox and Goswami (1961) clearly demonstrated in guinea pigs that the metabolic defect produced by tyrosine dosage is actually an enzymic imbalance, consisting of an adaptive increase in the activity of tyrosine transaminase, which forms pHPP and a reversible inactivation of pHPP oxidase that removes pHPP (Figure 4). Therefore, pHPP accumulates and is excreted in the urine unless sufficient ascorbic acid is available to prevent the inactivation of pHPP oxidase. The greater propensity of premature infants to hydroxyphenyluria could be due to pHPP oxidase deficiency because of immaturity of the liver, but in adults it seems to be a case of enzyme inactivation by substrate and usually occurs only when there is a gross overload of tyrosine or phenylalanine.

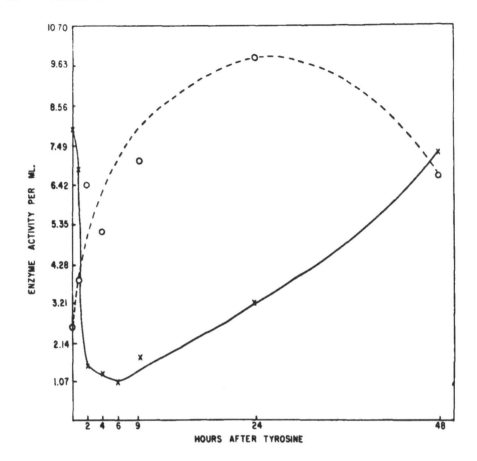

FIGURE 4   Time sequence of *in vivo* activity changes of liver tyrosine-α-ketoglutarate transaminase (-○-) and *p*-hydroxyphenylpyruvate (*p*HPP) oxidase (-×-) following an oral dose of 2.7 mmol *l*-tyrosine to normal guinea pigs. Enzyme activities are Δμ moles of *p*HPP per hour per milliliter of 25% liver homogenate. Each point is the average of two animals. (From Knox, W. E. and Goswami, M. N. D. [1961], *Ann. N.Y. Acad. Sci.*, 92, 192. With permission.)

Denson and Bowers (1961) found *p*HPA in the urines of 4 out of 25 normal adults and 7 out of 50 geriatric patients, but observed no correlation between the excretion of such abnormal phenolic acids and the white cell ascorbic acid (TAA)* levels of these subjects. Moreover, there was no consistent response to ascorbic acid therapy. However, Hyams and Ross (1963) made chromatograms of the urine of a 50-year-old woman who was admitted to hospital with megaloblastic anemia and osteoporosis due to scurvy. Large quantities of *p*HPA and *p*HPL acids were identified in her urine; these cleared as soon as treatment with ascorbic acid was commenced. "Plasma levels of vitamin C were low or nil before treatment, and tended to remain low when the hydroxyphenyluria had cleared. It was presumed that the degree of vitamin C saturation of the tissues, relative to the level of intake of tyrosine and phenylalanine, was the important factor in determining the abnormal excretion."

Light et al. (1966) observed elevated serum tyrosine levels in 32% and elevated phen-ylalanine levels in 20% of low-birth-weight infants on a cow milk formula diet. A prompt fall in both serum tyrosine and phenylalanine was noted within 2 to 4 h of a single intra-muscular dose of 50 to 100 mg of ascorbic acid (Figure 5). The authors pointed out the need to distinguish this ascorbate-responsive "tyrosinemia of prematurity" from the inborn metabolic abnormalities of phenylketonuria and tyrosinemia. However, they warned against

* TAA — total ascorbic acid, reduced and oxidized forms.

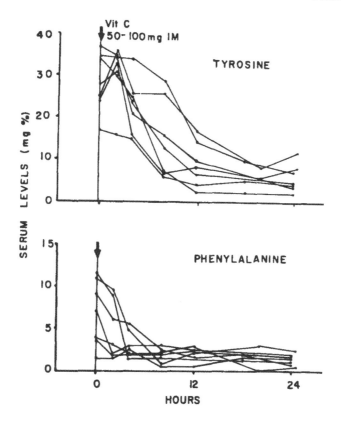

FIGURE 5   Response of low-birth-weight infants with abnormal blood amino acids to a single intramuscular dose of 50 to 100 mg of ascorbic acid. (From Light, I J., Berry, H. K., and Sutherland, J. M. [1966], *Am J. Dis. Child*, 112, 229. © 1966 American Medical Association. With permission.)

dismissing this disorder as a harmless biochemical anomaly of prematurity. Robinson and Warburton (1966) noted that tyrosyluria has been observed in association with idiopathic steatorrhea, thyrotoxicosis, rheumatoid arthritis, and parenchymatous liver disease, but they found no signs of tyrosyluria in five adults and one 6-year-old child with scurvy. However, they conceded that these scorbutic patients might have shown an abnormality of tyrosine metabolism if they had been challenged with a loading dose of tyrosine.

Dhatt et al. (1979) studied 30 children aged 6 months to 3 years, who were admitted to hospital in India weighing "less than 60 per cent of the 50th percentile of the Boston Standards." The plasma tyrosine and the urinary *p*HPA and *p*HPL levels were significantly increased in these children with marasmus, but were not affected by the provision of a high-protein diet. Administration of ascorbic acid caused a marked reduction in the urinary excretion of *p*HPL, but did not affect the excretion of *p*HPA. It was suggested that *p*HPL, which is derived from *p*HPP, is a good index of tyrosyluria and indicates ascorbic acid deficiency, whereas *p*HPA can also arise from tyramine by a pathway which is independent of ascorbic acid (i.e., tyrosine → tyramine → *p*HPA), so it is a poor indicator of ascorbic acid deficiency. It is worthy of note that *p*HPL was found in the urine of 2 out of 20 control subjects and in 27 of the 30 patients with marasmus, so it would seem that ascorbic acid deficiency may not have been entirely confined to the patient group.

Enwonwu and Okolie (1983) observed a highly significant increase in the liver tyrosine level (+118%, *p* <0.01) and a marked increase in the plasma phenylalanine to tyrosine ratio from 84 to 179 in infant monkeys on a low-ascorbic acid diet (20 mg/kg/d). Moreover, these workers reported a highly significant increase in the brain phenylalanine to tyrosine ratio from 102 to 192 (p <0.01) in the animals on the low-ascorbic acid diet. Incidentally, the brain histamine level was significantly increased even though ascorbic acid was still present in the blood at a level of 0.48 mg/100 ml (see Chapter 1 of this volume).

III. DOPA, DOPAMINE, NOREPINEPHRINE, AND EPINEPHRINE

There is evidence that ascorbic acid is involved in the conversion of phenylalanine and tyrosine into dopamine, norepinephrine, and epinephrine, but ascorbic acid does not seem to be a rate-limiting factor in the adrenal synthesis of these catecholamines, even in scurvy. Either there is enough ascorbic acid remaining in the adrenal in scurvy, or else other substances or other pathways most come into play in the scorbutic animal.

Early studies by McCarrison in 1919, by Ohata in 1930, and by Deutsch and Schlapp in 1935 suggested a decrease in the epinephrine content of the adrenal glands in scorbutic guinea pigs. Mouriquand and Leulier in 1927 and Guha in 1935 found no change in the epinephrine content of these glands in scurvy, but it now seems that these investigators were all using nonspecific assay methods. The works of Giroud and Martinet (1941), Banerjee (1945), and Banerjee and Ghosh (1946) all showed an increase in the epinephrine content of the adrenal glands in scurvy, even when the latter workers used a paired-feeding technique to rule out inanition as the cause of the change.

Nagatsu et al. (1964), Udenfriend (1966), and Petrack et al (1968) isolated tyrosine hydroxylase from brain, adrenal medulla, and sympathetically innervated tissues. This enzyme catalyzes the conversion of l-tyrosine to 3,4-dihydroxyphenylalanine (dopa) and requires tetrahydropteridines, such as folic acid, as well as Fe^{2+} and mercaptoethanol, for full activity *in vitro*. Petrack et al. found 2 μmol of ascorbic acid to be more effective than 50 μmol of mercaptoethanol in this reaction. However, greater concentrations of ascorbic acid were inhibitory.

It was already known, from the work of Holtz in 1939, that dopa is readily decarboxylated to form dopamine, and from the works of Levin et al (1950), Levin and Kaufman (1961), Friedman and Kaufman (1965), Goldstein et al. (1965), Kaufman and Friedman (1965), Friedman and Kaufman (1966), and Kaufman (1966) that dopamine can be hydroxylated to norepinephrine by the enzyme dopamine beta oxidase (beta hydroxylase) which is present in the adrenal medulla. Ascorbic acid acts as a cofactor for beta hydroxylase, but it seems that the hydroxylation of dopamine can occur to some extent even in the absence of ascorbic acid. This reaction is now thought to proceed via ascorbate free radical rather than the AA/DHAA system, as the enzyme dopamine beta hydroxylase undergoes a change of

$$Cu^+ \underset{+e}{\overset{-e}{\rightleftharpoons}} Cu^{++}$$

and a small ascorbate free radical signal has been demonstrated by Blumberg et al. (1965). They have suggested a reaction in which ascorbate free radical reacts with dopamine to produce a dopamine free radical which, in turn, reacts with an oxygen-complexed form of the enzyme to form norepinephrine.

Nagatsu et al. (1964) believed tyrosine hydroxylation to be the rate-limiting step in norepinephrine biosynthesis in the adrenal glands. They found the activity of dopa-decarboxylase and dopamine hydroxylase to be of a much better order than that of tyrosine hydroxylase. Petrack et al., on the other hand, found the three enzymes to be of comparable activity. They did, however, find that tyrosine hydroxylase absorption is inhibited approximately 50% by 5×10^{-4} M L-epinephrine or L-norepinephrine, so it would appear that endogenous epinephrine may control the rate of its own synthesis by feedback inhibition of tyrosine hydroxylase. Thoa et al. (1966) reported a reduction in the concentration and content of endogenous norepinephrine in the hearts of scorbutic guinea pigs, but this conclusion seems to have been based on the analysis of only four normal and four scorbutic animals. Nagatsu et al. (1968) found dopamine beta hydroxylase and epinephrine to be absent from the adrenal glands of newborn guinea pigs, but both reached adult levels within 5 d after birth. Neither the enzyme activity nor the epinephrine content of the adrenal glands was significantly affected by scurvy. Moreover, these authors mentioned earlier works in which they had found the norepinephrine levels of the brain, heart, and spleen to be unaffected, even when ascorbic acid was markedly decreased in the tissues. They therefore suggested that either ascorbic acid is not a specific cofactor for the last step in the synthesis of norepinephrine, or else an even greater degree of ascorbic acid deficiency is required to make dopamine beta hydroxylase rate limiting in the overall conversion of tyrosine to norepinephrine. Sears (1969) found the norepinephrine concentration in the iris of the guinea pig to be reduced to 50% of normal in scurvy. Nakashima et al. (1970) demonstrated that the tyrosine hydroxylase of guinea pigs with scurvy was reduced to half normal in the adrenal glands, to one third normal in the liver, and to one quarter normal in the brain.

Phillipson (1973) commented on the curious phenomenon described by Udenfriend, "that in scorbutic animals where tissue ascorbate levels were reduced to as little as one per cent of controls, the adrenal medullary cells could manufacture noradrenaline at a normal rate, in the face of a very low dopamine beta-hydroxylase activity." He quoted Kaufman as suggesting that in these situations the catechol grouping of dopamine may support the hydroxylation reaction. Phillipson also suggested the possibility that there may be enough ascorbate concentrated, where it is needed, at critical intracellular loci even in scurvy.

Dashman et al. (1973) cited the finding by Lokoshko and Lesnykh that the cardiac epinephrine level does not differ from normal in guinea pigs with hypovitaminosis C. Moreover, their own data showed no change in the norepinephrine and dopamine levels in the brains of guinea pigs fed unusually high levels of ascorbic acid. They did, however, report that high doses of ascorbic acid caused a 30% lower than normal level of norepinephrine in the hearts of these animals ($p < 0.005$). So if we are to accept the data of both Thoa et al. (1966) and Dashman et al. (1973), there would seem to be a reduced cardiac level of norepinephrine, both in scurvy and in hypervitaminosis C.

Saner et al. (1975) observed that the norepinephrine content of the hypothalamus of guinea pigs after 21 to 23 d on a scorbutogenic diet was significantly lower ($p < 0.01$) than that of both pair-fed and control animals. However, studies with the use of a tyrosine hydroxylase inhibitor (α-methyl-p-tyrosine) led these workers to conclude that there is an accelerated turnover of norepinephrine in the brain in ascorbic acid deficiency. This is puzzling in view of the fact that ascorbic acid has been shown to be a cofactor of dopamine beta hydroxylase. They suggested that perhaps ascorbic acid depletion of the brain was not complete.

Rate-limiting dopamine hydroxylase activity was suggested by the observations of Deana et al. (1975) who reported finding markedly increased levels of dopamine and markedly decreased levels of norepinephrine in the brains of guinea pigs with scurvy (Table 1), but Green et al. (1979) found no such change. They suggested that the administration of ascorbic acid interfered with the measurement of norepinephrine by the trihydroxyindole technique used by Deana.

Lewin (1976) has suggested that ascorbic acid has an important role in protecting the catecholamines by direct chemical interaction, and that it helps to remove the toxic product of oxidation, adrenochrome. It seems that the high concentration of ascorbic acid in the adrenal glands not only facilitates the conversion of dopamine to norepinephrine and epinephrine, it may also play a role in protecting these two catecholamines from oxidation (Figure 6). Lewin asserts that in the absence of ascorbate, both epinephrine and norepinephrine are oxidized to their respective adrenochromes, as shown for epinephrine in Figure

Table 1

**CATECHOLAMINE LEVELS IN LIVER AND BRAIN
OF NORMAL AND ASCORBIC ACID-DEFICIENT
GUINEA PIGS[a]**

	Control(8)	Deficient(12)	p
Norepinephrine			
Brain	0 353 ± 0.075	0.178 ± 0.055	<0.001
Liver	0 786 ± 0.195	0.557 ± 0.245	<0.005
B/L[b]	0 449	0.320	
Dopamine			
Brain	0.261 ± 0.177	0.539 ± 0.233	<0.001
Liver	0.778 ± 0.348	0.707 ± 0.456	<0.3
B/L	0.336	0.762	
Total catecholamines			
Brain	0 673 ± 0.155	0.740 ± 0.242	<0.3
Liver	1.477 ± 0.330	1.286 ± 0.477	<0.01
B/L	0.415	0.566	

Note: A highly significant decrease in norepinephrine and a highly significant increase in dopamine concentration were observed in the brain tissue of the scorbutic guinea pigs.

[a]   Values are expressed as μg/g of fresh tissue, and they are given as means ± SD for the number of animals, in parentheses.
[b]   B/L is the brain to liver ratio.

From Deana, R., Bharaj, B. S., Verjee, Z. H., and Galzigna, L. (1975), *Int. J. Vitam. Nutr Res.*, 45, 175. With permission.

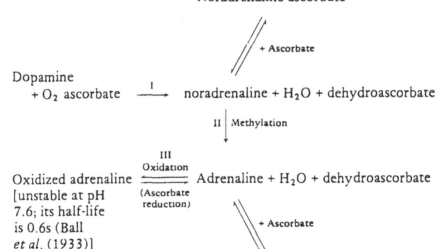

FIGURE 6.   Schematic representation of ascorbate-assisted formation of adrenaline from dopamine. (From Lewin, S. [1976], *Vitamin C: Its Molecular Biology and Medical Potential*, Academic Press, London With permission.)

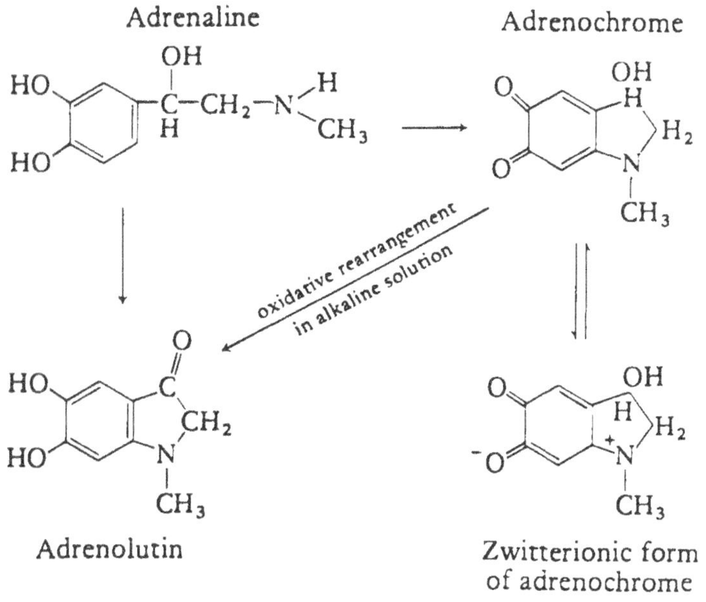

Adrenaline   Adrenochrome

Adrenolutin   Zwitterionic form of adrenochrome

FIGURE 7. Relationship between adrenaline, adrenochrome, and adrenolutin (From Lewin, S. [1976], *Vitamin C. Its Molecular Biology and Medical Potential,* Academic Press, London With permission.)

7. He states that these highly toxic substances can be inactivated by hydrogen bonding between 5,6-dihydrocatechol groups and ascorbate.

Subramanian (1977) has suggested that regional variations in tissue ascorbate levels may play an important role affecting the local levels of neurotransmitters, including epinephrine, norepinephrine, serotonin, and histamine in the brain and nerves. Certainly some studies of ascorbic acid deficiency in guinea pigs have demonstrated changes in tissue catecholamine levels, but there is no evidence that high doses of ascorbic acid affect the metabolism of catecholamines. Neither Behrens and Madère (1980), studying rats, nor Igisu et al. (1982), studying human volunteers, found any evidence of altered catecholamine metabolism following megadose ascorbic acid. Behrens and Madère studied the norepinephrine, epinephrine, and dopamine levels in the serum, adrenal glands, liver, and brain tissue of ascorbate-loaded rats. Igisu et al. studied the urinary excretion of these catecholamines in 11 students at Kyushu University after 1, 2, and 3 months of ascorbic acid supplementation (3 g daily) and found values that did not differ significantly from those of 8 control students receiving placebo tablets.

Hoehn and Kanfer (1980), working at the University of Manitoba, created chronic ascorbic acid deficiency in guinea pigs. They provided a diet which was totally devoid of ascorbic acid for 2 weeks to deplete the body stores of this vitamin and then gave the animals a diet containing ascorbic acid 0.5 mg/100 g body weight per day. In this way, it is possible to study guinea pigs after 2 to 12 weeks of marginal ascorbate deficiency, without the loss of weight and inanition that is inevitable in animals with frank scurvy. Depletion of ascorbic acid stores was found to cause a 25% decrease in the norepinephrine content of the brain and a corresponding 25% increase in the brain dopamine level (Figure 8), just as one would expect as a result of impaired dopamine beta hydroxylase activity. These catecholamine levels returned to normal when ascorbic acid was provided.

IV. TYROSINE OXIDATION TO DOPA AND TO MELANIN

Lerner and Fitzpatrick (1950), reviewing the biochemistry of melanin formation, reported

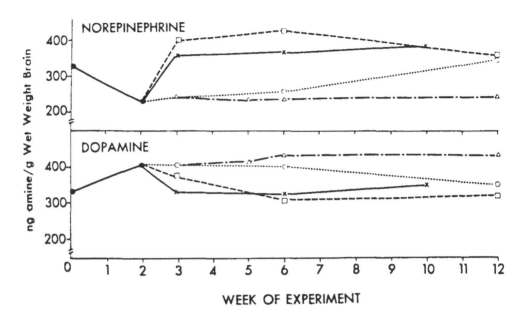

FIGURE 8.   Norepinephrine and dopamine concentrations in guinea pig brain tissue when various amounts of ascorbic acid (AA) were fed for 12 weeks During the first 2 weeks no ascorbic acid was provided in the diet (●—●); then animals received a stock diet containing AA 50 mg/100 g diet (×—×); AA 10 mg/d (□—□; 1 mg AA/100 g body weight (○—○); 0 5 mg AA/100 g body weight per day (△—△) Chronic subclinical ascorbic acid deficiency caused a 25% reduction in the brain norepinephrine and a 25% increase in the brain dopamine level. (From Hoehn, S. K. and Kanfer, J. N. [1980], *J. Nutr.*, 110, 2085. © American Institute of Nutrition With permission.)

that ascorbic acid has been shown to shorten the induction period of tyrosinase *in vitro*, but inhibits melanin formation from tyrosine and from DOPA and also decolourizes melanin from jet black to light tan. Large doses of ascorbic acid have been reported to decrease the skin pigmentation of patients with Addison's disease. Conversely, melanin hyperpigmentation in South African Bantu patients with hemosiderosis and toxic psychosis has been attributed to ascorbic acid deficiency by Robins (1972a). Moreover, Robins (1972b) has suggested that the melanosis described by Cawley et al. (1969) in association with hemochromatosis may also be due to ascorbic acid deficiency.

V. TYROSINE IODINATION AND CONVERSION TO THYROXINE

The ratio of reduced to oxidized ascorbic acid in the thyroid gland of the guinea pig is markedly reduced in scurvy from $\frac{6.9}{2.1}$ to $\frac{0.3}{2.0}$ mg/100 g of tissue, or from 3.3:1 to 0.15:1, according to the findings of Kassouny and Rivers (1972). This is undoubtedly associated with a big increase in the oxidation-reduction potential of the gland and most probably accounts for the decreased organic ^{131}I uptake by the thyroid in guinea pigs with acute scurvy. This effect on thyroid function may possibly account for the arrested growth of scorbutic animals.

VI. CONCLUSIONS CONCERNING CLINICAL RELEVANCE

Bates (1981), reviewing the effects of ascorbic acid on tyrosine metabolism, illustrated the dopamine beta hydroxylase and the *p*HPP hydroxylase pathways, as shown in Figure 9. The oxidation of tyrosine is frequently impaired in premature infants on cow milk formulae,

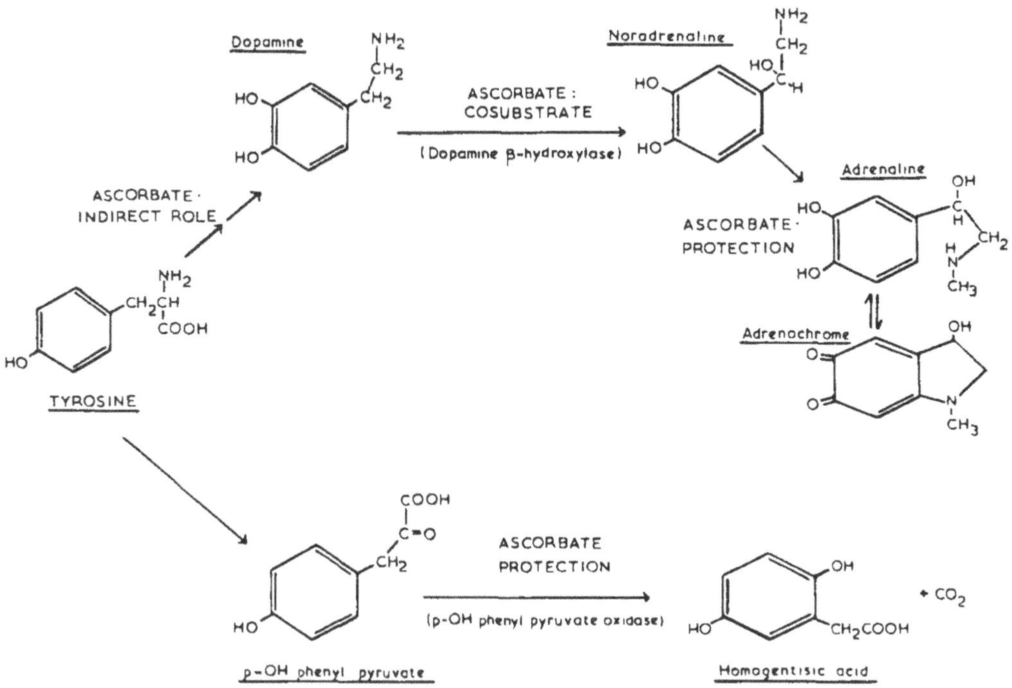

FIGURE 9.   Role of ascorbate in catecholamine synthesis and tyrosine degradation. (From Bates, C. J. [1981], in *Vitamin C (Ascorbic Acid)*, Counsell, J. N and Hornig, D. H., Eds., Applied Science, London, 1. With permission)

and this is associated with hydroxyphenyluria which responds to ascorbic acid. But it seems that ascorbic acid deficiency affects tyrosine metabolism in adults only under exceptional circumstances, such as malabsorption syndrome, hemosiderosis, and following a high tyrosine or phenylalanine load. There is still considerable doubt about the roles played by alterations of dopamine, melanin, and thyroxine metabolism in the mental depression, pigmentation, and thyroid deficiency which have been reported in scurvy.

REFERENCES

Abbott, L. D., Jr. and James, G. W., III (1950), Effect of vitamin B_{12} on the urinary phenol fractions in pernicious anemia, *J. Lab. Clin. Med* , 35, 35.

Aterman, K., Boscott, R. J., and Cooke, W. T. (1953), The significance of urinary p-hydroxyphenylacetic acid (pHPA), *Biochem. J.*, 55, XVII.

Banerjee, S. (1945), Relation of scurvy to the adrenalin content of the adrenal glands of guinea pigs, *J Biol. Chem* , 159, 327

Banerjee, S. and Ghosh, N. C. (1946), Adrenalin in scurvy, *J Biol. Chem.*, 166, 25.

Bates, C. J. (1981), The function and metabolism of vitamin C in man, in *Vitamin C (Ascorbic Acid)*, Counsell, J N. and Hornig, D. H., Eds., Applied Science, London, 1.

Behrens, W. A. and Madère, R. (1980), Effects of high ascorbic acid intake on the metabolism of catecholamines in the rat, *J. Nutr.*, 110, 720.

Blumberg, W. E., Goldstein, M., Lauber, E., and Peisach, J. (1965), Magnetic resonance studies on the mechanism of the enzymic beta-hydroxylation of 3,4-dihydroxyphenylethylamine, *Biochim. Biophys. Acta*, 99, 187.

Boscott, R. J. and Cooke, W. T. (1954), Ascorbic acid requirements and urinary excretion of p-hydroxyphenylacetic acid in steatorrhoea and macrocytic anaemia, *Q. J. Med. (New Ser)*, 23, 307.

Cawley, E. P., Hsu, Y. T., Wood, B. T., and Weary, P. E. (1969), Hemochromatosis and the skin, *Arch. Dermatol.,* 100, 1.

Crandon, J. H., Landau, B., Mikal, S., Balmanno, J., Jefferson, M., and Mahoney, N. (1958), Ascorbic acid economy in surgical patients as indicated by blood ascorbic acid levels, *N. Engl. J. Med.,* 258, 105

Darby, W. J., McGanity, W. J., Stockell, A., and Woodruff, C. W. (1953), Ascorbic acid, pteroylglutamates and other factors in scorbutic hydroxyphenyluria, *Proc. Nutr. Soc.,* 12, 329.

Dashman, T., Horst, D., Bautz, G., and Kamm, J. J. (1973), Ascorbic acid effect of high doses on brain and heart catecholamine levels in guinea pigs and rats, *Experientia,* 29, 832.

Deana, R., Bharaj, B. S., Verjee, Z. H., and Galzigna, L. (1975), Changes relevant to catecholamine metabolism in liver and brain of ascorbic acid deficient guinea pigs, *Int. J Vitam. Nutr. Res.,* 45, 175.

Denson, K. W. and Bowers, E. F. (1961), The determination of ascorbic acid in white blood cells. A comparison of w.b.c. ascorbic acid and phenolic acid excretion in elderly patients, *Clin. Sci.,* 21, 157

Dhatt, P. S., Saini, A. S., Gupta, I., Mehta, H. C., and Singh, H. (1979), Tyrosyluria in marasmus, *Br. J. Nutr.,* 42, 387.

Enwonwu, C. O. and Okolie, E. E. (1983), Differential effects of protein malnutrition and ascorbic acid deficiency on histidine metabolism in the brains of infant nonhuman primates, *J. Neurochem.,* 41, 230.

Friedman, S. and Kaufman, S. (1965), 3,4-Dihydroxyphenylethylamine beta-hydroxylase: a copper protein, *J. Biol. Chem.,* 240, PC552.

Friedman, S. and Kaufman, S. (1966), An electron paramagnetic resonance study of 3,4-dihydroxyphenylethylamine beta-hydroxylase, *J. Biol. Chem.,* 241, 2256.

Giroud, A. and Martinet, M. (1941), Modifications fonctionnelles de la medullo-surrenale en rapport avec les variations de l'acide ascorbique, *C R. Soc. Biol.,* 135, 1344.

Goldstein, M., Lauber, E., Blumberg, W. E., and Peisach, J. (1965), Dopamine-beta-hydroxylase-a copper enzyme, *Fed. Proc. Fed Am Soc. Exp. Biol.,* 24, 604.

Govan, C. D., Jr. and Gordon, H. H. (1949), The effect of pteroylglutamic acid on the aromatic amino acid metabolism of premature infants, *Science,* 109, 332.

Green, M. D., Bell, W., Kraut, C., and Omaye, S. (1979), Failure of ascorbic acid to influence brain catecholamines in the guinea pig, *Experientia,* 35, 515.

Hoehn, S. K. and Kanfer, J. N. (1980), Effects of chronic ascorbic acid deficiency on guinea pig lysosomal hydrolase activities, *J. Nutr.,* 110, 2085.

Hyams, D. E. and Ross, E. J. (1963), Scurvy, megaloblastic anaemia and osteoporosis, *Br. J. Clin. Pract.,* 17, 332.

Igisu, H., Fujino, T., and Dohnao, T. (1982), Effect of large dose of ascorbic acid on urinary excretion of catecholamines in man, *Int. J. Vitam. Nutr. Res.,* 52, 1464.

Johnson, B. C. and Dana, A. S. (1948), Ascorbic acid therapy of pteroylglutamic acid-deficient rats, *Science,* 108, 210.

Kassouny, M. E. and Rivers, J. M. (1972), Vitamin C depletion and in vitro uptake and organification of ^{131}I by guinea pig thyroid tissue, *J. Nutr.,* 102, 797.

Kaufman, S. (1966), Coenzymes and hydroxylases: ascorbate and dopamine-beta-hydroxylase; tetrahydropteridines and phenylalanine and tyrosine hydroxylases, *Pharmacol. Rev.,* 18, 61.

Kaufman, S. and Friedman, S. (1965), Dopamine-beta-hydroxylase, *Pharmacol. Rev.,* 17, 71.

Knox, W. E. and Goswami, M. N. D. (1961), Ascorbic acid in tyrosine metabolism, *Ann. N.Y. Acad. Sci.,* 92, 192.

Kretchmer, N., Levine, S. Z., McNamara, H., and Barnett, H. L. (1956), Certain aspects of tyrosine metabolism in the young. I. The development of the tyrosine oxidizing system in human liver, *J. Clin. Invest.,* 35, 236.

La Du, B. N., Jr. (1952), Effect of ascorbic acid on oxidation of tyrosine, *Fed. Proc. Fed. Am. Soc. Exp. Biol.,* 11, 244.

La Du, B. N., Jr. and Greenberg, D. M. (1953), Ascorbic acid and the oxidation of tyrosine, *Science,* 117, 111.

La Du, B. N. and Zannoni, V. G. (1961), The role of ascorbic acid in tyrosine metabolism, *Ann. N.Y. Acad. Sci.,* 92, 175.

Lan, T. H. and Sealock, R. R. (1944), The metabolism in vitro of tyrosine by liver and kidney tissues of normal and vitamin C-deficient guinea pigs, *J. Biol. Chem.,* 155, 483.

Lerner, A. B. and Fitzpatrick, T. B. (1950), Biochemistry of melanin formation, *Physiol. Rev.,* 30, 91.

Levin, E. Y. and Kaufman, S. (1961), Studies on the enzyme catalyzing the conversion of 3,4-dihydroxyphenylethylamine to norepinephrine, *J. Biol. Chem.,* 236, 2043.

Levin, E. Y., Levenberg, B., and Kaufman, S. (1960), The enzymatic conversion of 3,4-dihydroxyphenylethylamine to norepinephrine, *J. Biol. Chem.,* 235, 2080.

Levine, S. Z., Gordon, H. H., and Marples, E. (1941), A defect in the metabolism of tyrosine and phenylalanine in premature infants. II. Spontaneous occurrence and eradication by vitamin C, *J. Clin. Invest.,* 20, 209.

Levine, S. Z., Marples, E., and Gordon, H. H. (1939), A defect in the metabolism of aromatic amino acids in premature infants; the role of vitamin C, *Science,* 90, 620.

Levine, S. Z., Marples, E., and Gordon, H. H. (**1941**), A defect in the metabolism of tyrosine and phenylalanine in premature infants I. Identification and assay of intermediary products, *J Clin. Invest.*, 20, 199.

Lewin, S. (**1976**), *Vitamin C: Its Molecular Biology and Medical Potential*, Academic Press, London.

Light, I. J., Berry, H. K., and Sutherland, J. M. (**1966**), Amino-acidemia of prematurity. Its response to ascorbic acid, *Am. J. Dis. Child.*, 112, 229.

Lustberg, T. J., Schulman, J. D., and Seegmiller, E. J. (**1969**), Metabolic fate of homogentisic acid-1-^{14}C (HGA) in alkaptonuria and effectiveness of ascorbic acid in preventing experimental ochronosis, *Arthritis Rheum.*, 12, 678.

Morris, J. E., Harpur, E. R., and Goldbloom, A. (**1950**), The metabolism of *l*-tyrosine in infantile scurvy, *J. Clin. Invest.*, 29, 325.

Nagatsu, T., Levitt, M., and Udenfriend, S. (**1964**), Tyrosine hydroxylase. The initial step in norepinephrine biosynthesis, *J. Biol. Chem.*, 239, 2910.

Nagatsu, T., van der Schoot, J. B., Levitt, M., and Udenfriend, S. (**1968**), Factors influencing dopamine beta-hydroxylase activity and epinephrine levels in guinea pig adrenal gland, *J. Biochem. (Tokyo)*, 64, 39.

Nakashima, Y., Suzue, R., Sanada, H., and Kawada, S. (**1970**), Effect of ascorbic acid on hydroxylase activity. I. Stimulation of tyrosine hydroxylase and tryptophan-5-hydroxylase activities by ascorbic acid, *J. Vitaminol.*, 16, 276.

Nitowsky, H. M., Govan, C. D., Jr., and Gordon, H. H. (**1953**), Effect of hemopoietic and other agents on the hydroxyphenyluria of premature infants, *AMA Am. J. Dis. Child.*, 85, 462.

Painter, H. A. and Zilva, S. S. (**1947**), The influence of *l*-ascorbic acid on the rupture of the benzene ring of *l*-tyrosine consumed in high doses by guinea pigs, *Biochem. J.*, 41, 511.

Painter, H. A. and Zilva, S. S. (**1950**), The influence of L-ascorbic acid on the disappearance of the phenolic group of L-tyrosine in the presence of guinea pig liver suspensions, *Biochem. J.*, 46, 542

Peters, J. P. and Van Slyke, D. D. (**1946**), *Quantitative Clinical Chemistry Interpretations*, Vol. 1, Williams & Wilkins, Baltimore.

Petrack, B., Sheppy, F., and Fetzer, V. (**1968**), Studies on tyrosine hydroxylase from bovine adrenal medulla, *J. Biol. Chem.*, 243, 743.

Phillipson, O. T. (**1973**), in *Frontiers in Catecholamine Research*, Usdin, E. and Snyder, S. H., Eds., Pergamon Press, New York, 174.

Reinits, K. G. (**1950**), Metabolism of L-ascorbic acid and L-tyrosine in guinea pig liver, *J. Biol. Chem.*, 182, 11.

Robins, A. H. (**1972a**), Melanin hyperpigmentation in South African Bantu patients with toxic psychosis. The probable role of iron overload, *S. Afr. Med. J.*, 46, 1639.

Robins, A. H. (**1972b**), Hemochromatosis, melanosis and hypovitaminosis C, *Arch. Dermatol.*, 106, 768.

Robinson, R. and Warburton, F. G. (**1966**), Tyrosine metabolism in human scurvy, *Nature (London)*, 212, 1605.

Rodney, G., Swendseid, M. E., and Swanson, A. L. (**1947**), Tyrosine oxidation by livers from rats with a sulfasuxidine-induced pteroylglutamic acid deficiency, *J. Biol. Chem.*, 168, 395.

Rogers, W. F. and Gardner, F. H. (**1949**), Tyrosine metabolism in human scurvy, *J. Lab. Clin. Med.*, 34, 1491.

Salmon, R. J. and May, C. D. (**1950**), Metabolism of tyrosine in experimental megaloblastic anemia and in scurvy in the monkey, *J. Lab. Clin. Med.*, 36, 591.

Salmon, R. J. and May, C. D. (**1953**), Further studies on the metabolism of *l*-tyrosine in scorbutic monkeys. The effect of folinic acid, cortisone, adrenalectomy and sulfhydryl groups on urinary excretion of *p*-hydroxy-phenyl compounds, *J. Lab. Clin. Med.*, 41, 376.

Saner, A., Weiser, H., Hornig, D., Da Prada, M., and Pletscher, A. (**1975**), Cerebral monoamine metabolism in guinea pigs with ascorbic acid deficiency, *J. Pharm. Pharmacol.*, 27, 896.

Schepartz, B. and Nadel, E. M. (**1952**), Effect of ascorbic acid, folic acid and vitamin B$_{12}$ upon tyrosine oxidation in acetone powder preparations of guinea pig liver, *Fed. Proc. Fed. Am. Soc. Exp. Biol.*, 11, 425.

Sealock, R. R. (**1942**), The relation of vitamin C to the metabolism of the aromatic amino acids, *Fed. Proc. Fed. Am. Soc. Exp. Biol.*, 1, 287.

Sealock, R. R., Galdston, M., and Steele, J. M. (**1940**), Administration of ascorbic acid to an alkaptonuric patient, *Proc. Soc. Exp. Biol. Med.*, 44, 580.

Sealock, R. R., Goodland, R. L., Sumerwell, W. N., and Brierly, J. M. (**1952**), The role of ascorbic acid in the oxidation of L-tyrosine by guinea pig liver extracts, *J. Biol. Chem.*, 196, 761.

Sealock, R. R. and Lepow, J. P. (**1948**), Antipernicious anemia extracts and tyrosine metabolism in the scorbutic guinea pig, *J. Biol. Chem.*, 174, 763.

Sealock, R. R., Perkinson, J. D., and Silberstein, H. E. (**1940**), Ascorbic acid and the metabolism of the aromatic amino acids, phenylalanine and tyrosine, *J. Biol. Chem.*, 133, 1xxxvii.

Sealock, R. R. and Silberstein, H. E. (**1939**), The control of experimental alcaptonuria by means of vitamin C, *Science*, 90, 517.

Sealock, R. R. and Silberstein, H. E. (**1940**), The excretion of homogentisic acid and other tyrosine metabolites by the vitamin C-deficient guinea pig, *J. Biol. Chem.*, 135, 251.

Sears, M. I. (1969), Vitamin C as a requirement for the storage of norepinephrine by the iris, *Biochem. Pharmacol* , 18, 253.

Steele, B. F., Hsu, C.-H., Pierce, Z. H., and Williams, H. H. (1952), Ascorbic acid nutriture in the human. I. Tyrosine metabolism and blood levels of ascorbic acid during ascorbic acid depletion and repletion, *J. Nutr.*, 48, 49.

Subramanian, N. (1977), On the brain ascorbic acid and its importance in metabolism of biogenic amines, *Life Sci.*, 20, 1479.

Swendseid, M. E., Burton, I. F., and Bethell, F. H. (1943), Excretion of keto acids and hydroxyphenyl compounds in pernicious anemia, *Proc. Soc. Exp. Biol. Med.*, 52, 202

Swendseid, M. E., Wandruff, B., and Bethell, F. H. (1947), Urinary phenols in pernicious anemia, *J Lab. Clin. Med.*, 32, 1242.

Thoa, N. B., Wurtman, R. J., and Axelrod, J. (1966), A deficient binding mechanism for norepinephrine in hearts of scorbutic guinea pigs, *Proc. Soc. Exp. Biol. Med.*, 121, 267.

Udenfriend, S. (1966), Tyrosine hydroxylase, *Pharmacol. Rev.*, 18, 43.

Woodruff, C. W. (1950), Tyrosine metabolism in infantile scurvy, *J. Lab. Clin. Med.*, 36, 640.

Woodruff, C. W., Cherrington, M. E., Stockell, A. K., and Darby, W. J. (1949), The effect of pteroylglutamic acid and related compounds upon tyrosine metabolism in the scorbutic guinea pig, *J. Biol. Chem.*, 178, 861.

Woodruff, C. W. and Darby, W. J. (1948), An in vivo effect of pteroylglutamic acid upon tyrosine metabolism in the scorbutic guinea pig, *J. Biol. Chem.*, 172, 851.

Woodruff, C. W. and Darby, W. J. (1949), Influence of pteroylglutamic acid on tyrosine metabolism in the scorbutic guinea pig, *Am. J. Dis. Child.*, 77, 128.

Woolf, L. I. and Edmunds, M. E. (1950), The metabolism of tyrosine and phenylalanine in premature infants: the effect of large doses, *Biochem. J.*, 47, 630.

Zannoni, V. G. and La Du, B. N. (1960), Studies on the defect in tyrosine metabolism in scorbutic guinea pigs, *J. Biol. Chem.*, 235, 165.

Chapter 7

TRYPTOPHAN METABOLISM

The essential amino acid L-tryptophan is used in the body for many purposes besides the building of protein molecules and its complex conversion to nicotinic acid for the synthesis of NAD (coenzyme I, DPN) and NADP (coenzyme II, TPN). One interesting metabolic change that it undergoes is hydroxylation to 5-hydroxytryptophan, followed by decarboxylation to form 5-hydroxytryptamine or serotonin (Figure 1), which has an important role as a neurotransmitter in the brain, as originally outlined by Brodie et al. (1955). Harper (1961) noted that, "the serotonin produced in the rest of the body does not pass the blood-brain barrier to any significant degree;" he therefore suggested that serotonin must be produced within the brain itself from precursors which do gain access to the brain.

Cooper (1961) isolated an enzyme which catalyzed the hydroxylation of tryptophan to 5-hydroxytryptophan. He found it in the mucosa of the small intestine and in the kidney of both the rat and the guinea pig, but was unable to find any measurable quantities of the enzyme in extracts of the liver, brain, pineal gland, large intestine, or spleen. Cupric ions and ascorbic acid were found to be necessary for activity of this enzyme, and Cu^{++} could not be replaced by Fe^{+++}, Mo^{vi} or Mn^{++}. L-Xyloascorbic acid could be replaced by D-ascorbate, isoascorbate or dehydroascorbate, but cytochrome C and flavin adenine dinucleotide or mononucleotide were inactive. In studies of both the enzyme tryptophan-5-hydroxylase and also 5-hydroxytryptophan decarboxylase, the hydroxylation step was found to be the rate-limiting reaction in the formation of serotonin. The hydroxylation reaction was said to proceed equally well under either aerobic or anaerobic conditions, but this was not confirmed by subsequent workers. The hydroxylating enzyme was present in the particulate fraction of the tissue extracts and the soluble fraction was reported as possessing an inhibitor for this enzyme.

Sokoloff (1964) reported that ascorbic acid had been used successfully by Pukhalskaya to protect the hydroxyl group of serotonin in experiments with rats and mice. Polis et al. (1969) stated that 5-hydroxytryptophan penetrates the brain readily, but the isotope studies of Green and Sawyer (1965) provided evidence in rats that extracranial tryptophan and 5-hydroxytryptophan are not the precursors of cerebral serotonin. They suggested that brain proteins uniquely provide the tryptophan and showed that the brain is capable of hydroxylating tryptophan to 5-hydroxytryptophan. Nakamura et al. (1965) also isolated tryptophan hydroxylase from rabbit brain.

Nakashima et al. (1970) found tryptophan-5-hydroxylase in the adrenal glands, the liver, and the brain of guinea pigs. They studied one group of male guinea pigs after 18 d on a scorbutogenic diet and another group on the same diet, but having received ascorbic acid, 200 mg daily, by intraperitoneal injection for the last 3 d. In the ascorbic acid-treated guinea pigs, the ascorbic concentration in the adrenal gland was higher than that of other tissues, but in scurvy, the ascorbic acid concentration was highest in the brain, as indicated in Table 1. The effect of ascorbic acid on the tryptophan hydroxylase activity of the brain is shown in Table 2. The mean rate of serotonin formation in the nonscorbutic animals was 0.27 nmol/h/g of brain, whereas the rate in the scorbutic animals was 0.17 nmol/h/g. The brains of the ascorbic acid-treated guinea pigs showed a 59% greater tryptophan-5-hydroxylase activity than did the brains of those with scurvy (Figure 2). Both tyrosine hydroxylase and tryptophan-5-hydroxylase were reported as requiring 2-amino-4-hydroxyl-6,7-dimethyltetrahydropteridine (DMPH$_4$), NADP, and molecular oxygen. Adequate dietary ascorbic acid was also required, but addition of ascorbate *in vitro* did not increase the activity of the enzyme, so it was suggested that ascorbic acid has an indirect effect, possibly affecting the

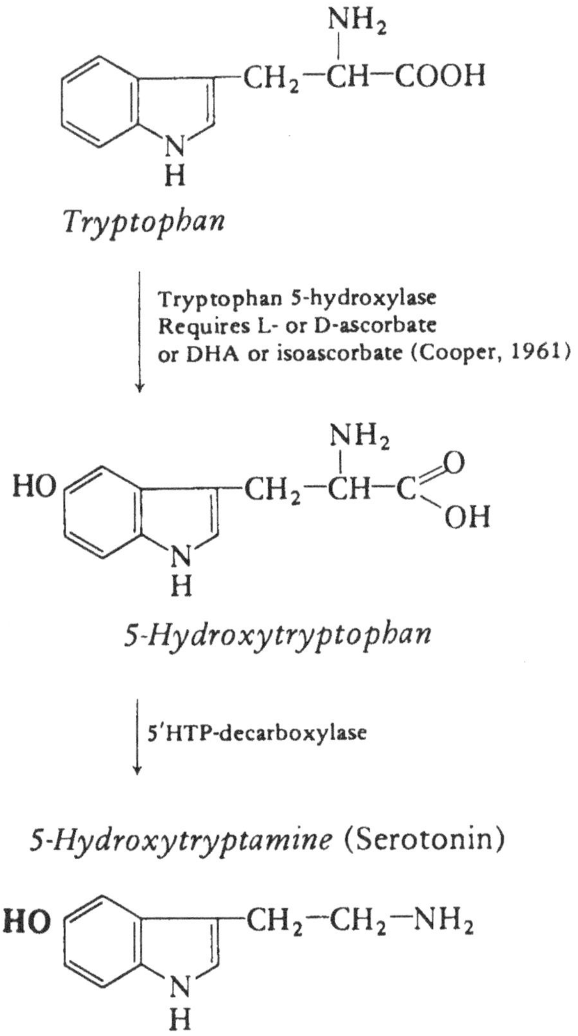

$$\text{Tryptophan} \quad \underset{\underset{\text{H}}{\overset{\displaystyle N}{|}}}{\text{indole ring}} \text{---CH}_2\text{---}\underset{\underset{\text{NH}_2}{|}}{\text{CH}}\text{---COOH}$$

Tryptophan

Tryptophan 5-hydroxylase
Requires L- or D-ascorbate
or DHA or isoascorbate (Cooper, 1961)

$$\text{HO---} \underset{\underset{\text{H}}{\overset{\displaystyle N}{|}}}{\text{indole ring}} \text{---CH}_2\text{---}\underset{\underset{\text{NH}_2}{|}}{\text{CH}}\text{---C}\underset{\text{OH}}{\overset{\text{O}}{\diagdown}}$$

5-Hydroxytryptophan

5'HTP-decarboxylase

5-Hydroxytryptamine (Serotonin)

$$\text{HO---} \underset{\underset{\text{H}}{\overset{\displaystyle N}{|}}}{\text{indole ring}} \text{---CH}_2\text{---CH}_2\text{---NH}_2$$

FIGURE 1   Conversion of tryptophan to serotonin, as described by Cooper (1961). (From Lewin, S. [1976], *Vitamin C: Its Molecular Biology and Medical Potential*, Academic Press, London. With permission.)

activation of a low-molecular-weight coenzyme or a metal ion such as Fe^{+++}, or by aiding in the synthesis of cytochrome P-450; these authors made no mention of the cupric ion.

Shimizu et al. (1960), using a silver nitrate staining technique to study the distribution of ascorbic acid in the brains of rats, mice, and guinea pigs, concluded that this vitamin is concentrated in the locus coeruleus, the dorsal nucleus of the vagus nerve, and the area postrema as well as the hypothalamus. This led them to suppose that ascorbic acid might be implicated in the function of the autonomic centers of the brain. Clearly, ascorbic acid may be important in these centers for the synthesis of dopamine, norepinephrine, and epinephrine from tyrosine and in the synthesis of serotonin from tryptophan. Shimizu et al. (1960) also suggested that the pronounced amine oxidase activity in the locus coeruleus might be related to an active aromatic amine metabolism in these nuclei. Giarmann and Freedman (1960) found very high levels of serotonin in the pineal glands of men and monkeys. In fact, they stated that the serotonin levels found in the majority of human and

Table 1
ASCORBIC ACID CONTENT IN TISSUES OF SCORBUTIC AND NONSCORBUTIC GUINEA PIGS

	Nonscorbutic	Scorbutic
Liver	0.255	0.006
Adrenal gland	1.105	0.015
Brain	0.266	0.079

Note: All values are expressed as milligrams per gram of tissue The scorbutic guinea pigs were fed ascorbic acid-deficient diet for 2 weeks. In normal guinea pigs, the adrenal glands had a higher concentration of ascorbic acid than the brain, but in the scorbutic animals the level of ascorbic acid in the brain was higher than that in the adrenals.

From Nakashima, Y , Suzue, R., Sanada, H., and Kawada, S (1970), *J. Vitaminol* , 16, 276. With permission.

Table 2
EFFECT OF ASCORBIC ACID ON BRAIN TRYPTOPHAN-5-HYDROXYLASE ACTIVITY

Nonscorbutic	Scorbutic
0 269 ± 0.035	0.169 ± 0 013
159 (%)	100 (%)

Note: The tryptophan-5-hydroxylase activities of normal and scorbutic guinea pig brain tissues were measured; the values are expressed as the number of nanomoles of serotonin synthesized per hour per gram of brain. The normal enzyme activity is also shown as a percentage of the scorbutic value.

From Nakashima, Y , Suzue, R , Sanada, H., and Kawada, S. (1970), *J. Vitaminol* , 16, 276. With permission

simian pineal glands were the highest for any neural structure of any species examined. Bovine pineal glands had low levels of serotonin, while psychotic patients showed widely varying levels.

Levine et al. (1964) reported the hydroxylation of tryptophan and phenylalanine by murine neoplastic mast cells, but did not report the effects of altering the ascorbate concentration. Schildkraut (1965), reviewing a catecholamine hypothesis of affective disorders, expressed the opinion that the observed stimulation of human behavior by monoamine oxidase inhibitors is better related to the rise in brain norepinephrine levels than to an elevation of serotonin. However, he pointed out that reserpine induces sedation which is associated with a decrease in brain levels of norepinephrine, dopamine, and serotonin. He cited several studies indicating a mood-elevating effect of tryptophan and possible synergism between the effects of tryptophan and monoamine oxidase inhibitors in the treatment of depression.

Rajalakshmi et al. (1967) measured the distribution of ascorbic acid in different areas of the rat brain. They obtained values ranging from a low of 45 mg/100 g in the dorsal cortex, to a high of 72 mg/100 g in the hypothalamus. It is interesting to note that the hypothalamus (the seat of the emotions), which was found to have the highest concentration of ascorbic acid, was also found to have a high concentration of serotonin.

During ascorbic acid depletion studies in guinea pigs, Hughes et al. (1971) found that loss of ascorbic acid from the brain and the lens of the eye was considerably less rapid than the loss from the adrenal glands, spleen, and aqueous humor. After 14 d on a scorbutogenic diet, the concentrations of ascorbic acid in the brain and the eye lens were 24 and 27% of their initial concentrations; the corresponding values for the aqueous humour, adrenal glands, and spleen were 3, 4 and 5%. Thus, the brain and its ascorbate-dependent enzyme systems are to some extent protected against ascorbic acid deficiency under ordinary circumstances, though this protection may be lost in such conditions as meningitis, aspirin intoxication, or protracted scurvy.

Cho-Chung and Pitot (1967) isolated and purified the enzyme tryptophan pyrrolase which catalyzes the first irreversible step in the degradative metabolism of tryptophan in mammalian liver to nicotinic acid and nicotinamide, a series of reactions which accounts for the niacin-sparing effect of tryptophan in the rat.

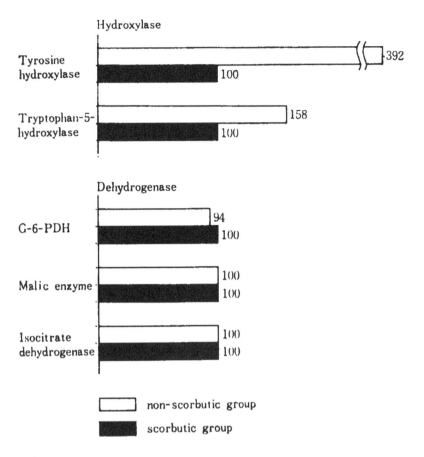

FIGURE 2.   The effect of ascorbic acid on brain hydroxylase and dehydrogenase activities. The specific activities of the enzymes were expressed as percentages of the scorbutic values. Both tyrosine hydroxylase and tryptophan-5-hydroxylase activities were decreased in scurvy. (From Nakashima, Y , Suzue, R., Sanada, H., and Kawada, S. [1970], *J. Vitaminol.*, 16, 276. With permission.)

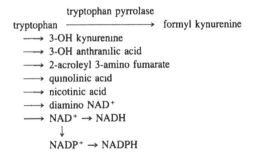

These workers noted that NADPH had a strong inhibitory action on tryptophan pyrrolase and that other nicotinyl derivatives, including NADH, nicotinamide mononucleotide, nicotinamide, and nicotinic acid, were also inhibitory to the enzyme, but only at considerably higher concentrations.

    In a tryptophan loading study of Bantu patients with hemosiderosis and chronic scurvy, Hankes et al. (1973) found significantly increased levels of anthranilic acid glucuronide, acetylkynurenine, kynurenine, and hydroxykynurenine in the urine. Hankes et al. (1974) suggested that the high iron levels of the tissues of these patients stimulate the formulation

Table 3
VITAMIN CONCENTRATIONS IN HUMAN PLASMA AND
CSF[a]

Vitamin	Plasma (μM)	CSF (μM)	CSF/plasma ratio
At low or normal plasma concentrations			
Ascorbic acid	24	78	3.3
Inositol	28	138	4.9
Folates	0.013	0.032	2.5
At higher plasma concentrations			
Ascorbic acid	52	106	2.0
Inositol	346	317	0.9
Folates	0.129	0.090	0.7

Note: Human cerebrospinal fluid ascorbic acid levels are twice as high as plasma levels
   when ascorbate is plentiful; they are three times as high when ascorbic acid supplies
   are low.

[a]   CSF — cerebrospinal fluid.

From Spector, R. (1977), *N Engl. J. Med.*, 296, 1393. With permission.

of monodehydroascorbate and reduce the level of available ascorbic acid. We may conjecture that the increased oxidation-reduction potential converts NADH to NAD, and NADPH to NADP, thus holding the levels of NADH and NADPH down so that there is insufficient NADPH for the feedback control of tryptophan pyrrolase, which regulates the quantity of tryptophan converted into the kynurenine pathway. Hankes et al. (1974) also surmised that, "this uncontrolled tryptophan metabolism can present other metabolic problems. If these patients (like the pellagrin) deplete their tryptophan supply by converting it into kynurenine and other excreted products, they have very little remaining for conversion into products of the serotonin-synthesis pathway." This could be the reason why a number of pellagra and scurvy patients eventually develop mental problems such as dementia and depression.

The finding of increased levels of tryptophan metabolites in the urine of ascorbic acid-deficient people is of great interest because at least three of them (3-hydroxyanthranilic acid [3-HOA], 3-hydroxykynurenine, and 2-amino-3-hydroxyacetophenone) were found by Allen et al. (1957) to be carcinogenic when implanted into the urinary bladders of mice. Moreover, Schlegel (1975) of Tulane University has pointed out that cancer of the bladder is most common in elderly male smokers who are known to have low ascorbic acid levels. This may be more than a coincidence, as Schlegel found that high doses of ascorbic acid prevented the carcinogenic effect of 3-HOA pellets implanted in the bladders of mice. Schlegel reported that 3-HOA is very unstable under simulated physiological conditions and is rapidly oxidized, but can be stabilized by ascorbic acid. Schlegel was also interested in the chemiluminescence of urine, which was found to be higher in the urine from smokers and bladder tumor patients than in urine from normal individuals. The origin of this chemiluminescence was not known, but it, too, could be suppressed by higher concentrations of ascorbic acid.

The existence of short-lived free radicals of noradrenaline and of serotonin were demonstrated by Borg (1965a, b). Lewin (1976) has suggested that the findings of Polis et al. (1969) can be interpreted as supporting a role for these free radical forms of noradrenaline and serotonin as neurotransmitters in the central nervous system.

Spector (1977), discussing the active transfer mechanism in the choroid plexus, which maintains a higher level of ascorbic acid in the cerebrospinal fluid than in the blood plasma (Table 3), stated that, "ascorbic acid is homogenously distributed throughout the mammalian brain, in which its only known role is to serve as a cofactor for the enzyme that converts dopamine into norepinephrine.''

However, there are reasons for believing that tryptophan conversion to serotonin by brain tissue may be reduced in scurvy, and that tryptophan conversion to kynurenine metabolites in the urine may be increased. It would therefore seem likely that the active transfer of ascorbic acid from blood plasma to the brain is also important for the conversion of tryptophan to serotonin in the brain.

Subramanian (1977) has reported an uneven distribution of ascorbic acid in the rat brain, the order being hypothalamus, 466; hippocampus, 410; cerebral cortex, 400; cerebellum, 355; striatum, 342; medulla-pons, 210; and the midbrain, 172 μg/g, respectively. He noted that ascorbic acid is closely linked with the synthesis of almost all the presumed neurotransmitters in the brain, including dopamine, norepinephrine, epinephrine, and serotonin. Moreover, he pointed out that ascorbic acid detoxifies histamine and may also be involved in the function of monoamine oxidase which destroys the pressor amines.

It is clearly tempting to try to link the depression which is associated with ascorbate deficiency to abnormalities of serotonin metabolism, but the exact state of affairs still seems to elude detection, and the data do not seem to support the hypothesis. Saner et al. (1975) found no significant change in the cerebral 5-hydroxytryptamine content either in pair-fed or ascorbic acid-deficient animals. The concentration of 5-hydroxyindoleacetic acid was significantly decreased in both these groups (compared with control animals fed *ad libitum*), but there was no significant difference between pair-fed and ascorbic acid-deficient animals.

Urba-Holmgren et al. (1979) gave their reasons for believing that serotonin may exert a positive modulating effect on yawning. Ogasahara et al. (1980) found higher levels of serotonin in the cerebral cortex and the midbrain of rats during slow-wave sleep and paradoxical sleep than during wakefulness. These findings stimulated interest in the possibility that tryptophan supplements might be helpful for inducing sleep. Pollet and Leathwood (1983) reported the effects of a 500-mg dose of L-tryptophan, concealed in a 10-g bar of bitter chocolate, vs. chocolate alone on human sleep during five test and five placebo nights. Volunteers took the chocolate 1 h before retiring and made subjective assessments of sleep latency, sleep quality, sleep depth, night awakenings and sleepiness the next morning. It was found that relative to placebo, L-tryptophan significantly decreased perceived sleep latency ($p < 0.05$) and was sensed as being calming ($p < 0.01$). The effect was apparently short lasting, as other scores for night awakenings and sleepiness next morning were not affected. A detailed analysis showed that women (n = 33) and the under 40s (n = 24) were the subgroups most sensitive to L-tryptophan.

Since L-tryptophan is the precursor of serotonin, a neurotransmitter implicated in the control of sleep, and since ascorbic acid is necessary for the 5-hydroxylation of tryptophan, it is interesting to note that women and young people, who tend to have higher ascorbic acid levels, were the ones who benefited most from the somnorific effect of the tryptophan supplement. However, ascorbic acid itself does not cause somnolence; rather it increases alertness, as shown by Kubala and Katz (1960), who found that orange juice supplements caused a significant improvement in the intelligence quotient (I.Q.) scores of children from kindergarten through college.

The observation by Cohen et al. (1983) that serotonin can mediate endothelium-dependent relaxation of coronary arteries may be relevant in connection with the negative correlation between dietary ascorbic acid intake and ischemic heart disease (see Table 1, Chapter 20 of this volume). However, the direct contractile and the indirect relaxing activities of serotonin on the smooth muscle of the coronary arteries present a complex problem.

REFERENCES

Allen, M. J., Boyland, E., Dukes, C. E., Horning, E. S., and Watson, J. G. (1957), Cancer of the urinary bladder induced in mice with metabolites of aromatic amines and tryptophan, *Br. J. Cancer*, 11, 212.

Borg, D. C. (1965a), Transient free radical forms of hormones EPR spectra from catecholamines and adrenochrome, *Proc Natl. Acad. Sci. U S A.*, 53, 633.

Borg, D. C. (1965b), Transient free radical forms of hormones. EPR spectra from iodothyronines, indoles, estrogens and insulin, *Proc Natl. Acad. Sci. U S A.*, 53, 829.

Brodie, B. B., Pletscher, A., and Shore, P. A. (1955), Evidence that serotonin has a role in brain function, *Science*, 122, 968.

Cho-Chung, Y. S. and Pitot, H. C. (1967), Feedback control of rat liver tryptophan pyrrolase. I. End product inhibition of tryptophan pyrrolase activity, *J. Biol Chem.*, 242, 1192.

Cohen, R. A., Shepherd, J. T., and Vanhoutte, P. M. (1983), 5-hydroxytryptamine can mediate endothelium-dependent relaxation of coronary arteries, *Am J Physiol.*, 245, H1077.

Cooper, J. R. (1961), The role of ascorbic acid in the oxidation of tryptophan to 5-hydroxytryptophan, *Ann. N Y Acad. Sci.*, 92, 208.

Giarman, N. J. and Freeman, D. X. (1960), Serotonin content of the pineal gland of man and monkey, *Nature (London)*, 186, 480.

Green, H. and Sawyer, J. L. (1965), Tryptophan hydroxylase of rat brain, *Fed. Proc. Fed. Am Soc Exp Biol*, 24, 604.

Hankes, L. V., Jansen, C. R., and Schmaeler, M. (1974), Ascorbic acid catabolism in the Bantu with hemosiderosis (scurvy), *Biochem Med.*, 9, 244

Hankes, L. V., Leklem, J., Brown, R. R., Mekel, R. C., and Jansen, C. R. (1973), Abnormal tryptophan metabolism in patients with scurvy-type skin, *Biochem. Med.*, 7, 184.

Harper, H. A. (1961), *Review of Physiological Chemistry*, Lange Medical, Los Altos, CA.

Hughes, R. E., Hurley, R. J., and Jones, P. R. (1971), The retention of ascorbic acid by guinea pig tissues, *Br. J. Nutr.*, 26, 433.

Kubala, A. L. and Katz, M. M. (1960), Nutritional factors in psychological test behavior, *J. Genet Psychol.*, 96, 343.

Levine, R. J., Lovenberg, W., and Sjoerdsma, A. (1964), Hydroxylation of tryptophan and phenylalanine by murine neoplastic mast cells, *Fed. Proc. Fed. Am. Soc. Exp. Biol.*, 23, 563

Lewin, S. (1976), *Vitamin C. Its Molecular Biology and Medical Potential*, Academic Press, London.

Nakamura, S., Ichiyama, A., and Hayaishi, O. (1965), Purification and properties of tryptophan hydroxylase in brain, *Fed. Proc. Fed. Am. Soc. Exp. Biol.*, 24, 604.

Nakashima, Y., Suzue, R., Sanada, H., and Kawada, S. (1970), Effect of ascorbic acid on hydroxylase activity. I. Stimulation of tyrosine hydroxylase and tryptophan-5-hydroxylase activities by ascorbic acid, *J. Vitaminol.*, 16, 276.

Ogasahara, S., Taguchi, Y., and Wada, H. (1980), Changes in serotonin in rat brain during slow-wave sleep and paradoxical sleep: application of the microwave fixation method to sleep research, *Brain Res.*, 189, 570.

Pollet, P. and Leathwood, P. D. (1983), The influence of tryptophan on sleep in man, *Int. J. Vitam. Nutr. Res.*, 53, 223.

Polis, B. D., Wyeth, J., Goldstein, L., and Graedon, J. (1969), Stable free radical forms of plasma proteins or simpler related structures which induce brain excitatory effects, *Proc. Natl. Acad. Sci. U S.A.*, 64, 755.

Rajalakshmi, R., Malathy, J., and Ramakrishnan, C. V. (1967), Effect of dietary protein content on regional distribution of ascorbic acid in the rat brain, *J. Neurochem.*, 14, 161.

Saner, A., Weiser, H., Hornig, D., Da Prada, M., and Pletscher, A. (1975), Cerebral monoamine metabolism in guinea pigs with ascorbic acid deficiency, *J. Pharm. Pharmacol.*, 27, 896.

Schildkraut, J. J. (1965), The catecholamine hypothesis of affective disorders: a review of supporting evidence, *Am. J. Psychiatry*, 122, 509.

Schlegel, J. U. (1975), Proposed uses of ascorbic acid in prevention of bladder cancer, *Ann. N.Y. Acad. Sci.*, 258, 432.

Shimizu, N., Matsunami, T., and Onishi, S. (1960), Histochemical demonstration of ascorbic acid in the locus coeruleus of the mammalian brain, *Nature (London)*, 186, 479.

Sokoloff, B. (1964), The biological activity of serotonin, *Growth*, 18, 113.

Spector, R. (1977), Vitamin homeostasis in the central nervous system, *N. Engl. J. Med.*, 296, 1393.

Subramanian, N. (1977), On the brain ascorbic acid and its importance in metabolism of biogenic amines, *Life Sci.*, 20, 1479.

Urba-Holmgren, R., Holmgren, B., Rodriquez, R., and Gonzalez, R. M. (1979), Serotonergic modulation of yawning, *Pharmacol. Biochem. Behav.*, 11, 371.

Chapter 8

ADRENAL CORTICOID METABOLISM

I. INTRODUCTION

When Szent-Györgyi (1928) isolated a strong reducing agent from the adrenal cortex of the ox and found it to be an acidic six-carbon sugar, he named it hexuronic acid. It was not until 4 years later that King and Waugh (1932) and Svirbely and Szent-Györgyi (1932a, b) discovered that this substance was identical to vitamin C which could be extracted from oranges, lemons, cabbages, or Hungarian red peppers. Harris and Ray (1932) and Zilva (1932) fed the cortex of ox adrenals to guinea pigs and thereby proved the antiscorbutic activity of these glands. They found that 2 g of raw ox adrenal corresponded to 6 ml of orange juice. Studying guinea pigs on a vitamin C-deficient diet, Harris and Ray (1933) estimated tissue hexuronic acid by its ability to reduce silver nitrate, causing a black stain of metallic silver. They found that in guinea pigs deprived of vitamin C, the hexuronic acid disappeared from the adrenals and from the liver as the animals became vitamin deficient.

Many workers have attempted to elucidate the relationship between adrenal ascorbic acid and adrenal cortical function. Lockwood and Hartman (1933) found that injection of an ascorbate-free extract of the adrenal cortex, containing "cortin", did not protect guinea pigs from scurvy, but did delay the onset of the symptoms of scurvy in animals receiving a half-protective daily dose of orange juice. They concluded that cortin, or some related substance, aids in the utilization of vitamin C. This concept is discussed at some length in Chapter 16, Volume I.

In the past, several authors, including Giroud and Santa (1939), Giroud (1940), Giroud and Ratsimamanga (1940), and Giroud et al. (1940), reported evidence suggesting decreased adrenal cortical function in guinea pigs with scurvy. They believed ascorbic acid to be necessary for the production of steroids by the adrenal and suggested that cortin and ascorbic acid should be given together in the treatment of Addison's disease. Assays of cortin by its special effect on the pigment cells of goldfish seemed to support their thesis of cortin deficiency in scorbutic guinea pig adrenals, but these observations have since proven to have been misleading. The adrenal steroid output of guinea pigs is now known to be normal in the early stages of vitamin C deficiency and to be markedly elevated in scurvy; it is probably similar in human scurvy. The apparent decrease of adrenal corticoid effects, observed by Giroud et al. (1939, 1940) in scorbutic guinea pigs, was therefore not due to decreased production of cortical hormones, but could have been due to a lack of ascorbate for peripheral AA*-cortisol synergism, especially as regards the redox potential of the tissues. Similarly, the benefits reported by some authors following the administration of cortin in scurvy may have been due to such peripheral synergism between cortin and traces of dehydroascorbate remaining in the tissues.

II. HUMAN OBSERVATIONS

Wilkinson and Ashford (1936) reported three patients with Addison's disease who had abnormally low ascorbic acid stores and suggested an association between ascorbic acid deficiency and adrenal insufficiency, but Witts (1936) found normal storage and excretion of ascorbic acid in a woman with Addison's disease under his care. Giroud and Ratsimamanga (1940) cited Laederich et al. as having reported the disappearance of the symptoms of a

* AA — ascorbic acid, reduced form.

patient with Addison's disease following the administration of ascorbic acid alone. Di Bartolini and Michetti (1937) reported that the ratio of dehydroascorbic acid (DHAA) to ascorbic acid was increased in the blood of patients with Addison's disease. This is of particular interest in view of the later observations by Stewart et al. (1953a) that both adrenocorticotropic hormone (ACTH) and cortisone decrease the ratio of DHAA to AA in the blood plasma of human subjects *in vivo*, as shown in Figures 1 and 2 of Chapter 13, Volume I, of this book. Lack of cortisone could therefore account for a tendency to decreased ascorbate storage in Addison's disease, as ascorbic acid is stored in the reduced form.

Daughaday et al. (1948) studied the formaldehydogenic steroids in the urine of three adult patients with scurvy and obtained inconclusive results. The administration of vitamin C was associated with a decrease in the excretion of these steroids, followed by an increase, which was above normal in one instance.

Smolyanskii (1963) studied 144 elderly men and women (aged 60 to 90 years) at the Leningrad Medical Institute of Sanitation and Hygiene for healthy old people. They reported that ascorbic acid loading (500 mg daily for 12 to 14 d), improved the functional state of the adrenal cortex; 70% of the subjects showed increased blood levels of 17-hydroxycorticoids after this vitamin supplementation. The ascorbic acid loading also raised their urinary output of 17-ketosteroids and intensified their response to ACTH, as judged by the eosinopenic effect and by the effect on their production of steroid hormones. However, Kitabchi and Duckworth (1970) studied two men, aged 62 and 73, with scurvy and found them to have normal plasma cortisol levels. Moreover, both of these patients had normal plasma and urinary steroid responses to ACTH and to metyrapone stimulation. In fact, the excretion of 17-ketogenic steroids in response to metyrapone was greater before treatment than after treatment with ascorbic acid. Likewise, Shilotri and Bhat (1977) found no elevation in the plasma cortisol levels of human volunteers taking large doses of vitamin C. Furthermore, Dubin et al. (1978) studying 19 elderly women with low leukocyte ascorbic acid (TAA)* levels (less than 15 $\mu g/10^8$ cells), found them to have normal plasma cortisol levels and normal responses to ACTH; so earlier suggestions that ascorbic acid deficient individuals had impaired adrenal function have not been confirmed by the use of more modern methods.

III. GUINEA PIGS

Many studies of adrenal function in scorbutic guinea pigs were conducted in the 1950s; none of them was truly comparable to another, but they eventually led to the conclusion that adrenal corticosteroid production is increased in late scurvy. Oesterling and Long (1951) observed that the ascorbic acid in the adrenals of guinea pigs fell from a normal level of 165 to 6 mg/100 g in early scurvy and to 1.9 mg/100 g in late scurvy. The adrenal ascorbic acid level in early scurvy was not further depleted by ACTH, but there was a normal (42%) adrenal cholesterol depletion response to ACTH in early scurvy. Eisenstein and Shank (1951) observed a progressive decrease in the number of circulating eosinophils and a further prompt reduction by 50% in the eosinophil count after ACTH administration in scorbutic guinea pigs. These observations, along with an inverse relationship between adrenal weight and dietary ascorbic acid, were interpreted as indicating that ascorbic acid deficiency functions as a nonspecific stress, and that adrenal corticoid production is not impaired in scurvy.

Clayton and Prunty (1952) reported that guinea pigs on a scorbutogenic diet showed a gradual but well-marked increase in urinary 17-ketosteroid excretion, which reached a peak in the terminal phases of scurvy. In contrast, Banerjee and Deb (1952), studying scorbutic and pair-fed female guinea pigs, reported that the scorbutic animals excreted significantly less 17-ketosteroids in their urine. Nadel and Schneider (1952), measuring urinary formal-

---

* TAA — total ascorbic acid, reduced and oxidized forms.

dehydogenic steroids in the urine of male guinea pigs, showed normal, then low, and eventually very high urinary steroid levels as the animals became more and more severely scorbutic. Stepto et al. (1952) assayed the ACTH content of the pituitary glands of guinea pigs and concluded that the ACTH content increases as the duration of ascorbate deficiency increases. Done et al. (1953) found that the plasma concentration of 17-hydroxycortico-steroids was increased approximately tenfold in scorbutic guinea pigs, but returned to normal within 5 d after treatment with ascorbic acid. Becker et al. (1953) observed a sixfold increase in the conversion of acetate 1-^{14}C to cholesterol in the adrenals of severely scorbutic guinea pigs.

Using the Sayers test for biological assay of ACTH, Clayton et al. (1957) reported high levels in the serum of guinea pigs with scurvy and found that this was blocked by the administration of cortisone. These workers also observed that the urinary 17-ketosteroid and 17-ketogenic steroid excretion rose steadily during the development of scurvy, but were little changed in pair-fed controls receiving ascorbic acid. They concluded that the increased adrenal cortical activity in scurvy was not due to starvation: they pointed out that usually when cortisol is high, ACTH is low, so their findings in scurvy provided an exception. Both ACTH and cortisol levels were increased together, so they concluded that the dynamics of ACTH release must be abnormal in scurvy. Nevertheless, the ACTH levels of scorbutic guinea pigs returned to normal within 1 $1/2$ h after the administration of ascorbic acid. Perhaps histamine is responsible for this effect on the pituitary or the hypothalamus; blood histamine levels are markedly increased in scurvy (Chapter 1 of this volume), and histamine is known to stimulate the release of ACTH (Figure 5, Chapter 16, Volume I). Moreover, histamine is a particularly potent stimulant of adrenocortical activity and, unlike other forms of stress, it cannot be completely blocked by adrenal corticoid administration, as shown by Sayers and Sayers (1947, 1949) in the rat. Jones et al. (1958) demonstrated that even guinea pigs with moderate ascorbic acid deficiency had somewhat elevated plasma 17-hydroxycortico-steroid levels. They discussed a possible "braking role" of ascorbic acid in the synthesis or release of adrenal steroids.

Studying guinea pigs on a scorbutogenic diet, Guirgis (1965) reported first a decrease in the urinary 17-hydroxycorticoid excretion, from a daily mean of 98 to 54 μg on day 12, then an increase to 245 μg daily by day 21. Coinciding with these changes, there was a drop in the resistance of the guinea pigs to histamine aerosol on the 12th day and a marked rise in their resistance on the 21st day. While both ascorbic acid and hydrocortisone were recognized as having antianaphylactic properties, this author observed that the resistance of the animals to histamine aerosol changed, as the functional state of the adrenal cortex changed, in vitamin C deficiency.

More extensive studies of adrenal function in scorbutic guinea pigs by Hodges and Hotston (1970a, b), involving analysis of both plasma and adrenal cortisone and corticosterone concentrations, gave clear-cut results. Adrenal ascorbic acid levels and urinary 17-keto and ketogenic steroids were also measured. Comparison of Figure 1, showing the adrenal ascorbic acid (TAA) concentrations, with Figure 2, showing the plasma corticosteroid levels of the same guinea pigs at succeeding stages in the development of scurvy, clearly demonstrates an enormous (tenfold) increase in both plasma cortisol and corticosterone levels in late scurvy. This seems to remove all doubt about the ability of the adrenal cortex to function in the virtual absence of ascorbic acid. However, the adrenal glands of the same animals (Figure 3) showed a pronounced increase, followed by an even greater decrease in cortisol and corticosterone concentrations. The terminal fall in the concentration of adrenal steroids was greater than could be explained by the dilutional effect of adrenal enlargement in scurvy. It may mean that adrenal corticosteroid synthesis could not keep pace with the rate at which these hormones were being released, as suggested by Hodges and Hotston, or it could theoretically have represented decreased utilization followed by decreased synthesis of cor-

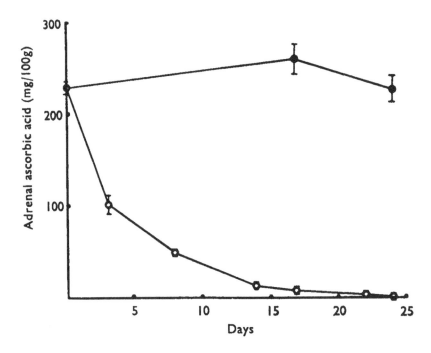

FIGURE 1.   Adrenal ascorbic acid (TAA) concentrations in guinea pigs on a diet deficient in vitamin C. Controls, ●—●; tests, ○—○; vertical bars indicate standard errors. (From Hodges, J. R. and Hotston, R. T. [1970b], *Br. J. Pharmacol.*, 40, 740. With permission.)

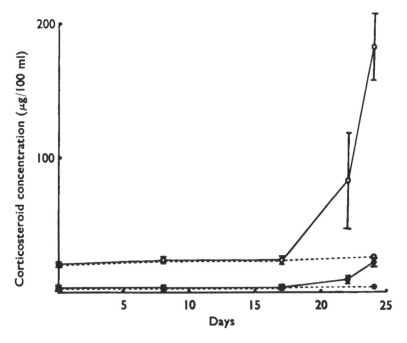

FIGURE 2.   Plasma corticosteroid concentrations in guinea pigs on a diet deficient in ascorbic acid. ○—○, cortisol; ●—●, corticosterone concentrations; dotted lines, control values; vertical bars, standard errors. (From Hodges, J. R. and Hotston, R. T. [1970b], *Br. J. Pharmacol.*, 40, 740. With permission.)

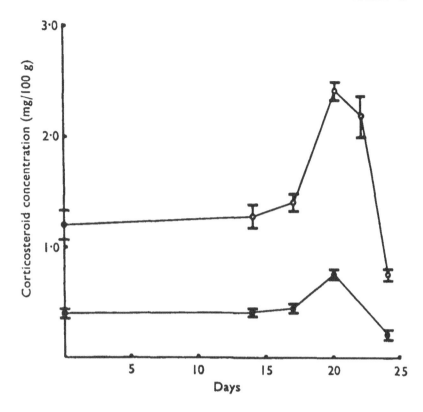

FIGURE 3.   Adrenal cortisol (o–o) and corticosterone (●–●) concentrations in guinea pigs on a diet deficient in ascorbic acid. The vertical bars indicate the standard errors. (From Hodges, J. R. and Hotston, R. T. [1970b], *Br. J. Pharmacol.*, 40, 740. With permission.)

ticosteroids. In the same study, the urinary 17-ketosteroid excretion showed a 20% increase on the 14th day and remained elevated for the next 7 d. In contrast, the excretion of 17-ketogenic steroids fell during the first 2 weeks of ascorbic acid deficiency and then increased considerably (Figure 4). Both histamine and corticotrophin increased the plasma corticosteroid concentrations when injected during the second week, but failed to change the high concentrations of the steroids in the third week of ascorbic acid deficiency.

Further studies by Hodges and Hotston (1971), using the "rat adrenal ascorbic acid depletion method" for the assay of ACTH, showed that the plasma ACTH level of vitamin C-deficient guinea pigs rose from a control value of 6.6 mU/100 ml on day 0, and 7.0 on day 14 to 245.5 mU/100 ml on day 21. Injection of ascorbic acid into scorbutic animals caused the ACTH levels to fall to undetectable levels in 90 min. The simultaneous elevations of ACTH and corticosteroid levels do suggest a blockage in the corticopituitary negative-feedback system in scurvy. This may be due to the inability of cortisone to promote the inactivation of histamine in the tissues when ascorbic acid is deficient, as shown in the scheme which is illustrated in Figure 5.

Encarnacion et al. (1974) reported that the mean plasma corticoid levels of guinea pigs which had received a scorbutogenic diet for 15 to 39 d was not significantly different from that of a control group receiving 2 g of ascorbic acid per kilogram of diet. However, they observed that all fasted guinea pigs and all guinea pigs stressed by massive dosage with ascorbic acid (86 g/kg of diet) had elevated plasma corticoid levels. These results are hard to reconcile with those of Fordyce and Kassouny (1977) who observed that a partial dietary ascorbic acid deficiency (0.1 mg/kg body weight) caused a marked increase in the adrenal calcium and plasma corticosteroid levels (from 85 to 130 μg/dl), but no change in the plasma

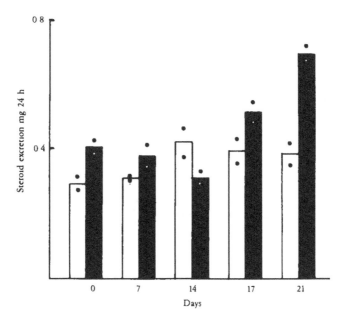

FIGURE 4.   Urinary 17-oxo (□) and 17-oxogenic (■) steroids excreted in 24 h by guinea pigs on a diet deficient in ascorbic acid. The dots indicate individual values. (From Hodges, J. R. and Hotston, R. T. [1970b], *Br. J. Pharmacol.*, 40, 740. With permission.)

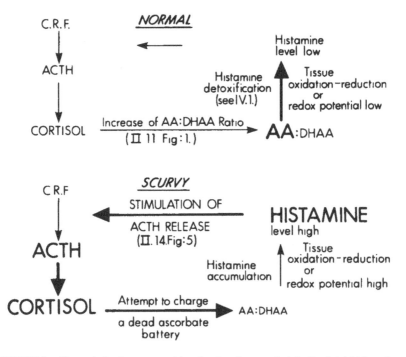

FIGURE 5.   Theoretical scheme to explain why the adrenocortical feedback inhibition of pituitary ACTH is blocked in scurvy, so that both ACTH and corticosteroid levels are increased in scorbutic guinea pigs.

Table 1[a]

Days on scorbutic diet	Plasma ascorbic acid (TAA) (mg/100 ml)	Leukocyte ascorbic acid (μg/10⁸ cells)	Adrenal ascorbic acid (mg/100 g)	Adrenal cortisol (mg/100 g)	Plasma cortisol (mg/100 ml)
O					
M	0.43	27 2	174.1	31.8	0.020
F	0.88	21 8	133.9	14.6	0.020
6					
M	0 50	22 8	167.5	30.1	0.040
F	1 03	23.2	130 4	13.1	0.020
12					
M	0 37	18.5	92.7	18 2	0.113
F	1 02	23.2	82.7	12 8	0.032
18					
M	0 21	12 7	23.3	7.8	0 267
F	0.26	12.6	27 6	9.2	0.177
24					
M	0.09	8.1	1 8	7.8	0.388
F	0.25	12.7	2.0	3.6	0.237
30					
M	0.09	9.2	0.9	10.2	0.587
F	0.08	4.6	1.2	2.0	0 296
36					
F	0.25	5.8	7.8	8.0	0 228

Note: Six male (M) and six female (F) guinea pigs were killed before and at 6-d intervals after subjection to a vitamin C-deficient diet As the animals became progressively more deficient in ascorbic acid, the adrenal cortisol levels fell, but the plasma cortisol levels showed a progressive increase.

[a] Data obtained and compiled from Odumosu (1982).

ACTH levels of guinea pigs. These authors felt that their findings supported the hypothesis of Kitabchi (1967a, b), that a high adrenal concentration of ascorbate, in the resting stage, exerts a "braking" influence on corticosteroid synthesis, and that a low adrenal ascorbate level releases this inhibition.

The work of Wilbur and Walker (1978) added further support to the "braking" theory, for these authors reported that excessive dietary ascorbic acid increased adrenal ascorbic acid levels and delayed cortisol release from the adrenals of guinea pigs. Detailed studies of adrenal functions in scorbutic guinea pigs by Björkhem et al. (1978) confirmed the existence of elevated blood ACTH (365 vs. 161 ng/l) and elevated plasma cortisol levels in scurvy. These workers also demonstrated that ascorbate inhibits adrenal mitochondrial side chain cleavage of cholesterol.

Odumosu (1982) confirmed the existence of a progressive rise of the 11-hydroxycorticosteroid (cortisol) levels in the plasma of guinea pigs subjected to a vitamin C-deficient diet, but he also observed a progressive fall in the adrenal cortisol levels in scurvy, as shown in Table 1. These data raise the question as to whether the high plasma cortisol levels of scurvy are due to increased adrenal production or to decreased peripheral utilization of the hormone. It is quite conceivable that ascorbic acid is necessary for the peripheral utilization of cortisone.

Doulas et al. (1987) studied the adenyl cyclase (ACL) activity of the adrenal tissues of guinea pigs receiving ascorbic acid 0, 0.1, 5, 20, and 100 mg/100 g body weight per day by intraperitoneal injection. They found no change in the basal ACL activity, but did observe a progressive decrease of sodium fluoride-stimulated ACL activity with increasing ascorbate dosage. Indeed, there was a highly significant negative correlation between NaF-stimulated ADL activity and plasma ascorbic acid concentration. Although these workers did not measure ACTH levels in that study, they concluded that the enlarged adrenals, the augmented ACL activity, and the increased plasma cortisol level of 257 μg/d (normal 42.5) which they found were strongly indicative of pituitary adrenal stimulation.

IV. RATS

Studies of ascorbic acid deficiency cannot be conducted in rats, as they synthesize ascorbic acid in the liver. However, Solomon and Stubbs (1961) demonstrated that hypophysectomy in rats caused an abrupt decrease of urinary excretion, an increase in the half-life, and a decrease in the size of the body pool of ascorbic acid in rats. Moreover, Civen et al. (1980) observed that a markedly increased dietary ascorbic acid (1%) intake caused a modest inhibition in both the adrenal (-22%) and plasma (-27%) corticosterone responses to stress.

V. *IN VITRO* STUDIES

Studies of beef adrenals by Cooper and Rosenthal (1962) led them to postulate that ascorbic acid inhibits, and catecholamines accelerate, certain steps in the synthesis of adrenal corticosteroids. They studied the effects of D-noradrenaline, D-adrenaline, and L-ascorbic acid on C-21 hydroxylation of 17α-OH-progesterone by bovine adrenocortical microsomes. Concentrations of ascorbate as low as 10^{-5} M inhibited C-21 hydroxylation by 30 to 60%. The inhibition could be relieved by similarly low concentrations of the catecholamines which, in the absence of ascorbate, accelerated C-21 hydroxylation by 30 to 60%. They concluded that ascorbate and catecholamines antagonistically affect the rate of side reactions that compete with the hydroxylase for the electrons of TPNH.

Kitabchi (1967a, b) renewed interest in this theory and reviewed the evidence that both 11β- and 21-hydroxylase of beef adrenal are inhibited by ascorbic acid. He postulated that this inhibition is released by the adrenal ascorbate-depleting activity of ACTH and that corticosteroid synthesis is thus accelerated.

Shimizu (1970) obtained soluble enzyme preparations capable of cholesterol side chain cleavage from bovine and porcine adrenal mitochondria. Studying the effect of ascorbic acid on these enzyme systems, he observed first stimulation (1 to 3 mM), and then, with higher concentrations (15 to 20 mM), inhibition of side chain cleavage by ascorbic acid. Since the average concentration of ascorbic acid in bovine and porcine adrenal glands has been reported to be about 10 mM, it was concluded that these effects of ascorbic acid were produced by concentrations of ascorbic acid within the physiological range (Figure 6).

Studies by Greenfield et al. (1980) confirmed that ascorbate inhibits the 21-hydroxylation system of bovine adrenal cortical microsomes. They also observed that both oxidized and reduced glutathione (GSSG and GSH) stimulated this enzyme; GSSG stimulated the enzyme fivefold and GSH stimulated it tenfold. They also isolated glutathione reductase and suggested a role for this enzyme affecting 21-hydroxylation. Bovine serum albumen and also EDTA were found to accelerate 21-hydroxylase activity, so it would seem to be inhibited by heavy metals and is accelerated when they are chelated.

Clearly more work is needed to obtain a fuller understanding of this subject.

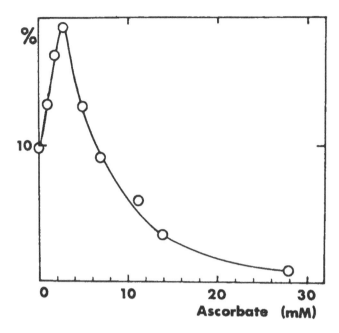

FIGURE 6. The rate of cholesterol side chain cleavage by a soluble enzyme preparation from porcine adrenal mitochondria, at different ascorbate concentrations, is indicated as the percentage of [^{14}C]isocaproic acid formed from [26-^{14}C]cholesterol The normal pig adrenal cortex is reported to contain approximately 10 mM of ascorbate, so adrenal ascorbate depletion by ACTH would presumably accelerate cholesterol side chain cleavage. (From Shimizu, K. [1970], *Biochim. Biophys. Acta,* 210, 333. With permission.)

REFERENCES

Banerjee, S. and Deb, C. (1952), Urinary excretion of 17-ketosteroids in scurvy, *J. Biol. Chem.,* 194, 575.

Becker, R. R., Burch, H. B., Salomon, L. L., Venkitasubramanian, T. A., and King, C. G. (1953), Ascorbic acid deficiency and cholesterol synthesis, *J. Am. Chem. Soc.,* 75, 2020.

Björkhem, I., Kallner, A., and Karlmar, K. E. (1978), Effect of ascorbic acid deficiency on adrenal mitochondrial hydroxylations in guinea pigs, *J. Lipid Res ,* 19, 695.

Civen, M., Leeb, J. E., Wishnow, R. M., and Morin, R. J. (1980), Effects of dietary ascorbic acid and vitamin E deficiency on rat adrenal cholesterol ester metabolism and corticosteroidogenesis, *Int. J. Vitam. Nutr. Res.,* 50, 70.

Clayton, B. E., Hammant, J. E., and Armitage, P. (1957), Increased adrenocorticotrophic hormone in the sera of acutely scorbutic guinea pigs, *J. Endocrinol.,* 15, 284.

Clayton, B. E. and Prunty, F. T. G. (1951), Relation of adrenal cortical function to scurvy in guinea pigs, *Br. Med. J.,* 2, 927.

Cooper, D. Y. and Rosenthal, O. (1962), Action of noradrenaline and ascorbic acid on C-21 hydroxylation of steroids by adrenocortical microsomes, *Arch. Biochem. Biophys.,* 96, 331.

Daughaday, W. H., Jaffe, H., and Williams, R. H. (1948), Adrenal corticol hormone excretion in endocrine and nonendocrine disease as measured by chemical assay, *J. Clin. Endocrinol.,* 8, 244.

Di Bartolini, B. and Giuseppe, M. (1937), Studio clinico del rapporto acido deidroascorbico/acido ascorbico, *Bull. Soc. Ital. Biol. Sper.,* 12, 307.

Done, A. K., Ely, R. S., Heiselt, L. R., and Kelley, V. C. (1953), Circulating 17-hydroxycorticosteroids in ascorbic acid-deficient guinea pigs, *Proc. Soc. Exp. Biol. Med.,* 83, 722.

Doulas, N. L., Constantopoulos, A., and Litsios, B. (1987), Effect of ascorbic acid on guinea pig adrenal adenylate cyclase activity and plasma cortisol, *J. Nutr.,* 117, 1108.

Dubin, B., MacLennan, W. J., and Hamilton, J. C. (1978), Adrenal function and ascorbic acid concentrations in elderly women, *Gerontology*, 24, 473.

Eisenstein, A. B. and Shank, R. E. (1951), Relationship of ascorbic acid to secretion of adrenocortical hormones in guinea pigs, *Proc Soc. Exp. Biol. Med.*, 78, 619.

Encarnacion, D., Devine, M. M., and Rivers, J. M. (1974), Influence of vitamin C nutriture and inanition on ACTH stimulated release of adrenal corticosteroids in guinea pigs, *Int. J. Vitam. Nutr. Res.*, 44, 309.

Fordyce, M. K. and Kassouny, M. E. (1977), Influence of vitamin C restriction on guinea pig adrenal calcium and plasma corticosteroids, *J Nutr*, 107, 1846.

Giroud, A. (1940), Les conditions de la fonction corticosurrénalienne; importance de l'acide ascorbique, *Presse Med*, 48, 841.

Giroud, A. and Ratsimamanga, A. R. (1940), L'insuffisance surrénalienne de l'avitaminose et de hypovitaminose C, *Presse Med.*, 48, 449.

Giroud, A. and Santa, N. (1939), Absence d'hormone corticale chez les animaux carencés en acide ascorbique, *C R. Soc. Biol.*, 131, 1176.

Giroud, A., Santa, N., and Martinet, M. (1940), Variations de l'hormone corticale en fonction de l'acide ascorbique, *C. R. Soc. Biol.*, 134, 23.

Greenfield, N., Ponticorvo, L., Chasalow, F., and Lieberman, S. (1980), Activation and inhibition of the adrenal steroid 21-hydroxylation system by cytosolic constituents: influence of glutathione, glutathione reductase, and ascorbate, *Arch. Biochem. Biophys.*, 200, 232.

Guirgis, H. M. (1965), The regulatory role of vitamin C on the adrenal function and resistance to histamine aerosol in the scorbutic guinea-pig, *J. Pharm. Pharmacol.*, 17, 674.

Harris, L. J. and Ray, S. N. (1932), Vitamin C and the adrenal cortex. I. Antiscorbutic activity of ox suprarenal, *Biochem. J.*, 26, 2067.

Harris, L. J. and Ray, S. N. (1933), Specificity of hexuronic (ascorbic) acid as antiscorbutic factor, *Biochem. J.*, 27, 580.

Hodges, J. R. and Hotston, R. T. (1970a), Pituitary-adrenocortical activity in the ascorbic acid deficient guinea pig, *Br. J. Pharmacol.*, 39, 193P.

Hodges, J. R. and Hotston, R. T. (1970b), Ascorbic acid deficiency and pituitary adrenocortical activity in the guinea pig, *Br. J. Pharmacol.*, 40, 740.

Hodges, J. R. and Hotston, R. T. (1971), Suppression of adrenocorticotrophic activity in the ascorbic acid deficient guinea pig, *Br. J. Pharmacol.*, 42, 595.

Jones, R. S., Perič-Golia, L., and Eik-Nes, K. (1958), Ascorbic acid deficiency and adrenocortical function in the guinea pig, *Endocrinology*, 63, 659.

King, C. G. and Waugh, W. A. (1932), The chemical nature of vitamin C, *Science*, 75, 357.

Kitabchi, A. E. (1967a), Inhibitory effect of ascorbic acid on steroid hydroxylase systems of beef adrenal cortex, *Fed. Proc. Fed. Am. Soc. Exp. Biol.*, 26, 484.

Kitabchi, A. E. (1967b), Ascorbic acid in steroidogenesis, *Nature (London)*, 215, 1385.

Kitabchi, A. E. and Duckworth, W. C. (1970), Pituitary adrenal axis evaluation in human scurvy, *Am. J. Clin. Nutr.*, 23, 1012.

Lockwood, J. E. and Hartman, F. A. (1933), Relation of the adrenal cortex to vitamins A, B and C, *Endocrinology*, 17, 501.

Nadel, E. M. and Schneider, J. J. (1952), Excretion of formaldehydogenic (FG) substances by normal and scorbutic guinea pigs, *Endocrinology*, 51. 5.

Odumosu, A. (1982), Ascorbic acid and cortisol metabolism in hypovitaminosis C guinea pigs, *Int. J. Vitam. Nutr. Res.*, 52, 175.

Oesterling, M. J. and Long, C. N. H. (1951), Adrenal cholesterol in the scorbutic guinea pig, *Science*, 113, 241

Salomon, L. L. and Stubbs, D. W. (1961), Some aspects of the metabolism of ascorbic acid in rats, *Ann. N.Y. Acad. Sci.*, 92, 128.

Sayers, G. and Sayers, M. A. (1947), Regulation of pituitary adrenocorticotrophic activity during the response of the rat to acute stress, *Endocrinology*, 40, 265.

Sayers, G. and Sayers, M. A. (1949), The pituitary adrenal system, *Ann. N.Y. Acad. Sci.*, 50, 522.

Shilotri, P. G. and Bhat, K. S. (1977), Effect of megadoses of vitamin C on bactericidal activity of leukocytes, *Am. J. Clin. Nutr.*, 30, 1077

Shimizu, K. (1970), Effects of ascorbic acid on the side-chain cleavage of cholesterol, *Biochim. Biophys. Acta*, 210, 333.

Smolyanskii, B. L. (1963), Effect of ascorbic acid on functional state of adrenal cortex in elderly persons, *Fed. Proc. Fed. Am. Soc. Exp. Biol.*, 22, T1173; and *Ter. Arch.*, 35 (1), 71, 1973.

Stepto, R. C., Pirani, C. L., Fisher, J. D., and Sutherland, K. (1952), ACTH content of pituitary at different levels of ascorbic acid intake, *Fed. Proc. Fed. Am. Soc. Exp. Biol.*, 11, 429.

Stewart, C. P., Horn, D. B., and Robson, J. S. (1953a), The effect of cortisone and adrenocorticotrophic hormone on the dehydroascorbic acid of human plasma, *Biochem. J.*, 53, 254.

Svirbely, J. L. and Szent-Györgyi, A. (1932a), Hexuronic acid as the antiscorbutic factor, *Nature (London),* 129, 576, 1932; and *Nature (London),* 129, 690, 1932.

Svirbely, J. L. and Szent-Györgyi, A. (1932b), The chemical nature of vitamin C, *Biochem. J.,* 26, 865.

Szent-Györgyi, A. (1928), Observations on the function of peroxidase systems and the chemistry of the adrenal cortex, *Biochem. J.,* 22, 1387.

Wilbur, V. A. and Walker, B. L. (1977), Dietary ascorbic acid and the time of response of the guinea pig to ACTH administration, *Nutr. Rep. Int.,* 16, 789.

Wilkinson, J. F. and Ashford, C. A. (1936), Vitamin-C deficiency in Addison's disease, *Lancet,* October 24, 967

Witts, L. J. (1936), Vitamin-C deficiency in Addison's diseases, *Lancet,* 2, 1184.

Zilva, S. S. (1932), The antiscorbutic activity of the cortex of the suprarenal gland of the ox, *Biochem. J.,* 26, 2182.

Chapter 9

URIC ACID CLEARANCE

Hypoxanthine (6-dioxypurine) is oxidized to xanthine (2,6,dioxypurine), and xanthine to uric acid (2,6,8-trioxypurine) by the enzyme xanthine oxidase, which has been found in liver and in milk. In the course of an investigation on the nature of the prosthetic group of xanthine oxidase, Feigelson (1952) observed that, "incubation of a purified cream xanthine oxidase preparation with trace amounts of ascorbic acid resulted in a marked loss of xanthine oxidase activity." The extent of the inhibition appeared to be a function of the concentration of ascorbic acid as well as the length of time that the enzyme and the ascorbate were incubated together in the absence of substrate.

A genetic lesion in primates and in Dalmatian dogs is responsible for loss of the enzyme uricase; this causes uric acid to be the end product of purine metabolism. Proctor (1970) has suggested that the resulting accumulation of uric acid may be of benefit to species which have lost the ability to synthesize ascorbic acid from simple sugars. He suggested the possibility that uric acid may act as an alternate electron donor when ascorbic acid levels are low and offered this as a possible explanation of the long time taken by primates to develop scurvy and the rapidity of its development in guinea pigs.

Stein et al. (1976) observed that ascorbic acid administration increased the urinary excretion of uric acid in human subjects and noted that this uricosuria was inhibited by acetyl salicylate. They concluded that ascorbic acid increases urinary excretion of uric acid by an action on the renal tubules; they also observed a decrease in the serum uric acid levels. Measuring the ratio of "uric acid clearance" to "creatinine clearance", they found that a single dose of ascorbic acid (4 g) given to nine subjects caused an increase to $202 \pm 41\%$ (SD) in the relative uric acid clearance 4 to 6 h later. Administration of 8 g of ascorbic acid daily for 10 d to three subjects resulted in an increase to $174 \pm 24\%$ in the relative uric acid clearance compared with control values. The serum uric acid values of the three subjects fell from 6.4 to 4.8, from 9.8 to 7.7, and from 4.1 to 2.9 mg/100 ml during the 10 d of ascorbic acid supplementation.

Berger et al. (1977) studied the urinary clearance rates of both ascorbic acid and uric acid in human subjects. They confirmed that ascorbic acid administration increased the excretion of uric acid and suggested that ascorbic acid probably competes with urate for renal tubular reabsorption, probably in the proximal tubule. This uricosuric effect was observed both in gouty and in nongouty men when the plasma ascorbic acid level rose above 6 mg/100 ml.

Studying four groups of guinea pigs receiving low (3 mg daily) and high (37.5 mg daily) ascorbic acid intakes, with and without 0.16% cholesterol in their food, Hanck and Weiser (1979) noted that the plasma uric acid levels decreased 20 to 30% in the cholesterol-loaded as well as in the nonloaded groups with higher ascorbic acid intakes, but the differences were not statistically significant. These authors remarked that, "gout is another entity associated with hypercholesterolemia and atherosclerosis which led to the recognition of hyperuricemia as a coronary risk factor." Not only was there this tendency for lower uric acid levels in the guinea pigs receiving the higher ascorbic acid diet, they also found decreased plasma total and free cholesterol and triglycerides in the ascorbate supplemented groups. Moreover, the cholesterol contents of both guinea pig and human aortas were found to be negatively correlated with plasma ascorbate levels. They concluded that, "ascorbic acid seems to be the only substance which depresses both elevated uric acid levels and hyperlipoproteinemia."

Elliot (1982) studied the effects of ascorbic acid loading on serum constituents in 26 volunteers (22 men and 4 women) whose ages ranged from 25 to 76 years. They took sodium ascorbate (3 g daily) for 12 weeks. Decreases in the serum levels of uric acid occurred in 22 of the 26 subjects and there was no change in the other 4. These subjects had a high normal mean serum ascorbic acid level of 1.2 mg/100 ml before loading and 2.1 mg/100 ml after loading. Nevertheless, the mean uric acid level of the group fell from 6.2 ± 0.24 (SE) to 5.7 ± 0.22 mg/100 ml.

It would seem that ascorbic acid may decrease plasma uric acid levels in two ways: by inhibiting the synthesis of uric acid from xanthine and by increasing the renal excretion of uric acid. If so, uric acid levels will tend to rise in ascorbic acid deficiency. This would theoretically have the effect of facilitating the alternate electron donor mechanism of the Proctor hypothesis.

REFERENCES

Berger, L., Gerson, C. D., and Yu, T. (1977), The effect of ascorbic acid on uric acid excretion with a commentary on the renal handling of ascorbic acid, *Am. J. Med*, 62, 71.

Elliott, H. C. (1982), Effects of vitamin C loading on serum constituents in man, *Proc. Soc Exp. Biol. Med.*, 169, 363

Feigelson, P. (1952), Inhibition of xanthine oxidase by ascorbic acid, *Fed. Proc. Fed. Am. Soc. Exp. Biol.*, 11, 210.

Hanck, A. and Weiser, H. (1979), The influence of vitamin C on lipid metabolism in man and animals, *Int. J. Vitam. Nutr Res.*, 19, 83.

Proctor, P. (1970), Similar functions of uric acid and ascorbic acid in man?, *Nature (London)*, 228, 868.

Stein, H. B., Hasan, A., and Fox, I. H. (1976), Ascorbic acid induced uricosuria: a consequence of megavitamin therapy, *Ann. Intern. Med.*, 84, 385.

Clinical Conditions Associated with Disorders of Ascorbic Acid Metabolism

Chapter 10

RHEUMATIC FEVER

I. EXPERIMENTAL PRODUCTION IN THE GUINEA PIG

The role of ascorbic acid in rheumatic fever has been the subject of much dispute since Rinehart and Mettier (1933), working at the University of California Medical Center in San Francisco, discovered that they could produce degenerative and proliferative lesions of the heart valves, the endocardium, and the myocardium in guinea pigs, which resemble those of human rheumatic carditis, by infecting vitamin C-deficient guinea pigs with beta hemolytic streptococci. This led them to conduct a series of experiments to determine the effects upon the heart valves and muscle of acute and chronic scurvy, of scurvy combined with infection, and of infection alone (Rinehart and Mettier, 1934).

A. Controls

Eight control guinea pigs were maintained on a basal vitamin C-deficient diet, supplemented with adequate amounts of orange juice. None of these animals showed any gross pathology. Macroscopically the heart valves of these control animals had a smooth, glistening surface and microscopically they had a compact, rich, fibrous stroma consisting of abundant tightly packed, wavy, uninterrupted collagen fibrils. All but one had a normal endocardium. In one animal a very mild proliferation of the endocardial layer was seen overlying the insertion of one of the chordae tendineae.

B. Infection

A total of 20 animals receiving the basal ration supplemented with orange juice were infected with beta hemolytic streptococci or with *B. aertrycke* by intracutaneous injection of a pure culture of the organisms into the skin of the thigh below the groin. One animal showed a necrotizing mitral valvulitis, a second showed an accumulation of a few polymorphonuclear leukocytes near the base of the tricuspid valve. Another developed uremia due to obstruction of the urinary tract and was found to have mild atypical mitral valvulitis. No lesions were seen in the myocardium.

C. Scurvy

Eight out of nine adequately examined hearts from guinea pigs with uncomplicated scurvy revealed thinning, fragmentation, and disorganization of the regular axial arrangement of the collagen fibers of the heart valves. A loss of the normal wavy contours and, at times, a hyaline degenerative change of the fiber substance was observed. In two animals there was also noted a mild but definitely proliferative reaction of the endothelial and subendothelial cells.

D. Combined Scurvy and Infection

Of 31 animals subjected to one or another combination of scurvy and infection with beta hemolytic streptococci or *B. aertrycke*, reasonably adequate examination of the heart was secured in 24 instances. All of these showed recognizable degenerative and/or proliferative changes in one or more of the heart valves. The mitral and aortic valves were most commonly and most severely involved, showing a mucoid degenerative appearance of the stroma and a distinct nodular proliferative reaction at the lines of closure. Multinucleate cells were seen in some of the proliferative verrucous lesions. Elsewhere homogeneous eosinophilic hyaline material, paler mucoid material, and proliferating cells made up the lesion. Mostly the proliferation was subendothelial, but in places the surface layer was eroded. The valvular pathology produced in the guinea pig by this combination of scurvy and infection is virtually identical to that classically described in rheumatic fever, where the verrucous lesions are formed by an edematous, mucoid, hyaline, fibrinoid swelling of the subendothelial layer, which pushes the endothelium before it into a warty nodule.

Judging their own work in a critical manner, Rinehart and Mettier stated that they could not fairly claim to have reproduced Aschoff nodes in the heart muscle. However, they pointed out that proliferative lesions were seen in the myocardium and beneath the mural endocardium in animals subjected to a combination of scurvy and infection, which bear a strong resemblance to and are believed to be fundamentally similar to reactions seen in rheumatic fever.

Although rheumatic fever develops in susceptible individuals following an upper respiratory infection due to group A beta hemolytic streptococci, Rinehart and Mettier found that beta hemolytic streptococci of group C and other natural guinea pig pathogens were capable of causing these myocardial and valvular lesions when combined with vitamin C deficiency in the guinea pig.

Rinehart et al. (1934), studying the knee joints of guinea pigs, demonstrated that chronic scurvy with infection or, to a lesser extent, chronic scurvy alone produces an arthropathy which has striking pathological similarities to rheumatic fever and also to rheumatoid arthritis. They found a subcutaneous nodule in one guinea pig, similar to those seen in rheumatic fever and drew attention to the widespread evidence of fibrinoid degeneration, which has been considered the fundamental lesion of rheumatic fever. They observed degenerative changes in the skeletal muscle, focal necrosis in the liver, fibrosis of the Malpighian bodies of the spleen, erythrophagocytosis in the lymph nodes, and focal lymphocytic accumulation in the kidneys, all of which are frequently seen in rheumatic fever.

E. Scurvy and Exotoxin

Stimson et al. (1934), who had been unable to reproduce rheumatic fever in other animals, soon confirmed the experimental work of Rinehart and Mettier by infecting vitamin C-deficient guinea pigs. Moreover, they reported that it was possible to produce degenerative and proliferative myocardial lesions resembling Aschoff nodes by injecting streptococcal exotoxin into scorbutic guinea pigs without the introduction of living organisms. Schultz (1936a) repeated the experiments of Rinehart and Mettier, with only partial confirmation. He found that chronic scurvy and chronic infection, acting synergistically, may induce nonpurulent carditis in guinea pigs. Valvulitis with fibrinoid degeneration and an intense proliferative reaction constituted the most prominent lesion. However, he concluded that the changes only slightly resembled those seen in rheumatic fever.

Taylor (1937) also observed valvulitis, myocarditis, and occasionally pericarditis, but no gross valvular vegetations, in scorbutic guinea pigs and observed Gram-positive organisms where none had been injected. The lesions were most common in the mitral valve, the atrioventricular junction, the perivascular areas in the myocardium, and the papillary muscles. Once the heart lesions have developed, he found that curing the scurvy would not cure the lesions, though it did prevent the development of congestive heart failure. He stated that this scorbutic carditis is not specific for hemolytic streptococci and concluded, "If scorbutic carditis is related to rheumatic fever, then ascorbic acid will be of greatest use in prevention and not treatment." This seems to be true.

McBroom et al. (1937) found myocardial and valvular lesions, both degenerative and proliferative, to be equally evident in scorbutic guinea pig hearts, with or without infection. They therefore concluded that these lesions represented scorbutic carditis and had nothing to do with rheumatic carditis. If only they had known that hemolysis causes destruction of ascorbic acid (see Chapter 15, Volume I) they might have suspected (as does the present writer) that the beta hemolytic streptococcus can cause local hemolysis and thus local scurvy whenever it is active in a tissue with borderline ascorbic acid deficiency. That is what is so special about the beta hemolytic streptococcus in causing the human disease, and that explains why these special bacteria are not necessary when studying guinea pigs on an ascorbic acid-deficient diet. All are agreed that infection alone does not cause this carditis.

II. RHEUMATIC PURPURA

Rinehart (1937) suggested that a scorbutic state may be the basis of the hemorrhagic features commonly seen in the acute phases of rheumatic fever. As evidence that hemorrhagic manifestations are of frequent occurrence in rheumatic fever, he cited Poynton and Paine (1914) who described blood-stained synovial fluid in rheumatic fever, Coburn (1933) who stated that hemorrhagic lesions were widespread in all patients dying during marked activity of the rheumatic process, Chester and Schwartz (1934) who recorded the rather frequent occurrence of purpuric skin lesions on the legs and arms in rheumatic fever and considered this to be evidence of recurrent activity of the disease, Holtz and Friedman (1934) who noted a hemorrhagic eruption in the mucosa of the mouth and throat in rheumatic patients, and Van der Sande (1935) who also recorded the occurrence of hemorrhagic manifestations.

Chester and Schwartz observed that no common medication had been administered to the 10 children with purpura among 21 with rheumatic fever under his care at the Montefiore Hospital in New York in 1931. He therefore concluded that the purpuric lesions were the result of the rheumatic state.

Coburn (1933) reported on 320 patients who died with rheumatic fever, who were observed by him during life, and who, in most instances, were examined post mortem. He stated that, "when it had been possible to study the tissues of patients during the first two weeks of a rheumatic attack, hemorrhagic lesions have been conspicuous." Indeed, "in all the patients dying during marked activity of the rheumatic process, hemorrhagic lesions were found to be widespread . . . " Hemorrhagic lesions in the skin, brain, kidneys, and thoracic and abdominal viscera suggested to Coburn an alteration in the vascular permeability with diapedesis and damage to the mesodermal tissues. He concluded that, "the anatomic observations on these patients with acute rheumatism indicate that, in addition to the well recognized swelling of endothelium and fragmentation of collagen, diffuse hemorrhagic changes are characteristic of activity of the rheumatic process." Coburn observed that a marked alteration of the permeability of the walls of the blood vessels is suggested by the occurrence of epistaxis, purpuric erythemas, and hemorrhages from the urinary tract, the respiratory tract, and the intestinal tract in acute rheumatism. He stated, "The relative frequency of these hemorrhages is illustrated by the incidence of 96 nose-bleeds among 15 of 30 convalescent rheumatic children." Hemoptysis, hematemesis, melena, hemorrhagic pneumonia, and hemorrhagic edema of the larynx were also observed in his patients. He found that, "The course of hematuria minima (microscopic haematuria) ran parallel to the apparent activity of the rheumatic process. Its onset preceeded the recognized manifestations of acute rheumatism by 24 hours."

Pelner (1942) reported the development of severe ringing in the ears due to salicylate intolerance, as well as several nosebleeds, a positive tourniquet test, and a low plasma ascorbic acid level (0.4 mg/100 ml) in a 16-year-old boy during salicylate treatment for rheumatic fever. The salicylates were discontinued and he was given ascorbic acid, 100 mg three times a day. The nosebleeds and the tinnitus ceased after 48 h of vitamin C therapy and did not return, although the salicylates were increased to the initial high level and were continued at that dosage for 10 additional days.

The combination of purpura, defective collagen, and swollen painful joints is characteristic of ascorbic acid deficiency, so it is readily understandable that a beta hemolytic streptococcal infection can create the pathology of rheumatic fever in an ascorbic acid deficient subject.

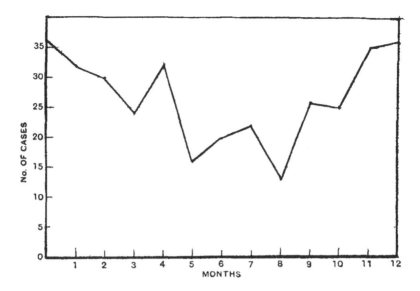

FIGURE 1.   Of children above the age of 3 years, who attended Guy's Hospital Outpatient Clinic, 15% were found to have rheumatic fever, rheumatic carditis, or chorea. Their parents were carefully questioned as to the actual onset of the disease. The months of occurrence of first attacks or recurrences are plotted here for each month of the year. 1 denotes January, 2 denotes February, etc. It is evident that there was a markedly increased incidence of these diseases in the winter months. (From Campbell, M. and Warner, E. C [1930], *Lancet*, 1, 61. With permission.)

III. EPIDEMIOLOGY OF RHEUMATIC FEVER

An association between ascorbic acid deficiency and rheumatic fever certainly helps to explain the geographic, socioeconomic, and seasonal distribution of the disease. Rinehart et al. (1934) pointed out that the debilitated state of the prerheumatic child has been emphasized clinically and that rheumatic fever is a disease of the poor and undernourished. Campbell and Warner (1930) and Swift (1930) found that rheumatic fever was from 15 to 20 times more frequent in the laboring classes than in the middle and upper classes, and most frequent among poor children of elementary school age. Typically the disease occurs in a growing child of a poor family living in a city, particularly in the North Temperate Zone; it has its greatest incidence in the winter months, as shown in Figure 1. It has always been more common in the poorly housed and poorly nourished, especially in the late winter and early spring, and for some reason, girls are somewhat more commonly affected than boys. A family history of rheumatic fever is much more common in patients than in controls, but one does not know whether this is due to genetic factors, to dietary deficiency, or to other social and environmental factors such as crowding. According to Hedley (1939), the annual death rate from rheumatic heart disease in persons aged 5 to 24, was lower in all regions of the U.S. during the period of 1930 to 1936 than in the period of 1922 to 1929. In Philadelphia, where the study was initiated, the reduction was 25%, and this occurred before the introduction of either sulfonamides or penicillin. It did, however, coincide with a period of increased knowledge of nutrition and better distribution of garden and dairy products.

Warner et al. (1935) stated that, "it is noticeable how the better-fed children and adults seem to escape rheumatic fever." However, in a study of the foods available in the homes of rheumatic fever patients and their families, in a poor law institution and at Christ's Hospital, they found no history of a shortage of potatoes or fruits and other vegetables in

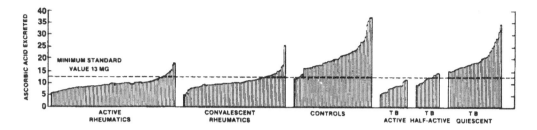

FIGURE 2. The amounts of ascorbic acid (AA) excreted in the urine per day by active and convalescent rheumatic fever patients are arrayed for comparison with control subjects. The results obtained in patients with active, half active, and quiescent surgical tuberculosis are also shown. Ascorbic acid in milligrams per 10 stones of body weight. (From Abbasy, M. A., Gray Hill, N., and Harris, L. J [1936], *Lancet*, 2, 1413. With permission.)

the rheumatic fever patients or their families. The only possible correlation was with the supply of fresh milk. The incidence of rheumatic fever in the Lent terms at Christ's Hospital (School) in London fell from 2.4% in 1918 to 1922, to 1.1% in 1923 to 1927, and to 0.75% in 1928 to 1932, as the average daily fresh milk supply increased from 0.48 to 0.73 to 0.86 pt per boy.

IV. ASCORBATE STATUS IN RHEUMATIC FEVER

A. Urinary Excretion Tests

Abbasy et al. (1936) studied the urinary excretion of ascorbic acid (AA)* by 107 patients with active rheumatic fever, 8 patients convalescing from rheumatic fever, and 64 normal individuals on the same diet, who served as controls. While control subjects excreted an average of 20 mg of ascorbic acid per 24 h and none less than 13 mg, those with active rheumatic fever excreted an average of 9 mg and convalescent rheumatics, 10 mg/ 24 h. Very few patients with rheumatic fever had an excretion rate exceeding 13 mg/24 h. Thus, the data showed very little overlap between rheumatic fever and normal (Figure 2). Other infections such as osteomyelitis and tuberculosis, had been found to be associated with decreased urinary ascorbic acid levels, but Abbasy et al. observed that an unusual feature of rheumatic fever was that the patients continued to show a disturbance of ascorbic acid metabolism during convalescence and continued to give a poor response to a standard test dose of ascorbic acid (700 mg/10 stone**/d), as shown in Figure 3.

However, these observations were disputed. Perry (1935), using the urinary excretion test, concluded from studies on a small group of five active and six quiescent cases of rheumatic fever in children that vitamin C deficiency is not an important factor in the causation of acute rheumatism, but that mild degrees of vitamin C deficiency are not uncommon in rheumatic children. Unfortunately he did not study any normal children for comparison.

Sendroy and Schultz (1936) found an apparent ascorbic acid deficiency in 8 out of 13 cases of rheumatic fever, but only 2 of these could be ascribed to poor diet. They stated, "Through digestive disturbances, patients with rheumatic fever evidently may develop a real hypovitaminosis on an ordinarily adequate diet." They concluded that their findings did not support the concept that a condition of ascorbic acid deficiency is a predisposing factor in the causation of rheumatic fever.

Parsons (1938) studied 18 children with acute rheumatism and found that they excreted more ascorbic acid in their urine on average than did nonrheumatic children. This was probably due to the fact that some of them were already being treated with salicylates on

* AA — ascorbic acid, reduced form.
** 1 stone = 63.5 kg or ≈14 lb.

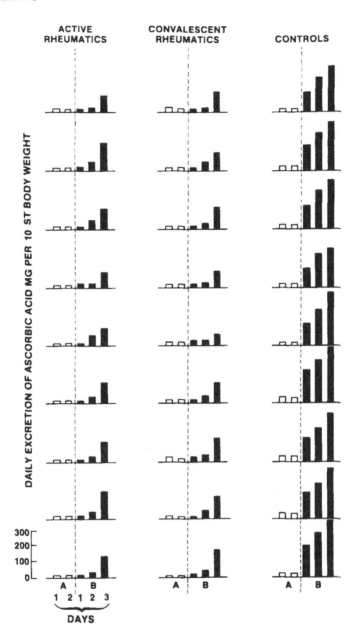

FIGURE 3.   Urinary excretion of ascorbic acid (AA) by active and convalescent rheumatic fever patients and by control subjects, (A) before and (B)
after the daily administration of test doses of ascorbic acid (700 mg/10 stones
body weight). Active and convalescent rheumatic fever patients gave little
response until the third day, and then generally less than the controls gave on
the first day. (From Abbasy, M. A., Gray Hill, N., and Harris, L. J. [1936],
Lancet, 2, 1413. With permission.)

admission to hospital; salicylates are known to cause a pronounced increase in urinary
ascorbic acid excretion (Chapter 22, Volume I).

B. Blood Values

Rinehart et al. (1936) found the plasma ascorbic acid (AA) levels to be uniformly lowered
in 21 cases of acute rheumatic fever. The mean level was 0.39 mg/100 ml, as compared to

a mean of 0.81 in 19 children admitted to the hospital for tonsillectomy. The low levels rose in response to extra supplements of vitamin C, but in many this rise was refractory; that is, the intake required to maintain an adequate level of vitamin C in the plasma was much above the average requirement for normal persons.

Rinehart et al. (1938) reported an average plasma ascorbic acid (AA) level of 0.30 (range 0.11 to 0.68) mg/100 ml in 30 patients with acute rheumatic fever and 0.8 (range 0.22 to 1.57) mg/100 ml in 19 control subjects. These findings were disputed by Abt et al. (1942) who found relatively low ascorbic acid levels in 76 children with scarlet fever (mean 0.49 mg/100 ml), but not in 17 patients with acute rheumatic fever (mean 0.5 mg/100 ml) nor in 26 with acute rheumatic heart disease. However, the ascorbic acid levels of the two children with rheumatic carditis who died, one receiving an ascorbic acid supplement and one not, both were exceptionally low.

Kaiser and Slavin (1938) investigated the incidence of hemolytic streptococci in the tonsils of 123 children, as related to the ascorbic acid (AA) content of the tonsils and the blood. They found a higher incidence of streptococci in the tonsils of the children who had the lower ascorbic acid values in the blood and tonsils. They also performed virulence tests on the streptococci isolated from the tonsils and noted not only that the streptococci are less likely to be found in the tonsils of children who have high blood ascorbic acid values, but also that the streptococci isolated from such children were seldom virulent.

In a study of the whole blood ascorbic acid (AA) levels of 13 children with acute rheumatic fever and 12 with chronic rheumatic fever, Kaiser (1938) found the blood levels of the rheumatic children to be lower than the levels in a group of normal children in good health. However, he found that children suffering from other acute febrile illnesses, such as pneumonia, nephritis, and tonsillitis, also had low blood ascorbic acid levels. His studies did not support the contention that vitamin C deficiency is a significant etiological factor in rheumatic infection; nevertheless, he stressed the importance of including adequate amounts of vitamin C in the diets of all persons suffering from rheumatic fever.

V. ASCORBIC ACID SUPPLEMENTS IN RHEUMATIC FEVER

Faulkner (1935) studied the effects of administering vitamin C (200 to 300 mg/d) in 27 patients with rheumatic fever. He reported no specific therapeutic effect on the course of the rheumatic fever, but observed an increase in the reticulocyte count in the majority of patients, which he likened to that which follows the administration of vitamin C to patients with the anemia of scurvy.

Schultz et al. (1935) failed to observe any demonstrable improvement in a therapeutic trial, but they did not state the dose of ascorbic acid, the number of patients, or the duration of the trial in their brief communication.

Schultz (1936b) studied 56 patients, aged 4 to 19 years, each of whom had experienced one or more attacks of rheumatic fever in previous years and was living at home. Of these, 28 were assigned to Group A and received ascorbic acid, 100 mg daily by mouth; 28 carefully matched patients constituted Group B and received lactose placebo capsules. Capillary fragility, as tested by the suction method, showed a decrease in the mean capillary strength of the control group from January to April, during which time the mean capillary strength of the supplemented group was increased. Thus it would seem that many of these patients with quiescent rheumatic fever had subclinical ascorbic acid deficiency and that Group A patients were protected from this by the ascorbic acid supplement. However, this dose of ascorbic acid (100 mg daily) did not prevent recurrence of rheumatic activity, nor did 250 mg of ascorbic acid daily seem to be of any benefit in the treatment of 17 patients with active rheumatic fever.

Perla and Marmorston (1937) pointed out that, "prolonged insufficiency of the vitamin may irreparably injure the organism to the point at which restoration of the normal state may never occur." They further added, "It is possible that, once a rheumatic infection is established, the restoration of normal availability of vitamin C will not affect the carditis or prevent recurrences of the infection which is entrenched in the organism."

Abt et al. (1942) reported that the blood ascorbic acid levels of patients with rheumatic fever rose as a result of supplementation with 300 to 600 mg of ascorbic acid a day, but the course of the disease did not appear to be affected in any way by the dietary supplements given. However, a report by Massel et al. in the *New England Journal of Medicine* (1950) described seven patients who all appeared to benefit from daily supplementation with ascorbic acid at a dose of 4 g/d. McCormick (1955) also reported successful treatment of rheumatic fever in several patients using massive doses (1 to 10 g daily) of ascorbic acid.

VI. USE OF ASCORBIC ACID IN PROPHYLAXIS AGAINST RHEUMATIC FEVER

Roff and Glazebrook (1939), caring for a large number of 16-year-old boys at a Royal Naval training establishment, observed that many of the new recruits had gingivitis. All received full dental care; local causes of chronic hypertrophic gingivitis, such as calculus, were removed, and they were all instructed in dental hygiene. Local causes of gingivitis were, however, few compared with the lesion more often observed, which appeared to be without obvious cause, and differed clinically in being a gingivostomatitis rather than a gingivitis. In these cases the gums were congested and spongy, the surfaces having a gelatinous feel. Bleeding did not occur on simple palpation, but if one pierced with a probe, the hemorrhage was more copious than usual. The congestion was uniform, from the gums into the sulci on to the buccal mucous membrane, extending backwards and involving the tonsils and the pharyngeal wall. In some cases, the lips were of a deeper red, suggesting an increased vascularity. In all of these cases, vitamin C deficiency was found, using the "Test-Dose" method of Abbasy et al. (1935). On a dosage of 200 mg per boy per day, saturation, as determined by urinary excretion, was not achieved until 22 doses had been given. This would appear to indicate that the deficiency of vitamin C was in the neighborhood of 4 g. It was probably less because of vitamin losses and utilization during the 22 d it took to saturate them. Of the boys, 300 were saturated with vitamin C, as above, and 300 boys living under exactly the same conditions on the same diet served as a control group. In the supplemented group, the gingivostomatitis was reduced from 17.6 to 4.9%, and in the control group from 16.3 to 12.6%. Those who failed to respond to treatment with vitamin C had marginal gingivitis of the type which responds to oral hygiene. The incidence of gingivostomatitis was greatest in boys from an area of poverty on Tyneside; recruits from that area were also found to be more prone to develop rheumatic fever.

A subsequent study at the same Naval training establishment by Glazebrook and Thomson (1942) showed that a supplement of ascorbic acid, 200 mg daily, had only a slight effect on the incidence of common colds and tonsillitis. It reduced the duration of illness from streptococcal tonsillitis, but afforded complete protection against the subsequent development of rheumatic fever or pneumonia. There were 16 cases of rheumatic fever and 17 cases of pneumonia among 1100 controls, and no case of either disease among 335 youths receiving ascorbic acid supplements.

Since the discovery of penicillin and so many other antibiotics capable of killing the beta hemolytic streptococcus, these antibacterial agents undoubtedly play the leading role in prophylaxis against recurrences of rheumatic fever, but it is wrong to forget the lessons of the past.

Rinehart continued to work and to write on the role of ascorbic acid deficiency in the etiology of rheumatic fever and the role of ascorbic acid and bioflavonoids in the treatment of this disease for many years (Rinehart, 1935, 1936, 1943, 1944, 1945, 1953, 1955), but his work was soon forgotten. In 1965, two editorials reviewing the natural history, prevention, and treatment of rheumatic fever appeared in the *British Medical Journal* (on June 26 and September 11) which did not even mention diet.

VII. OTHER PROPHYLACTIC AND THERAPEUTIC MEASURES

A cooperative Anglo-American clinical trial initiated in 1950, enrolled 497 children with rheumatic fever in 18 months and compared the relative merits of adrenocorticotropic hormone (ACTH), cortisone, and aspirin in their treatment (Rutstein and Densen, 1965). After 10 years of observation, the authors reported that they found no evidence that one of these treatments was better than another. In fact, the results were remarkably good in all three groups, probably because of the use of penicillin and sulfadiazine and the availability of ascorbic acid supplements. None of these substances was discussed in the final report, as they were available to all.

Keith (1960) discussed the decreasing incidence of morbidity and mortality from rheumatic fever in Canada, the U.S. and Great Britain. He mentioned a possible decrease in the virulence of the group A beta hemolytic streptococcus and also improved social and economic conditions of the population, with less crowding in the homes, but gave reasons for believing that the use of penicillin for children with nasopharyngitis and tonsillitis was probably of greatest importance. Keith advocated monthly intramuscular injections of long-acting benzathine penicillin to prevent recurrences of rheumatic fever. There used to be extensive debates on the relative merits of oral penicillin G, and oral penicillin V, both of which had been recommended by the Ministry of Health of the U.K. Loss of penicillin G by the action of gastric acid was discussed in the *British Medical Journal* leader of June 26, 1965. This undoubtedly reduces the amount of penicillin available for absorption, but it may serendipitously have a beneficial effect, by forming penicillamine (dimethyl cysteine) which is a strong reducing and chelating agent belonging to the group of sulfhydril amino acids which reduce dehydroascorbic acid (DHAA) to ascorbic acid *in vitro* and *in vivo*, thereby aiding tissue storage of ascorbic acid. In other words, oral penicillin G may convey benefits beyond its antibiotic capabilities. Nevertheless, Stollerman (1960) reports that monthly injections of 1.2 million U of long-acting benzathine penicillin is much more active than oral penicillin or sulfadiazine in preventing recurrences of rheumatic fever.

VIII. THE HYPERIMMMUNE RESPONSE

Hypersensitivity to hemolytic streptococcal nucleoproteins was reported by Swift (1930) in 88% of children with rheumatic fever between 6 and 10 years of age, compared with only 12% of nonrheumatic children. However, he pointed out that we do not know definitely whether the relationship between streptococcal hypersensitiveness and rheumatic fever is causal or merely concomitant.

The theory that local antigen-antibody reactions cause the lesions of rheumatic fever received renewed interest when Vazquez and Dixon (1957, 1958), using a fluorescent antibody technique, demonstrated localization of human gamma globulin in the endocardium and in the perivascular connective tissue of the hearts of three children who had died of active rheumatic heart disease; this was nonspecific gamma globulin and not necessarily related to any antigen-antibody reaction. Nevertheless, such localization is consistent with the report of Stimson et al. who produced degenerative and proliferative myocardial lesions resembling Aschoff bodies by injecting streptococcal exotoxin into scorbutic guinea pigs and valvular lesions by injecting streptococci into scorbutic guinea pigs.

Elevated titers of antistreptolysin O and/or antistreptodornase B provide evidence of the

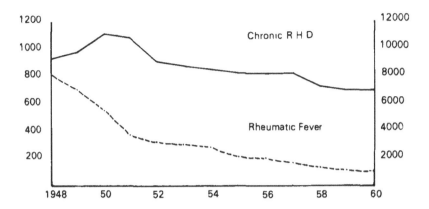

FIGURE 4.   Actual deaths from rheumatic heart disease and from rheumatic fever. England and Wales, all ages 1948 to 1960. (From Bywaters, E. G. L. [1978], in *Copeman's Textbook of Rheumatic Diseases*, 5th ed., Scott, J. T., Ed., Churchill Livingstone, London. With permission.)

streptococcal infection which usually precedes rheumatic fever by 2 or 3 weeks. Certainly rheumatic fever is an abnormal response to infection, but it seems to be the preexisting state of the child that determines the occurrence of rheumatic disease following a streptococcal throat infection. Campbell and Warner observed as early as 1930 that it is the debilitated child who develops rheumatic fever. They listed a series of symptoms observed by parents as occurring months or years before the onset of rheumatic fever: restlessness in sleep, night terrors, unaccountable crying and bad temper, irritability, stammering, fainting, twitching and blinking, nocturnal enuresis, and loss of weight. The incidence of rheumatic fever following streptococcal tonsillitis is closely related to the extent of the rise in the antistreptolysin O (ASO) titer. Stollerman (1961) reported that the rise in the ASO titer was related to the virulence of the streptococcus. High bacterial virulence and low host resistance are clearly important factors in the development of any infection. We are interested here in the evidence that ascorbic acid deficiency may be a crucial factor not only in decreasing resistance to infection, but also in causing hypersensitivity to one or more of 20 distinct extracellular antigens which are now known to be produced by group A beta hemolytic streptococci.

Kaplan et al. (1961) found that the serum of patients with rheumatic fever contained antibodies reacting both with streptococci and with myocardial sarcolemma; these belong to the class of so called "auto immune" antibodies. Their existence is strong evidence that the streptococcus is concerned with the development of heart damage. However, Bywaters (1978) pointed out that cross-reacting antibodies had been found in a number of patients with uncomplicated sore throats, in a high proportion of patients following myocardial infarction, and also patients following cardiac surgery, so they are not specific for rheumatic fever. This does not cast any doubt on the fact that rheumatic fever is due to group A beta hemolytic streptococcal infection and that rheumatic heart disease involves immunologically mediated muscle damage, but does suggest that other factors must be involved: poverty, crowding, debility, and poor nutrition seem to be necessary for the development of the full-blown disease.

Detailed studies have shown cross-reactivity between various streptococcal antigens and substances identified in human myocardial sarcolemmal membranes, myocardium, heart valves, caudate nucleus neurons, and even structures associated with the cardiac conduction system. Nevertheless, Williams et al. (1980), studying T cell binding to streptococcal membranes, consider that rheumatic fever is still an enigma.

In recent years, with improved nutrition and better housing, the incidence of rheumatic fever has been markedly reduced (Figure 4) in all but the poorest areas of the world, but post-mortem findings continue to show lesser degrees of cardiac valvular damage. This suggests that "miniature" rheumatic fever still occurs at a subclinical level in many people following streptococcal throat infections.

IX. THE EFFECT OF CORTISONE ON ASCORBIC ACID METABOLISM IN RHEUMATIC FEVER

While conducting a study of the reduced and oxidized ascorbic acid levels in the blood plasma of women with menorrhagia at the University of California Medical Center in San Francisco in 1963, the writer saw a 19-year-old Irish woman who was critically ill with a recurrence of rheumatic fever and who also had excessive vaginal bleeding. The results of analysis of five blood samples from this patient by a method similar to that used by Stewart et al. (1953), at intervals over a period of 6 weeks, before, during, and after the institution of prednisone therapy (and reinstitution of multivitamin therapy) gave the results shown in Table 1 and Figure 5 (Clemetson, 1969). Her plasma reduced ascorbic acid (AA) levels were low, in spite of the vitamin supplements which she had received intermittently during her stay in hospital. However, her total ascorbic acid (TAA)* levels were quite within the normal range. The striking abnormality was the very low percentage of her plasma ascorbic acid which was in the reduced (AA) form, when compared with results obtained by the same method in other individuals with similar TAA levels. The percentage of the plasma ascorbic acid in the reduced form tends to be low when the total ascorbic acid is low. Such percentage values are to be found in patients who have low total ascorbic acid (TAA) levels and have a capillary fragility state bordering on scurvy, but not in people with TAA levels within the normal range.

This patient with rheumatic fever and severe menorrhagia had a capillary fragility state, with a prolonged bleeding time of 30 minutes (normal 2 to 4 min), but adequate platelets and normal coagulation time and clot retraction. The first plasma sample from this patient with rheumatic fever showed slight hemolysis and the high percentage of ascorbic acid in the oxidized form (at 1 h after drawing blood) was therefore thought to be an artifact, as Kellie and Zilva (1935) have shown that hemolysis can cause a loss of ascorbic acid *in vitro*. Therefore, another sample was drawn 1 week later and analysis of this sample, which showed no hemolysis, demonstrated a lower total ascorbic acid level and a high percentage in the oxidized form. There was immense clinical improvement within 5 d of starting prednisone therapy, and she recovered from the acute attack.

Rinehart (1937) found persistently low plasma ascorbic acid levels in an 11-year-old child with severe acute rheumatic carditis, in spite of daily oral supplementation of 200 to 400 mg of ascorbic acid. Her plasma ascorbic acid levels remained low for 40 d before she died.

Thus, it would seem that ascorbic acid alone will not rectify the deficiency once the disease is active, perhaps because the patients are unable to keep ascorbic acid in the reduced form. In this respect, is is interesting that Stewart et al. (1953) have shown that cortisone administered to normal individuals has the effect of reducing DHAA to ascorbic acid *in vivo*, just as penicillamine does *in vitro*.

X. CONCLUSIONS

The writer has not had the opportunity to conduct further studies on rheumatic fever, but clearly much work is needed to increase our understanding of this disease. In the meantime it would seem reasonable to ensure an adequate supply of ascorbic acid, as well as penicillin, for all patients with acute streptococcal tonsillitis and to use cortisone with ascorbic acid in

---

* TAA — total ascorbic acid, reduced and oxidized forms.

Table 1
19-YEAR-OLD GIRL WITH RHEUMATIC FEVER

Date	Notes	Max. temp. (°C)	Daily penicillin dosage	ASO titer[a]	Plasma Ascorbic Acid (μg/ml)			DHAA (as % of TAA)	Reduced/oxidized ascorbic acid ratio
					TAA[b]	AA[c]	DHAA[d]		
Feb. 18	Mitral & aortic valvulitis	39.0	Oral 600,000	125					
Mar. 6	External cardiac massage for ventricular fibrillation	37.7	i.v. 50 million						
Mar. 27	Specimen hemolyzed	38.7	None		4.2	2.0	2 2	52	0.9:1
Apr. 3	Still critically ill; no hemolysis	39.2	None	333	2.3	1.0	1 3	56	0 8:1
Apr. 11	5th day on prednisone; 7th day on multivitamins; immense improvement.	37.9	None		5 4	4.1	1.3	24	3 2.1
Apr. 25	Afebrile	37.0	None		11.7	7.5	4.2	36	1.8.1
May 8	Convalescent	37.2	None		7.2	5 0	2.2	30	2.3:1
15 samples from 5 healthy nurses:			Mean values		12.3	10.8	1.5	12.8 ± 5.6	7.2:1

Note: Clinical and laboratory data concerning a 19-year-old woman who was critically ill with acute rheumatic carditis Treatment with prednisone and ascorbic acid orally began on April 4. Numerous other medications were used in the treatment of this patient. It may be noted that only 13% of the ascorbic acid in the plasma of five healthy nurses was in the oxidized form, but more than 50% of the ascorbic acid in the plasma of this patient was in the oxidized form before treatment.

[a] ASO titer — antistreptolysin titer in Todd units.
[b] TAA — total ascorbic acid, reduced and oxidized forms.
[c] AA — ascorbic acid, reduced form
[d] DHAA — TAA minus AA = dehydroascorbic acid + diketogulonic acid, but mainly dehydroascorbic acid (DHAA) or ascorbone. (See Stewart et al., 1953.)

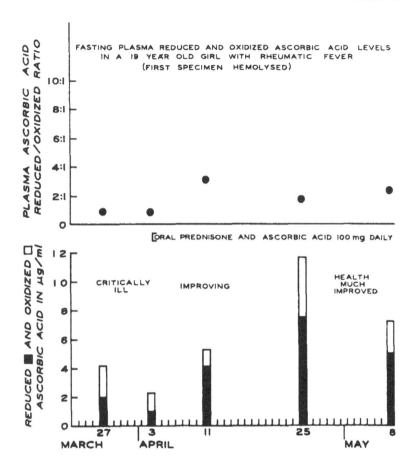

FIGURE 5. Showing the plasma reduced and oxidized ascorbic acid levels in the blood plasma of a 19-year-old woman while she was critically ill with acute rheumatic fever with carditis, and during her response to treatment with prednisone and ascorbic acid. The plasma ascorbic acid reduced/oxidized ratio, which is normally 7.1, had fallen below 1:1 when she was most severely ill with rheumatic fever. The AA/DHAA ratio rose as her health improved, but it did not return to normal.

the treatment of acute rheumatic fever. A capsule containing ascorbic acid (200 mg) and (+)catechin (200 mg) may be suitable for use, once a day, in routine prophylaxis against rheumatic fever. A capsule containing ascorbic acid (500 mg) and (+)catechin (200 mg) may be used twice or three times a day in the hope of preventing recurrences.

Actually the amount of (+)catechin needed to chelate heavy metal catalysts will depend on the amount of heavy metals in a glass of drinking water, but since this will vary from house to house and from time to time, one can only advise an excess of this natural nontoxic food fiber (see Chapter 11, Volume I).

Although aspirin has been shown to be useful in rheumatic fever, it seems to the writer like a poor substitute for vitamin C. Many substances with acidic hydroxyl groups, including ascorbic acid, bioflavonoids, estrogens, and aspirin decrease capillary fragility (or increase capillary strength); conversely, some amines, like histamine, weaken the capillaries. Daniels and Everson (1936) showed that small amounts of aspirin (2.5 gr) almost doubled the urinary excretion of ascorbic acid by children 4 to 6 years of age. Moreover, Sahud and Cohen (1971) showed that aspirin decreases the platelet ascorbic acid level in patients with rheumatoid arthritis. Thus, it would seem that aspirin may take the place of ascorbic acid in the tissues; aspirin does admittedly possess capillary strengthening, anti-inflammatory, and platelet antisludging activities in addition to its antipyretic and analgesic effects, which no doubt account for its observed benefits, but it cannot perform the essential functions of vitamin C, such as the removal of histamine and the conversion of proline to hydroxyproline, so aspirin seems like a poor substitute for the vitamin which it displaces and which the patient really needs.

Cortisone increases capillary strength and exerts its anti-inflammatory action by a more complex means, depending on seven structural requisites, as shown by Kramar et al. (1956). It seems highly probable that cortisone exerts its anti-inflammatory activity by acting on oxidation-reduction systems including ascorbic acid metabolism.

XI. FUTURE RESEARCH

Research is needed to ascertain the whole blood histamine levels of patients with rheumatic fever and to study the blood histamine response following the administration of ascorbic acid. In all probability, the blood histamine level will prove to be closely related to the activity of the disease. It will be important to find out whether elevated blood histamine levels can be restored to normal by ascorbic acid alone, as in normal subjects. More likely, both ascorbic acid and cortisone will be needed in acute rheumatic fever and ascorbic acid and (+)catechin in the chronic disease. Blood histamine levels may prove useful as a guide to the needs of the patient.

REFERENCES

Abbasy, M. A., Harris, L. J., Ray, S. N., and Marrack, J. R. (1935), Diagnosis of vitamin-C subnutrition by urine analysis, *Lancet,* 2, 1399.

Abbasy, M. A., Gray Hill, N., and Harris, L. J. (1936), Vitamin C and juvenile rheumatism, with some observations on the vitamin-C reserves in surgical tuberculosis, *Lancet,* 2, 1413.

Abt, A. F., Hardy, L. M., Farmer, C. J., and Maaske, J. D. (1942), Relation of vitamin C to scarlet fever, rheumatic infections and diphtheria in children, *Am. J. Dis. Child.,* 64, 426.

Bywaters, E. G. L. (1978), Rheumatic fever (including chorea), in *Copeman's Textbook of Rheumatic Diseases,* 5th ed., Scott, J. T., Ed., Churchill Livingstone, London, chap. 30.

Campbell, M. and Warner, E. C. (1930), A study of rheumatic disease in children, *Lancet,* 1, 61.

Chester, W. and Schwartz, S. P. (1934), Cutaneous lesions in rheumatic fever. Predominating signs of active rheumatic fever during a ward epidemic, *Am. J. Dis. Child.,* 48, 69.

Clemetson, C. A. B. (1969), Menorrhagia rheumatica, *Vie Med.,* (No. Hors Ser.), 8.

Coburn, A. F. (1933), Relationship of the rheumatic process to the development of alterations in tissues, *Am. J. Dis. Child.,* 45, 933.

Daniels, A. L. and Everson, G. J. (1936), Influence of acetylsalicylic acid (aspirin) on urinary excretion of ascorbic acid, *Proc. Soc. Exp. Biol. Med.,* 35, 20.

Faulkner, J. M. (1935), The effect of administration of vitamin C on the reticulocytes in certain infectious diseases. A preliminary report, *N. Engl. J. Med.,* 213, 19.

Glazebrook, A. J. and Thomson, S. (1942), The administration of vitamin C in a large institution and its effect on general health and resistance to infection, *J. Hyg.,* 42, 1.

Hedley, O. F. (1939), Trends, geographical and racial distribution of mortality from heart disease among persons 5-24 years of age in the United States during recent years (1922—1936) A preliminary report, *Public Health Rep.,* 54, 2271.

Holtz, E. and Friedman, G. (1934), A hemorrhagic eruption of the mouth and throat in the rheumatic state, *Am. J. Med. Sci.,* 187, 359.

Kaiser, A. D. (1938), Rheumatic infection: is vitamin C deficiency a factor?, *N.Y. State J. Med.,* 38, 868.

Kaiser, A. D. and Slavin, B. (1938), The incidence of hemolytic streptococci in the tonsils of children as related to the vitamin C content of tonsils and blood, *J. Pediatr.,* 13, 322.

Kaplan, M. H., Meyerserian, M., and Kushner, I. (1961), Immunological studies of heart tissue. IV. Serological reactions with human heart tissue as revealed by immunofluorescent methods: isoimmune, Wasserman, and autoimmune reactions, *J. Exp. Med.,* 113, 17.

Keith, J. D. (1960), Modern trends in acute rheumatic fever, *Can Med. Assoc. J*, 83, 789.

Kellie, A. E. and Zilva, S. S. (1935), Catalytic oxidation of ascorbic acid, *Biochem. J.*, 29, 1028.

Kramar, J., Kramar, M. S., and Levine, V. E. (1956), Correlation between chemical constitution and capillary activity of adrenal-cortical hormones, *Proc. Soc. Exp. Biol. Med.*, 92, 282.

Massel, B. F., Warren, J. E., Patterson, P. R., and Lehmus, H. J. (1950), Antirheumatic activity of ascorbic acid in large doses. Preliminary observations on seven patients with rheumatic fever, *N. Engl. J. Med.*, 242, 614.

McBroom, J., Sunderland, D. A., Mote, J. R., and Jones, T. D. (1937), Effect of acute scurvy on the guinea pig heart, *Arch. Pathol.*, 23, 20.

McCormick, W. J. (1955), The rheumatic diseases. Is there a common etiologic factor?, *Arch Pediatr.*, 72, 107.

Parsons, L. G. (1938), Nutrition and nutritional diseases, *Lancet*, 1, 123.

Perla, D. and Marmorston, J. (1937), Role of vitamin C in resistance, *Arch. Pathol.*, 23, 543 and 683.

Perry, C. B. (1936), Rheumatic heart disease and vitamin C, *Lancet*, 2, 426.

Poynton, F. J. and Paine, A. (1914), *Researches on Rheumatism*, Macmillan, New York.

Rinehart, J. F. (1935), Studies relating vitamin C deficiency to rheumatic fever and rheumatoid arthritis; experimental, clinical, and general considerations, *Ann. Intern. Med.*, 9, 586.

Rinehart, J. F. (1936), An outline of studies relating to vitamin C deficiency in rheumatic fever, *J. Lab. Clin. Med.*, 21, 597.

Rinehart, J. F. (1937), Vitamin C and rheumatic fever, *Int. Clin.*, II (Ser. 47), 22.

Rinehart, J. F. (1943), Rheumatic fever and nutrition, *Ann. Rheum. Dis.*, III (3), 1.

Rinehart, J. F. (1944), Treatment of rheumatic fever with crude hesperidin (vitamin P), *Calif. Health*, 1 (22), 163.

Rinehart, J. F. (1945), Observations on the treatment of rheumatic fever with vitamin P, *Ann. Rheum. Dis.*, 5, 11.

Rinehart, J. F. (1953), Histochemical observations in rheumatic fever, *Ann. Rheum. Dis.*, 12, 338.

Rinehart, J. F. (1955), Rheumatic fever: observations on the histogenesis, pathogenesis, and use of ascorbic acid and bioflavonoids, *Ann. N.Y. Acad. Sci.*, 61, 684.

Rinehart, J. F., Connor, C. L., and Mettier, S. R. (1934), Further observations on pathologic similarities between experimental scurvy combined with infection, and rheumatic fever, *J. Exp. Med.*, 59, 97.

Rinehart, J. F., Greenberg, L. D., and Christie, A. U. (1936), Reduced ascorbic acid content of blood plasma in rheumatic fever, *Proc. Soc. Exp. Biol. Med.*, 35, 350.

Rinehart, J. F., Greenberg, L. D., Olney, M., and Choy, F. (1938), Metabolism of vitamin C in rheumatic fever, *Arch. Intern. Med.*, 61, 552.

Rinehart, J. F. and Mettier, S. R. (1933), The joints in experimental scurvy and in scurvy with superimposed infection. With a consideration of the possible relation of scurvy to rheumatic fever, *Am. J. Pathol.*, 9, 952.

Rinehart, J. F. and Mettier, S. R. (1934), The heart valves and muscle in experimental scurvy with superimposed infection. With notes on the similarity of the lesions to those of rheumatic fever, *Am. J. Pathol.*, 10, 61.

Roff, F. S. and Glazebrook, A. J. (1939), Therapeutic application of vitamin C in peridental disease, *J. R. Nav. Med. Serv.*, 25, 340.

Rutstein, D. D. and Densen, E. (1965), Natural history of rheumatic fever and rheumatic heart disease. Ten year report of a co-operative clinical trial of A.C.T.H., cortisone, and aspirin, *Br. Med. J.*, September, 607.

Sahud, M. A. and Cohen, R. J. (1971), Effect of aspirin ingestion on ascorbic-acid levels in rheumatoid arthritis, *Lancet*, May 8, 937.

Schultz, M. P. (1936a), Cardiovascular and arthritic lesions in guinea-pigs with chronic scurvy and hemolytic streptococci infections, *Arch. Pathol.*, 127, 472.

Schultz, M. P. (1936b), Studies of ascorbic acid and rheumatic fever. II. Test of prophylactic and therapeutic action of ascorbic acid, *J. Clin. Invest.*, 15, 385.

Schultz, M. P., Sendroy, J., and Swift, H. F. (1935), The significance of latent scurvy as an etiologic factor in rheumatic fever, *J. Clin. Invest.*, 14, 698.

Sendroy, J., Jr. and Schultz, M. P. (1936), Studies of ascorbic acid and rheumatic fever. I. Quantitative index of ascorbic acid utilization in human beings and its application to study of rheumatic fever, *J. Clin. Invest.*, 15, 369.

Stewart, C. P., Horn, D. B., Robson, J. S. (1953), The effect of cortisone and adrenocorticotropic hormone on the dehydroascorbic acid in human plasma, *Biochem. J.*, 53, 254.

Stimson, A. M., Hedley, O. F., and Rose, E. (1934), Notes on experimental rheumatic fever, *Public Health Rep.*, 49, 361.

Stollerman, G. H. (1960), Factors that predispose to rheumatic fever, *Med. Clin. North Am.*, 44, 17.

Stollerman, G. H. (1961), Factors determining the attack rate of rheumatic fever, *JAMA*, 177, 823.

Swift, H. F. (1930), Factors favoring the onset and continuation of rheumatic fever, *Am. Heart J.*, 6, 625.

Taylor, S. (1937), Scurvy and carditis, *Lancet*, 1, 973.

Van der Sande, D. (1935), Rare exanthem in acute polyarthritis, *Ned. Tijdschr. Geneeskd.*, 79, 5846.

Vazquez, J. J. and Dixon, F. J. (1957), Immunohistochemical study of lesions in rheumatic fever, systemic lupus erythematosis, and rheumatoid arthritis, *Lab. Invest.,* 6, 205.

Vazquez, J. J. and Dixon, F. J. (1958), Immunohistochemical analysis of lesions associated with "fibrinoid change", *AMA Arch. Pathol.,* 66, 504.

Warner, E. C., Winterton, F. G., and Clark, M. L. (1935), A dietetic study of cases of juvenile rheumatic disease, *Q. J. Med.,* 208, 227.

Williams, R. C., Jr., Van de Rijn, I., Mahros, Z. H., and Reid, A. H. (1980), Lymphocytes binding C-reactive protein and streptococcal membranes in acute rheumatic fever, *J. Lab. Clin. Med ,* 96, 803

Chapter 11

MENORRHAGIA

Excessive menstrual bleeding (menorrhagia), excessive irregular menstrual bleeding (menometrorrhagia), unduly frequent bleeding (polymenorrhea), and bleeding between the periods (intermenstrual bleeding) have many causes. In attempting to divine the cause of abnormal bleeding in any woman, a gynecologist has to consider such common things as complications of early pregnancy and must rule out the much less frequent but more ominous causes, such as cancer of the cervix, especially when there is irregular, intermenstrual, or postcoital bleeding. A pregnancy test and a study of vaginal and cervical cytology are helpful, but careful inspection and palpation of the pelvic organs are essential.

Do-it-yourself "Pap smears" are dangerous as they can provide a false sense of security. Depending on the age of the patient and other factors, a diagnostic curettage may be necessary to allow histological study of the endometrium. Some writers have stressed the importance of irregular maturation and irregular shedding of the endometrium. Nevertheless, having ruled out pregnancy, cancer, pelvic infections, polyps, and fibroids, and having found no enlargement of the uterus to suggest adenomyosis, there is no obvious local cause for the bleeding in about half of the patients, who are then classified as having "dysfunctional uterine bleeding".

Many gynecologists think of dysfunctional uterine bleeding as being almost synonymous with nonovulation. This endocrine dysfunction is seen in its extreme form as metropathia hemorrhagica, with a cystic ovary producing high and fluctuating levels of estrogen, unopposed by progesterone, and the endometrium showing cystic hyperplasia. Classically this condition occurs at the two extremes of reproductive life and is characterized by 6 weeks of amenorrhea, followed by several weeks of heavy irregular bleeding, but many nonovulatory women do not fit this pattern. Obesity and hypothyroidism are also well-recognized causes of dysfunctional uterine bleeding.

The present author has observed another very common form of dysfunctional uterine bleeding in which there are ovulatory menstrual cycles, but excessive and prolonged menstrual bleeding is associated with easy bruising and sometimes bleeding gums or nosebleeds. These women sometimes give a history of having had rheumatic fever and have a capillary fragility state which persists in them except when they are receiving treatment with high doses of vitamin C and bioflavonoids. Indeed, for want of a better name, Clemetson (1969) described these women as having "menorrhagia rheumatica". They have normal platelet counts and their blood coagulation studies are nearly always normal. It is true that menorrhagia can be caused by thrombocytopenia or thrombasthenia; it can result from hereditary coagulation defects and is seen in association with Minot von Willebrand syndrome, as discussed by Quick (1966), but these are all relatively rare causes. In contrast, the capillary fragility of nonthrombocytopenic purpura is common in women with regular ovulatory menorrhagia, and is not associated with any abnormality of the blood clotting mechanism. Menorrhagia rheumatica may not be the most appropriate name for this condition, as many women with regular ovulatory menorrhagia and capillary fragility give no history of having had rheumatic fever and have no joint symptoms. Just as Henoch's purpura and Schonlein's purpura cause abdominal pains and joint pains in children and young people, so nonthrombocytopenic purpura can cause menorrhagia in women of reproductive age.

There is no suggestion that these women have scurvy, but they do seem to have an abnormality of ascorbic acid metabolism for they often need 2 months of treatment with vitamin C (600 mg) and citrus bioflavonoids (600 mg) daily to control their capillary fragility and their excessive menstrual blood loss. Moreover, they usually revert to heavy bleeding again within about a month when they are given lactose placebo capsules.

It is interesting to note that Mme Randoin (1923) recorded that those who had suffered from scurvy were subjects following chronic rheumatism, with pains and stiffness in the joints, clearly suggesting a predisposition to this deficiency disease in "rheumatic" subjects. She also noted that their gums never return to their normal state and that they have a special predisposition to recurrence of scurvy.

> *"Enfin, on notait également que ceux qui avait souffert du scorbut étaient sujets par la suite à des rhumatismes chroniques, à des douleurs et à des raideurs dans les articulations; que leurs gencives ne revenaient jamais à leur état naturel et qu'ils manifestaient une disposition toute spéciale à être affectés de nouveau par cette maladie."*

Studying guinea pigs with scurvy produced by feeding them on a diet of alfalfa hay, rolled barley and water, Meyer (1928) found that, "Although the uterus sometimes seemed congested, gross haemorrhages were never encountered in it. Fresh blood on the vulva had a vesical or intestinal origin." But of course, guinea pigs do not menstruate, so perhaps one would not expect them to show uterine bleeding.

Stander, Javert and Kuder (1942) were among the first to draw attention to simple purpura as a cause of dysfunctional uterine bleeding. They reported a mean plasma ascorbic acid (AA)* level of 0.43 mg/100 ml in 20 patients with dysfunctional uterine bleeding — 70% of them having ascorbate levels below the normal range, which was given as 0.5 to 1.2 mg/100 ml. Some showed improvement after curettage alone, but in those who did not, these authors obtained encouraging results by treating them with vitamin C and orange juice, along with vitamin K.

Subsequently, Morris (1953) reported four patients with excessive prolonged menstrual bleeding associated with capillary fragility, all of whom were markedly improved by treatment with vitamin C and orange juice. Blood vitamin C levels before treatment were 0.4, 0.55, 1.4, and 0.9 mg/100 ml, so none of them was even approaching scurvy. Nevertheless, they all experienced a reduction in both the amount and the duration of menstrual bleeding following treatment. One woman had 8-d menses which were reduced to 4 d, one had 6- or 7-d menses which were reduced to $3^1/_2$ to 4 d, one had 8- to 9-d menses which were reduced to 5 d, and the other woman reported that her menstrual periods had been reduced by only 1 d, but there was less bleeding than in any previous period in 19 years. Since her blood ascorbic acid level was 0.9 mg/100 ml before treatment, which is a perfectly normal level, and 2.0 mg/100 ml after 2 months of treatment, it is not at all clear how this treatment works. Perhaps it is the ratio of reduced to oxidized ascorbic acid that is altered by these dietary supplements.

Javert (1955) found plasma ascorbic acid levels below 0.5 mg/100 ml in 57% of women with menorrhagia and in 29% of women without menorrhagia, as shown in Table 1, and he reported that some of them responded to treatment with high-dosage vitamin C, K, and hesperidin. Pearse and Trisler (1957) reported 10 patients with menometrorrhagia treated with vitamin C and citrus bioflavonoids (C.V.P.®), 600 mg of each, daily for periods ranging from 2 weeks to 7 months. All were brought under control with this treatment.

Prueter (1961) reported 26 patients with heavy irregular bleeding who ranged in age from 14 to 44 years. Clearly many of them had nonovulation, but he reported excellent results, even in them, and also in one woman whose bleeding was due to a submucous myoma. He noted that ascorbic acid alone had not been found very useful, but that a combination of ascorbic acid and citrus bioflavonoids (C.V.P.®), given in large doses, was very effective. For some women he prescribed one capsule containing 200 mg of ascorbic acid and 200 mg of bioflavonoids 4 times a day, but he doubled that dose for patients with very heavy

---

* AA — ascorbic acid, reduced form.

Table 1
BLOOD PLASMA VITAMIN C CONCENTRATION IN NONPREGNANT PATIENTS WITH AND WITHOUT MENORRHAGIA

	No. of cases	Plasma vitamin C concentration		
		Average values (mg/100 ml)	Deficiency	
			No.	%
With menorrhagia	28	0.48	16	57.1
Without menorrhagia	75	0.79	22	29.3
Total	103		38	36.8

Note: Although the plasma ascorbic acid levels of women with menorrhagia do tend to be lower than normal, they do not even approach scorbutic levels

From Javert, C. T. (1955), *Ann. N.Y. Acad. Sci.*, 61, 700. With permission.

bleeding. One patient found she could control her bleeding by taking the capsules only when she felt the imminent approach of menstruation. She took one capsule every hour for the first day of her cycle, which gave her the most relief from her flooding that she had experienced in 10 years.

Three of Prueter's patients complaining of infertility, in association with menorrhagia, were able to conceive during treatment, and three other patients who had been pregnant before, became pregnant again during treatment. This improvement in fertility was also witnessed by the present writer, for three of his patients became pregnant while taking vitamin C and citrus bioflavonoids for menorrhagia, and one of them, a 40-year-old mother of three grown children, expressed anger at becoming pregnant again "as a result of the treatment," saying she had not been taking any precautions for years. She felt she should have been warned about this effect of the treatment.

While working in central Canada, at the University of Saskatchewan, Clemetson et al. (1962) studied the capillary strength of normal women during the menstrual cycle, using a 2-cm-diameter suction cup applied to the inner aspect of the upper arm. The typical pattern was found to be a short postovulatory and a longer premenstrual decrease in capillary strength (Figure 1). The capillary strength was found to fall at times when the estrogen levels would be falling, and rose when estrogens were administered to menopausal women, as shown by Clemetson et al. (1962). Daily studies of the capillary strength of a normal woman for 9 months revealed that her skin capillary strength increased and her menstrual bleeding decreased when the tomatoes in her garden became ripe in August 1959, and she was consuming them, two to six a day, so that they would not go to waste (Figure 2). Subsequently her menstrual bleeding became heavier again, but treatment with vitamin C and citrus bioflavonoids (Duo C.V.P.®) had the same effect as the tomatoes, reducing her menstrual blood flow from 15 pads in December to 6, 3, and 4 pads, respectively, in her next 3 menses (Figure 2). These findings prompted our studies of women with menorrhagia and the finding that they tend to have a capillary fragility state (Figure 3). Treatment with vitamin C and citrus bioflavonoids caused an appreciable increase in capillary strength and a decreased blood loss in about eight out of ten of them, as reported by Clemetson and Blair (1962). Some of the women were poor and might have had an inadequate intake of fresh fruits and vegetables during the long months of winter, but others were quite wealthy and clearly had access to all that they might choose.

An unexpected feature of this treatment was the length of time needed to provide relief from the bleeding. Scurvy can be cured in 1 week or 10 d, but these women, who were in

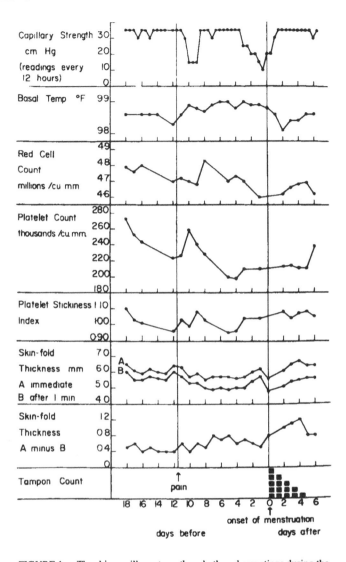

FIGURE 1.   The skin capillary strength and other observations during the same menstrual cycle of a normal woman. The short postovulatory and longer premenstrual decrease in capillary strength are quite typical. (From Clemetson, C. A. B., Blair, L. M., and Brown, A. B. [1962], *Ann. N.Y. Acad. Sci.*, 93, 277. With permission.)

no way scorbutic, often required 2 months of treatment with both ascorbic acid (200 mg) and bioflavonoids (200 mg) three times a day before their excessive menstruation and their capillary fragility could be brought under control (Figure 4). Moreover, one could not speak of cure, as both their capillary fragility and their excessive bleeding tended to return within a month or so when they were given lactose placebo capsules, or when treatment was discontinued (Figure 5).

Extensive studies revealed no coagulation defects in these women. They had no gross pelvic pathology and most had no evident hormonal disturbance. There were, however, two teenage girls with heavy irregular bleeding typical of nonovulation, who, like Prueter's patients, showed a prompt response to this treatment after prolonged and varied hormone therapy by other physicians had failed. In all, 32 out of 37 women showed decreased blood loss when treated with the test capsules (Duo C.V.P.®) for 2 months, while only 1 out of 13 showed a similar improvement with lactose placebo capsules.

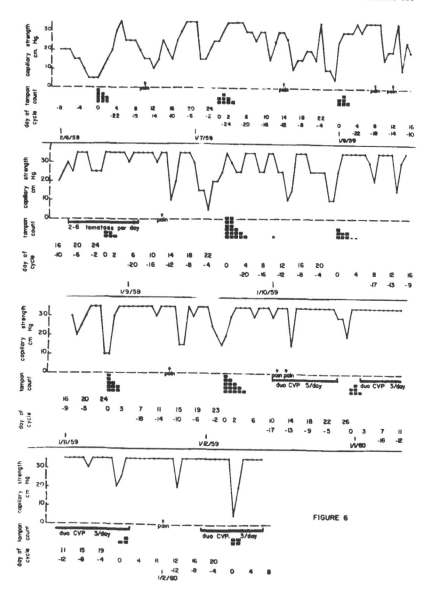

FIGURE 2. Daily capillary strength measurements for 9 months on a normal, fertile 24-year-old woman who had a fairly regular 24- to 26-d menstrual cycle. See text. (From Clemetson, C. A. B., Blair, L. M., and Brown, A. B. [1962], *Ann. N.Y. Acad. Sci.*, 93, 277. With permission.)

Similar results were reported by Arcangeli (1967) and several other speakers at a conference on bioflavonoids held at Stresa on Lago Maggiore in April of 1966. All reported this combination of ascorbic acid and bioflavonoids to be beneficial, not only in the treatment of menorrhagia, but also in the treatment of threatened or recurrent abortion.

Hauer and Schleicher (1966) used a mixture of hydroxyethyl rutosides* (bioflavonoids) in the treatment of 100 patients at the age of puberty because of premenstrual syndromes or dysmenorrhea. Of these patients, 61 were chosen at random and observed over a period of at least 4 cycles. In 24 of these young women there was relief of symptoms, and in a

* Venoruton® P4, Zyma SA, Nyon, Switzerland.

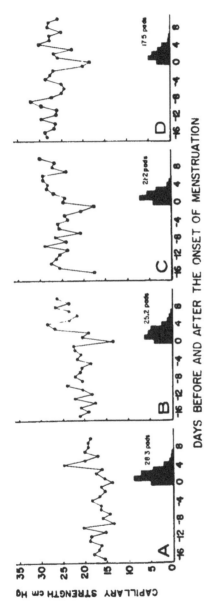

DAYS BEFORE AND AFTER THE ONSET OF MENSTRUATION

FIGURE 3.   Twelve women with menorrhagia whose capillary strength averaged less than 20 cmHg before treatment. (A) Before treatment, (B) first month on treatment with vitamin C and citrus bioflavonoids, (C) second month, (D) third month of treatment. Note that the premenstrual drop of capillary strength was absent before treatment, but became evident when the average capillary strength rose as a result of treatment. (From Clemetson, C. A. B. and Blair, L. M. [1962], *Am. J. Obstet. Gynecol.*, 83, 1269. With permission.)

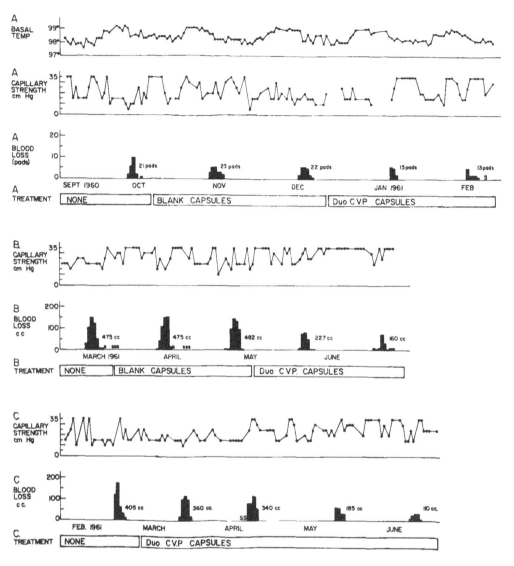

FIGURE 4   (A) Blood loss, capillary strength, and basal body temperature of a patient before and during treatment with vitamin C and citrus bioflavonoids. (B and C) Menstrual blood loss of two other patients measured by rubber intravaginal cup. (From Clemetson, C. A. B. and Blair, L. M. [1962], *Am. J. Obstet Gynecol.*, 83, 1269. With permission.)

further 24 there was some alleviation of the symptoms; so two thirds obtained some relief and side effects were not worth mentioning.

Clemetson (1969) reported the plasma reduced ascorbic acid (AA) and total ascorbic acid (TAA)* levels in the blood plasma of six normal women and six women with menorrhagia, using the same technique as Stewart et al. (1953). While 88% of the ascorbic acid was found to be in the reduced form (AA) in normal women, only 62% was found to be in the reduced form in women with menorrhagia (Figure 6). Unfortunately, such small differences between the levels of reduced and total ascorbic acid are difficult to measure accurately. Moreover, some oxidation of the vitamin occurs even when the samples are analyzed within the hour. However, the results do suggest an abnormality of ascorbic acid metabolism in such women with menorrhagia and capillary fragility.

---

* TAA — total ascorbic acid, reduced and oxidized forms.

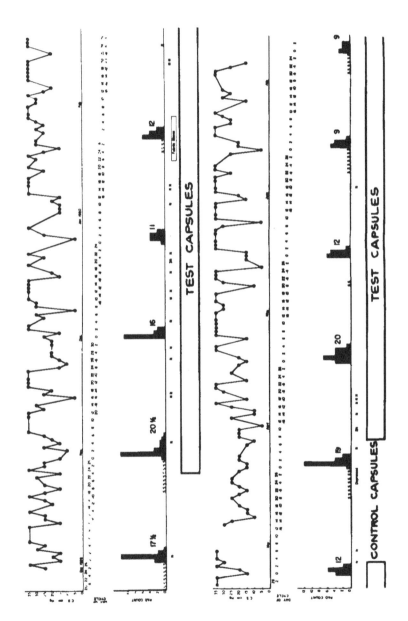

FIGURE 5. A record of the daily capillary strength measurements on a 34-year-old woman with excessive menstrual bleeding, indicated by the black blocks. She also had premenstrual spotting and nosebleeds. The bleeding came under control after 2 months of treatment with vitamin C and citrus bioflavonoids, but returned within 1 month when lactose placebo capsules were substituted. Reinstitution of treatment again produced a slow but satisfactory reduction of her menstrual flow, but her premenstrual spotting continued. N indicates nose bleeds; S indicates spotting. (From Clemetson, C. A. B and Blair, L. M. [1962]. *Am J Obstet. Gynecol.*, 83, 1269. With permission)

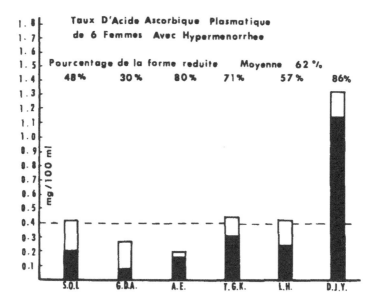

A

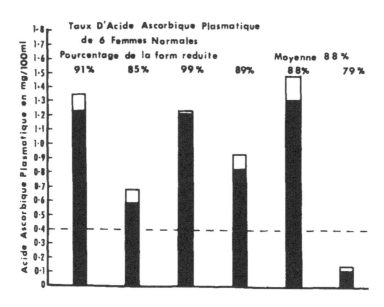

B

FIGURE 6. Plasma reduced ascorbic acid (AA; black blocks) and total ascorbic acid (TAA) analyses of blood plasma from six normal women (A) and in 6 women with menorrhagia (B) revealed a lower percentage of the vitamin in the reduced form (62%) in women with menorrhagia than in normal women (88%). (From Clemetson, C. A. B. [1969], *Vie Med.* (No. Hors Ser.), December, 8. With permission.)

Studies using carbon monoxide to prevent oxidation and the Hughes (1956) homocysteine method for analysis of blood plasma samples would be useful. Moreover, the future may see the development of methods for direct measurement of the redox potential of the blood; such studies could perhaps help to elucidate this problem.

Among 16,125 women of reproductive age, Federl and Streli (1967) found 1500 women to have excessive bleeding associated with decreased capillary resistance. Out of this group, 310 women with menorrhagia or menometrorrhagia were treated with a combination of ascorbic acid, 500 mg, and rutin, 50 mg (Rutiscorbin®tablets), three times a day. Very good results were obtained, but women who had had menorrhagia for many years reacted only slowly to the treatment and their menorrhagia returned when the treatment was discontinued. With this treatment, 205 patients showed improvement in 4 weeks, 77 patients in 8 weeks and 14 in 2 weeks; 14 women did not show any improvement as a result of the treatment.

These results with rutin and ascorbic acid are very similar to those of Clemetson and Blair (1962), using citrus bioflavonoids and ascorbic acid, in that the full effect was often delayed for 2 or 3 months, and in that the bleeding often recurred when the treatment was discontinued.

Lithgow and Politzer (1977) made some very interesting observations of vitamin A deficiency in women with menorrhagia. The mean serum vitamin A level of 191 women with normal menstruation was 166 IU/100 ml, and the mean level in 71 patients with menorrhagia was 67 IU/100 ml. In fact, 16% of the women with menorrhagia had markedly deficient serum levels of vitamin A (0 to 50 IU/100 ml), 39% were moderately deficient (50 to 100 IU), and 14% were found to be mildly deficient (100 to 115 IU) in vitamin A in that study. Moreover, treatment with retinol (25,000 IU twice daily for 15 d) provided complete relief from menorrhagia in 23 out of 40 (57.5%) for a period of at least 3 months and some reduction in the amount or the duration of the menstrual flow for a further 14 (35%). These workers also observed pyridoxine (vitamin B_6) deficiency in 10% of women with menorrhagia. Lithgow and Politzer also found vitamin A deficiency in six patients with amenorrhea, and observed a marked rise in their mean 17β-estradiol levels from 78 to 308 pg/ml over periods ranging from 8 to 59 d following treatment with retinol.

Dietary deficiency, malabsorption, recent infections, overexposure to sunlight, and excessive intake of alcohol were given as causes of vitamin A deficiency. Clearly, menorrhagia is a highly complex nutritional and endocrinological problem, and deficiency of several vitamins may be involved. Indeed, Biskind et al. (1944) obtained excellent results in the treatment of women with menometrorrhagia, cystic mastitis, and premenstrual tension by oral and parenteral administration of vitamin B complex. Most of these women showed clinical evidence of vitamin B deficiency such as glossitis, stomatitis, or cheilosis. They were believed to have nonovulatory bleeding associated with excessive estrogen levels due to failure of estrogen conjugation, resulting from hepatic vitamin B complex deficiency.

One can readily understand that FAD (derived from dietary riboflavin) activates glutathione reductase in the presence of NADP (derived from dietary nicotinic acid) and that the resulting intracellular reduced glutathione promotes tissue storage of ascorbic acid. Likewise, pyridoxine, folic acid, and α-tocopherol are all cogs in the essential metabolic machinery of the liver, and a deficiency of any one can affect the activity of the whole.

A recent *in vitro* study of hamster pituitary glands in a complete culture medium by Wun, Wang-Song (1984) at the University of Texas in Houston has demonstrated that the addition of $10^{-3} M$ ascorbic acid buffered to pH 7.4 caused an immediate release of follicle-stimulating hormone and leuteinizing hormone from the gland. The significance of this observation is difficult to interpret because the human plasma ascorbate level rarely exceeds $10^{-4} M$ (1.76 mg/100 ml), even after ingestion of 1 g of ascorbic acid. Moreover, one does not know whether it was ascorbic acid or its oxidation products that caused the release of gonadotrophins from the hamster pituitary.

REFERENCES

Arcangeli, A. (1967), Sindromi emorragiche ed endocrinopatie manifestazioni emorragiche in soggetti con obesita cura con bioflavonoidi,, in *Bioflavonoidi*, Zambotti, V., Ed., Scuole Griafiche Artigianelli Pavoniani, Milano, 293.

Biskind, M. S., Biskind, G. R., and Biskind, L. H. (1944), Nutritional deficiency in the etiology of menorrhagia, metrorrhagia, cystic mastitis, and premenstrual tension, *Surg. Gynecol Obstet.*, 78, 49.

Clemetson, C. A. B. (1969), Menorrhagia rheumatica, *Vie Med.* (No. Hors Ser.), December, 8

Clemetson, C. A. B. and Blair, L. M. (1962), Capillary strength of women with menorrhagia, *Am. J. Obstet. Gynecol.*, 83, 1269.

Clemetson, C. A. B., Blair, L. M., and Brown, A. B. (1962), Capillary strength and the menstrual cycle, *Ann. N.Y. Acad. Sci.*, 93, 277.

Clemetson, C. A. B., Blair, L. M., and Reed, D. H. (1962), Estrogens and capillary strength, *Am. J. Obstet. Gynecol.*, 83, 1261.

Fegerl H. von and Streli, E. (1967), Meno-Metrorrhagien bei verminderter Kapillarresistenz, *Wien. Med. Wochenschr.*, 117, 763.

Hauer, E. and Schleicher, K. (1966), Behandlung der funktionellen Dysmenorrhoe und des prämenstruellen Syndroms, *Muench. Med. Wochenschr.*, 108, 2466.

Hughes, R. E. (1956), The use of homocysteine in the estimation of dehydroascorbic acid, *Biochem. J.*, 64, 203.

Javert, C. T. (1955), Decidual bleeding in pregnancy, *Ann. N.Y Acad. Sci.*, 61, 700.

Lithgow, D. M. and Politzer, W. M. (1977), Vitamin A in the treatment of menorrhagia, *S. Afr. Med. J.*, 51, 191.

Meyer, A. W. (1928), The symptomatology and gross microscopy of experimental scurvy in the guinea pig, in Studies on Scurvy by Meyer and McCormick'', *Stanford Univ. Publ. Univ. Ser. Med Sci.*, 2(2), 133.

Morris, G. E. (1953), Hypermenorrhea due to scurvy, *Postgrad. Med.*, 14, 443.

Pearse, H. A. and Trisler, J. D. (1957), A rational approach to the treatment of habitual abortion and menometrorrhagia, *Clin. Med.*, 4, 1081.

Prueter, G. W. (1961), A treatment for excessive uterine bleeding, *Appl. Ther.*, 3, 351.

Quick, A. J. (1966), Menstruation in hereditary bleeding disorders, *Obstet. Gynecol.*, 28, 37.

Randoin, L. (1923), La question des vitamines. II. Le facteur antiscorbutique, *Bull. Soc. Chim. Biol.*, 5, 806.

Stander, H. J., Javert, C. T., and Kuder, K. (1942), The management of abnormal vaginal bleeding, *Surg. Gynecol. Obstet.*, 75, 759.

Stewart, C. P., Horn, D. B., and Robson, J. S. (1953), The effect of cortisone and adrenocorticotropic hormone on the dehydroascorbic acid in human plasma, *Biochem. J.*, 53, 254.

Wun, Wang-Song, A. (1984), Personal communication, unpublished work.

Chapter 12

WOUND DEHISCENCE

The failure of wounds to heal in scurvy was well documented by James Lind (1753) in his *Treatise of the Scurvy*, where he cited the account by Richard Walter (1748) of Lord Anson's voyage round the World from 1740 to 1744 as follows:

"...But a most extraordinary circumstance, and what would be scarcely credible upon any single evidence, is, that the scars of wounds which had been for many years healed, were forced open again by this virulent distemper. Of this there was a remarkable instance in one of the invalids on board the Centurion, who had been wounded above fifty years before at the battle of the Boyne: for though he was cured soon after, and had continued well for a great number of years past; yet, on his being attacked by the scurvy, his wounds, in the progress of his disease, broke out afresh, and appeared as if they had never been healed. Nay, what is still more astonishing, the callous of a broken bone, which had been compleatly formed for a long time, was found to be hereby dissolved; and the fracture seemed as if it had never been consolidated."

Studies of scorbutic guinea pigs by Wolbach and Howe (1926) demonstrated that the histological basis of the failure of wounds to heal in scurvy lies in the inability of the fibroblasts, osteoblasts, and odontoblasts to produce and maintain the intercellular supporting substances — collagen, bone, and dentin. Only weak precollagen is formed in scorbutic wounds; strong fibrous tissue does not form until ascorbic acid is provided.

Sokolov (1932) of Leningrad analyzed 730 cases of wound rupture after laparotomy and observed that wound disruption occurred more frequently in Northern peoples during winter and spring. Colp (1934), of New York City, with a much smaller series of 29 cases, found no seasonal relationship; but Fallis (1937), of Detroit, noted an increased incidence of wound disruption in March, April, and May. Indeed his seasonal analysis of 49 abdominal wound disruptions at the Henry Ford Hospital showed the following distribution: winter 11, spring 20, summer 5, and fall 13. The incidence in spring was four times the summer rate, which is interesting because spring is the season when the ascorbic acid stores of the body are often at their lowest (Chapter 19, Volume I).

The causes of death of patients following partial gastrectomy at St. Bartholemew's Hospital in London were reported by Payne (1936) and were reviewed by Archer and Graham (1936). Many of those who died following surgery for gastric or duodenal ulcer showed peritonitis due to leakage at the site of anastomosis, and there was an almost complete absence of any fibrinous response or healing. The stitches stood out as though the operation had just been completed. These findings and Payne's discussion of the part played by malnutrition prompted Archer and Graham (1936) to conduct ascorbic acid saturation tests. They found that six out of nine patients with gastric or duondenal ulcers at that hospital were in a subscurvy state.

Plasma ascorbic acid studies by Ingalls and Warren (1937) at the Peter Bent Brigham Hospital in Boston led them to conclude that subclinical or asymptomatic scurvy is not uncommon and is probably a major cause of postoperative wound dehiscence, especially in patients with gastric or duodenal ulcers. They found that 14 out of 20 peptic ulcer patients had deficient plasma ascorbic acid levels of 0.2 mg/100 ml or below (normal) 0.7 to 1.0 mg/100 ml), requiring treatment before surgery. They also described the pathological findings in another patient, a 54-year-old man with carcinoma of the pylorus who died 8 d after surgery and 2 d following resuture of his abdominal wound. "At autopsy the surgical incision was open throughout its entire extent. The edges of the wound were greatly thickened, edematous and necrotic. The skin for approximately 4 cm on either side showed a deep purplish black discoloration. The subcutaneous tissues were stained dark greenish black and were extremely soft and friable. There was considerable exudate, though no frank pus was seen. The gastrojejunostomy and the sutured portion of the duodenum had also broken open, and there was a definite peritonitis.'' These writers observed that this objective description by the pathologist is similar to the note by Lanman and Ingalls (1937) on wounds experimentally produced in guinea pigs — ''the scar tissue of the wounds of the normal animals after rupture was pink and firm, while that of the wounds of the scorbutic animals was livid and of soft consistency.''

Working at the Children's and Infants' Hospital in Boston, Lanman and Ingalls (1937) reported that plasma analysis disclosed a much higher incidence of asymptomatic scurvy than could be recognized by clinical examination. They reported wound dehiscence, 3 d after a second operation and congenital atresia of the bowel, in a 6-week-old infant. The wound was resutured, but the child died on the following day. There had been no clinical or radiological signs of scurvy during life, but at autopsy the diagnosis was scurvy. Microscopic sections of the costochondral junctions were described as follows: ''There is a definite lattice formation and many of the cartilage spicules are fractured. Osteoblasts are present along the cartilage spicules but there is no osteoid formation.'' This is evidence of early scurvy. Sections of the skin at the operation site showed numerous fibroblasts without formation of collagen. As a result of their studies, these workers stated, ''A large body of evidence is being assembled to show that the conception of vitamin C deficiency should include not only absolute depletion of ascorbic acid stores, but also such low levels of supply as place the functioning of connective tissue cells at a physiological disadvantage.'' Their studies of partially scorbutic guinea pigs showed that they had markedly inferior tensile strength in experimental wounds when compared with controls at 10, 20, and 30 d postoperatively. The abdominal wounds of the subscorbutic group ruptured at a pressure averaging approximately one third of that required to rupture the wounds of normal animals.

Taffel and Harvey (1938) reported that the tensile strength of stomach wounds in partially scorbutic guinea pigs is markedly decreased from the eighth to the tenth days, which is when most human wound dehiscences occur.

Lanman (1938) reiterated his belief that scurvy was far more common in infants and children than was generally known, and that a partial vitamin C deficiency is of much greater importance in the healing of human wounds than had been recognized.

Hartzell and Winfield (1939) reported that their own observations, and also reports of others in the literature, showed a preponderance of wound disruptions in men over women, the ratio being roughly 2:1. This is of interest, as men tend to have lower ascorbic acid levels (Chapter 6, Volume I). They gave details of a 68-year-old man who underwent surgery for volvulus at the Receiving Hospital in Detroit. His abdominal wound broke down completely, with evisceration on the seventh postoperative day; he was found to have a plasma ascorbic acid level of 0.13 mg/100 ml (normal range cited at 0.65 to 2 mg/100 ml).

Wolfer and Hoebel (1939) and Bartlett et al. (1940) stressed the importance of ascorbic acid for proper wound healing in surgical patients. Holman (1940), working at Stanford University in California, found that 44% of his ''run of the mill'' surgical patients had low blood ascorbic acid levels, and 13% were on the verge of scurvy. As a result of these findings, he recommended nutritional preparation of all surgical patients. Whenever possible, he gave a high-protein, high-caloric diet with vitamin supplements for several days before surgery. He concluded, ''Even a few days, therefore, spent in nutritional preparation of the patient should reduce the operative hazard and diminish the length of convalescence.'' Needless to say, he also advocated attention to postoperative nutritional support for all patients.

Lund and Crandon (1941b) studied 58 patients having operations on the biliary tract and found the incidence of postoperative hernia to be greatest in patients with low plasma ascorbic acid levels. Moreover, in a detailed analysis of 12 complete abdominal wound disruptions,

Table 1
SUMMARY OF ANALYSES IN 20 CASES OF WOUND DISRUPTION: AVERAGE VALUES FOR PLASMA ASCORBIC ACID AND SERUM PROTEIN CONCENTRATIONS

Group	Number of patients	Plasma ascorbic acid (mg/100 cc)	Range (mg/100 cc)	Serum protein (g/100 cc)
Normal values students	20	0.96	0 64—1.86	7.19
Hospitalized patients with benign conditions	30	0.87	0.49—1.61	6.93
Wound disruption (infected wounds)	6	0 38	0.28—0.58	5 84
Wound disruption (clean wounds)	14	0.36	0 13—0.66	5.83

Note: The average values of the plasma ascorbic acid and serum protein concentrations in the patients whose wounds disrupted were decidedly lower than normal

From Hartzell, J. B., Winfield, J. M., and Irvin, J. L (1941), *JAMA*, 116, 669. ©1941 American Medical Association. With permission.

following general surgical procedures, they found that 3 had leukocyte ascorbic acid levels below 4 mg/100 g (normal 30 to 40 mg/100 g); so those patients were close to scurvy. Moreover, 10 of the 12 had plasma ascorbic acid levels below 0.2 mg/100 ml at the time of wound dehiscence (normal 0.4 to 1.4 mg/100 ml). Although Lund and Crandon at that time did not consider plasma ascorbic acid levels as being indicative of tissue stores of the vitamin, later work reported by Crandon et al. (1952) showed that plasma ascorbate levels are more closely related to ascorbate levels in human tissue biopsies than are leukocyte ascorbate levels.

Hunt (1941), working at St. Bartholemew's Hospital in London, reported the post-mortem findings in 28 patients who died following abdominal surgery. In eight, the wounds showed a deficiency in the formation of collagen resembling that in subscurvy guinea pigs. In these eight cases, there were five instances of abdominal disruption and three of leakage of gastrointestinal suture lines. He reported that a 75% reduction in wound disruption occurred following the routine administration of ascorbic acid for all major operations, but he was careful to point out that other factors may have played a part in this improvement and that ascorbic acid is not a panacea.

Hartzell et al. (1941) reported the results of investigations that they had conducted on 20 patients with wound discruption at the Harper and Grace Hospitals in Detroit. They reported that the serum protein level was below normal in every instance and that the plasma ascorbic acid (AA)* levels were low (below 0.6 mg/100 ml) in all but one. Certainly the mean ascorbic acid levels were lower than normal (Table 1), but only one of these patients had a plasma ascorbic acid level below 0.2 mg/100 ml. Thus, it would seem to the writer that most of the wound disruptions in this series may have been due to causes other than vitamin C deficiency.

Studying plasma and buffy coat ascorbic acid levels of normal individuals, patients with scurvy, and preoperative surgical patients, Crandon et al. (1953) detected traces of ascorbic acid (TAA)** in the plasma and in the buffy coat (leukocytes and platelets) of some patients with frank scurvy, while other patients had no detectable ascorbic acid in either plasma or buffy coat, yet showed no clinical evidence of scurvy. They stated that their findings suggested a thesis consistent with the observations of Lind (1753) and postulated by Hess

---

* AA — ascorbic acid, reduced form.
** TAA — total ascorbic acid, reduced and oxidized forms.

Table 2

**INFORMATION ABOUT SURGICAL
PATIENTS WITH NO ASCORBIC ACID IN
BLOOD BUFFY COAT, BUT WITHOUT
MANIFESTATION OF CLINICAL
SCURVY**

Total patients studied (1136 samples)	561
No ascorbic acid in buffy coat	19
Followed up	16
Subjected to laparotomy	8
Eviscerated between 9th and 12th postoperative day	4
Developed persistent draining sinus in wound	1
Dead within 2 years	9

Note: Showing the high rate of evisceration (four/eight) in
patients who had no detectable ascorbic acid (TAA) in
their plasma or buffy coat

From Crandon, J. H., Mikal, S., and Landeau, B. R. (1953),
Proc. Nutr. Soc., 12, 273. With permission.

(1920) that clinical scurvy is the result not of vitamin C deficiency alone, but of the vitamin deficiency plus local tissue stress. In light of the present knowledge (Chapter 1 of this volume), we can readily concede that local tissue damage may cause a local release of histamine which could add to the scorbutic histaminemia and might precipitate bleeding leading to local hemolysis and further ascorbate depletion (Chapter 15, Volume I). Undoubtedly there can be patients who are severely deficient in ascorbic acid, who show no signs of scurvy, but are nutritionally unfit for surgery; they are surgical catastrophes waiting to happen.

Careful studies of 1136 blood samples from 561 different selected surgical patients by Crandon et al (1953) included both plasma and buffy coat ascorbic acid (TAA) levels. Of these, 19 were found with no ascorbic acid in the buffy coat and, for the most part, only a trace in their blood plasma; none of whom showed any clinical evidence of scurvy at the time of blood sampling. "Eight of these nineteen were subjected to laparotomy, and of these, four had evisceration of their abdominal wounds requiring resuture between the 9th and 12th postoperative days and one other developed a persistent draining sinus in his wound" (Table 2). None of these 19 patients had bleeding gums or perifollicular hemorrhages, perhaps because of lack of stress in these areas, but when subjected to local tissue stress in the form of a laparotomy wound, five of the eight showed defects of wound healing consistent with scurvy.

In 70 patients, these same workers obtained specimens of fascia and muscle, as well as blood, at the time of surgery. Tissue ascorbic acid levels below 1.4 mg/100 g were considered as low, and above 6 mg/100 g as high. Of the 70 patients, 24 had low tissue ascorbic acid levels, and all but 3 of these also had low plasma ascorbic acid levels, below 0.2 mg/100 ml; these 3 had been receiving an ascorbic acid supplement. Against expectation, the correlation between the tissue and plasma ascorbate levels was found to be much better than that between the tissue and buffy coat ascorbate levels. Thus, plasma ascorbate levels were considered more reliable than buffy coat levels, probably because of the greater accuracy of technique. Of the 24 patients with low tissue ascorbic acid levels, 18 were followed up, and Table 3 shows that 12 of them suffered wound complications of one kind or another. The authors stated that lack of ascorbic acid was only one of several adverse factors in these patients, who were in poor physical condition, as evidenced by their high mortality rate,

Table 3
INFORMATION ABOUT SURGICAL PATIENTS WITH LOW FASCIA AND BLOOD ASCORBIC ACID AT OPERATION (ROE-KUETHER METHOD)

Total in a low group		24
Followed up		18
Total wound complications		12
Undergoing laparotomy		13
Eviscerated	1	
Massive bleeding into wound	1	
Dehisced down to peritoneum	2	9
Developed incisional hernia	3	
Developed persistent draining sinus	2	
Herniorrhaphy, amputation, or sympathectomy		5
Poor or no healing	2	3
Draining sinus lasting 2 years	1	
Of 18, total dead within 2 years		11

Note: Showing the high incidence of wound complications (12/18) in patients who had low tissue (fascia and muscle) ascorbic acid (TAA) levels (below 1.4 mg/100 g).

From Crandon, J. H., Mikal, S., and Landeau, B. R. (1953), *Proc. Nutr. Soc.*, 12, 273. With permission.

but ascorbic acid deficiency was undoubtedly a major factor in the development of their wound complications.

Further studies by Crandon et at. (1958) showed that 19 out of 47, or over one third, of patients suffering wound dehiscence at the Boston City Hospital had serious ascorbic acid deficiency, as indicated by plasma ascorbic acid (TAA) levels below 0.2 mg/100 ml and buffy coat levels of less than 8 mg/100 ml.

Crandon et al. (1961) stated that, "among surgical patients subjected to laparotomy, there was an eight fold higher rate of wound dehiscence in those patients with deficient blood levels of the vitamin at primary operation compared to those with adequate levels;" the wound dehiscence rates were 13.9 and 1.7%, respectively. At the time of writing in 1961, this group of workers had studied 63 patients suffering wound dehiscence, and 36, or 57%, of this total showed serious deficiency of blood ascorbic acid at the time of dehiscence. This was significantly greater than the overall rate of ascorbic acid deficiency among 287 surgical patients, which was 40%. They concluded that, "a low vitamin C level is associated with wound dehiscence and that maintenance of the patient's blood ascorbic acid at an adequate level preoperatively appears to be a potent means of preventing wound dehiscence." They did not suggest that ascorbic acid deficiency was the only cause of wound dehiscence, but their data do suggest that it is a major factor.

In studies of the force required to cause dehiscence of experimental skin wounds in the forearms of volunteers, Lindstedt and Sandblom (1975) observed that patients with serum protein levels below 6.5 g/100 ml had weaker wounds at 5 d than did those with higher protein levels. They also noted somewhat weaker wounds in patients over 80 years of age, but there were no studies of younger people with comparable ascorbic acid levels.

Many factors, including the proper choice of incision, accurate approximation of all layers, maintenance of blood supply, good hemostasis, removal of devitalized tissues, avoidance of contamination, use of appropriate suture material, and accurate apposition of the wound edges are all very important in the closure of abdominal wounds, but ascorbic acid deficiency can prevent healing no matter how much care is taken over surgical technique. Moreover, ascorbic acid deficiency is much more common than is generally appreciated. Only when plasma ascorbic acid determination becomes a routine part of hospital admission will the full extent of this deficiency be recognized.

REFERENCES

Archer, H. E. and Graham, G. (1936), The subscurvy state in relation to gastric and duodenal ulcer, *Lancet*, 2, 364.

Bartlett, M. K., Jones, C. M., and Ryan, A. E. (1940), Vitamin C studies on surgical patients, *Ann. Surg.*, 111, 1.

Colp, R. (1934), Disruption of abdominal wounds, *Ann. Surg.*, 99, 14.

Crandon, J. H., Landau, B., Mikal, S., Balmanno, J., Jefferson, M., and Mahoney, N. (1958), Ascorbic acid economy in surgical patients as indicated by blood ascorbic acid levels, *N. Engl. J. Med.*, 258, 105.

Crandon, J. H., Lennihan, R., Mikal, S., and Reif, A. E. (1961), Ascorbic acid economy in surgical patients, *Ann. N.Y. Acad. Sci.*, 92, 246.

Crandon, J. H., Mikal, S., and Landau, B. S. (1952), Ascorbic acid in surgical patients with particular reference to their blood buffy layer, *Surg. Gynecol Obstet.*, 95, 274.

Crandon, J. H., Mikal, S., and Landeau, B. R. (1953), Ascorbic acid deficiency in experimental and surgical subjects, *Proc. Nutr. Soc.*, 12, 273.

Fallis, L. S. (1937), Postoperative wound separation: review of cases, *Surgery*, 1, 523.

Hartzell, J. B. and Winfield, J. M. (1939), Disruption of abdominal wounds. Collective review. International Abstract of Surgery, *Surg. Gynecol. Obstet.*, 68 (Suppl.), 585

Hartzell, J. B., Winfield, J. M., and Irvin, J. L. (1941), Plasma vitamin C and serum protein levels in wound disruption, *JAMA*, 116, 669.

Hess, A. F. (1920), *Scurvy Past and Present*, J. B. Lippincott, Philadelphia.

Holman, E. (1940), Vitamin and protein factors in preoperative and postoperative care of the surgical patient, *Surg. Gynecol. Obstet.*, 70, 261.

Hunt, A. H. (1940/1941), The role of vitamin C in wound healing, *Br. J. Surg.*, 28, 436.

Ingalls, T. H. and Warren, H. A. (1937), Asymptomatic scurvy. Its relation to wound healing and its incidence in patients with peptic ulcer, *N. Engl. J. Med.*, 217, 443.

Lanman, T. H. (1938), Relation of vitamin C deficiency to wound healing, *J. Pediatr.*, 12, 416.

Lanman, T. H. and Ingalls, T. H. (1937), Vitamin C deficiency and wound healing. An experimental and clinical study, *Ann. Surg.*, 105, 616.

Lind, J. (1753), *A Treatise of the Scurvy*, Reprinted (1953) as *Lind's Treatise on Scurvy*, Stewart, C. P. and Guthrie, D., Eds., Edinburgh University Press, Edinburgh.

Lindstedt, E. and Sandblom, P. (1975), Wound healing in man: tensile strength of healing wounds in some patient groups, *Ann. Surg.*, 181, 842.

Lund, C. C. and Crandon, J. H. (1941b), Ascorbic acid and human wound healing, *Ann. Surg.*, 114, 776.

Payne, Sir R. (1936), The post-mortem findings after partial gastrectomy, *St Bartholemew's Hosp. Rep.*, 191.

Sokolov, S. (1932), Das Aufplatzen der Bauchwunde nach Laparotomie mit Eventration bzw. Freiliegen der Eingeweide, *Ergeb. Chir. Orthopaed.*, 25, 306.

Taffel, M. and Harvey, S. C. (1938), Effect of absolute and partial vitamin C deficiency on healing of wounds, *Proc. Soc. Exp. Biol. Med.*, 38, 518.

Walter, Richard, (1748), *A Voyage Around the World, in the Years 1740, 41, 42, 43, 44*, by George Anson Esq., now Lord Anson, Commander in Chief of a squadron of his Majesty's ships, sent upon an expedition to the South Seas, Compiled from his papers and materials; as cited in **James Lind (1753)**, *A Treatise of the Scurvy*, Reprinted (1953) as *Lind's Treatise on Scurvy*, Stewart, C. P. and Guthrie, D., Eds., Edinburgh University Press, Edinburgh.

Wolbach, S. B. and Howe, P. R. (1926), Intercellular substances in experimental scorbutus, *Arch. Pathol*, 1, 1.

Wolfer, J. A. and Hoebel, F. C. (1939), The significance of cevitamic acid deficiency in surgical patients, *Surg. Gynecol. Obstet.*, 69, 745.

Chapter 13

HABITUAL ABORTION

The terms "recurrent abortion" or "habitual abortion" are usually used to describe the misfortune of three consecutive spontaneous abortions or miscarriages. Most spontaneous abortions are believed to be due to defective ova. Indeed, Hertig et al. (1956), studying hysterectomy specimens, found morphological abnormalities in 4 out of 8 preimplanted embryos and in 8 out of 26 implanted embryos. Moreover, Boué et al. (1975) reported chromosome abnormalities in more than 50% of first-trimester abortion specimens.

However, habitual abortion may belong to a different subset, for the prognosis is much less ominous than these figures would suggest. Warburton and Fraser (1964) showed that after one spontaneous abortion, the risk of recurrence rises from 12 to 25%, but does not increase appreciably thereafter. The known causes of habitual abortion include uterine hypoplasia, uterine anomalies, submucous uterine myomata, incompetent internal os of the cervix, hypothyroidism, chronic nephritis, systemic lupus erythematosus, and syphilis. The roles of *Toxoplasma gondii, Listeria monocystogenes,* Cytomegalovirus, *Chlamydia trachomatis,* and T-strain Mycoplasma (or pleuropneumonia-like organisms) remain to be determined. Other more subtle conditions such as "inadequate luteal phase" have been studied, but in the vast majority of instances, no obvious cause can be found. Chromosome rearrangements are found in one or other parent in 5 to 10% of couples, according to Simpson (1980).

In the past, some physicians placed great faith in progesterone injections to hold the pregnancy, on the assumption that there was a deficiency of this hormone, especially when the urinary pregnanediol levels were falling; but of course, all placental hormone levels will fall when the placenta is dying or is partially separated, whatever the reason; so hormone injections for habitual abortion are used little today. While some authors reported 80% of successful pregnancies following habitual abortion, using bed rest, vitamins, reassurance, and progesterone injections, other reported 80% success with bed rest, vitamins, and reassurance, without any hormone administration. Clearly, these figures relating to patients treated for habitual abortion are somewhat better than the figures for all patients following habitual abortion, for the simple reason that some women abort again even before they reach the physician. For this and other reasons it is difficult to know the meaning of any figures concerning habitual abortion.

Öhnell (1928) cited Westin as having noted a tendency to abortion in guinea pigs, in which scurvy had been produced experimentally. Only three among his own patients with scurvy were married women of reproductive age. Two of them had spontaneous abortions and the third gave birth to a stillborn fetus, so he suggested the need for further investigation of this association.

Some important observations were made by Javert and Stander (1943) who reported the results of analysis of 376 blood samples obtained from 246 pregnant women of average or high economic status in New York City. They noted a tendency for the plasma ascorbic acid (AA)* levels to fall as pregnancy advanced (see Figure 1, Chapter 14, Volume I). They also noted a tendency for low vitamin C and prothrombin levels in women with threatened, spontaneous, or habitual abortion (Table 1). Moreover, they made the clinical observation that, "some of the patients with deficient values in vitamin C and pro-thrombin complained of bruising easily, epistaxis, bleeding gums, and vaginal bleeding."

King (1945), on the other hand, found very littly evidence of vitamin C deficiency in patients with threatened, inevitable, or complete abortion in Montgomery, WV. Only 21%

---

* AA — ascorbic acid, reduced form.

Table 1
PLASMA VITAMIN C AND PROTHROMBIN CONCENTRATIONS IN PATIENTS WITH THREATENED, SPONTANEOUS, AND HABITUAL ABORTIONS

Group	Number of cases	Vitamin C (mg%)		Deficiency (0.5 mg or less)		Prothrombin concentration (% of normal)		Deficiency (70% or less)	
		Average	Range	Number	Percentage	Average	Range	Number	Percentage
Threatened abortion	20	0.46	0.08—0.81	14	70	65	22—100	13	65
Spontaneous abortion	37	0.30	0.00—0.98	27	73	46	23—100	30	81
Habitual abortion	22	0.36	0.10—0.98	14	64	61	15—100	14	68
Total	79	0.35	0.00—0.98	55	69	54	15—100	57	72
Control patients 4 to 16 weeks pregnant	50	0.55	0.00—1.6	25	50	92	70—100	9	15

From Javert, C. T. and Stander, H.J. (1943), *Surg. Gynecol. Obstet.*, 76, 115. With permission

Table 2
CONDITIONS FOUND IN THE DECIDUA OF 1334 ABORTION SPECIMENS AS COMPARED WITH 361 CONTROL SPECIMENS

	Spontaneous number	Abortions (%)	Control group[a] Number	%
Hemorrhage	810	60 7	31	8 6
Degeneration	680	50 9	75	20.7
Infection	572	42 8	13	3.6
Normal	83	6 2	268	71.5

Note. Decidual hemorrhage was seen much more frequently in spontaneous than in therapeutic abortion specimens

[a] Therapeutic and unintentional abortion specimens.

From Javert, C. T. (1955), *Ann. N.Y. Acad Sci.*, 61, 700 With permission

of the abortion patients and 13% of normal control subjects had plasma ascorbic acid levels below 0.5 mg/100 ml. Nevertheless, they recommended vitamin E, vitamin K, and vitamin C supplements for patients with threatened or habitual abortion.

Crampton (1947) discovered that not only vitamin C and other vitamins, minerals, and nutrients, but also roughage is necessary for successful pregnancy in the guinea pig. When his vitamin-supplemented ration was fed *ad libitum* along with fresh or dried, long-stored grass clippings, reproduction (80% successful pregnancies) and the growth of the young were normal; whereas when the roughage was omitted, only about 66% of pregnancies were successful, and there was some slowing of the growth of the young to maturity. The dried herbage clippings were not antiscorbutic; nevertheless, the abortion rate was 13% without it and 0% with it. Today, this roughage is known as fiber. The present writer suggests that it may be the chelating property of the catechins, tannins, and bioflavonoids of the fiber (Chapter 11, Volume I) that are so important in heavy metal balance and, hence, vitamin C metabolism.

Dill (1954), reporting 28 pregnancies in 15 patients who had had two or more late fetal losses prior to the study, obtained 50% of successful pregnancies using bed rest and 400 to 600 mg each of ascorbic acid and the bioflavonoid hesperidin daily. However, other treatments, including thyroxine and progesterone, were also used in some of these patients.

Javert (1954a, b, 1955) gave reasons for believing that decidual hemorrhage was responsible for a large number of spontaneous and repeated abortions. He demonstrated decidual hemorrhage in 61% of spontaneous and only 9% of induced abortions (Table 2). He also reported that decidual hemorrhage was considerably more frequent in the spontaneous abortuses of women with low plasma ascorbic acid (AA) levels (Table 3).

Greenblatt (1953, 1955) performed capillary fragility tests on his patients with habitual abortion and obtained positive tests in over 80%, a much higher incidence than was found in a control group. He, too, noted that many of these women bruised easily and had nasal, rectal, or gingival bleeding; treatment with vitamin C and the flavonoid hesperidin usually corrected the capillary defect. Among those with a positive petechial test, he reported 84.6% success for those who had aborted twice and 66.6% for those who had aborted three to eight times, but hormones and other treatments had been used where appropriate. Taylor (1956) reported success in the treatment of multiple and habitual aborters with vitamin C and citrus bioflavonoids.

Table 3
DECIDUAL HEMORRHAGE IN 100 CASES OF SPONTANEOUS ABORTION CORRELATED WITH MATERNAL BLOOD PLASMA VITAMIN C DEFICIENCY[a]

Maternal blood plasma	No. of cases	Decidual hemorrhage Number	%
Vitamin C deficiency, average value 0.22 mg%	45	30	66.6
Vitamin C sufficiency, average value 0.95 mg%	55	22	40.0
Total	100	52	52.0

Note. Incidence of decidual hemorrhage in 100 spontaneous abortuses correlated with maternal blood plasma vitamin C deficiency.

[a] Values below 0.50 mg%.

From Javert, C. T. (1955), *Ann. N.Y. Acad. Sci.*, 61, 700. With permission.

Summarizing his studies of habitual abortion, Javert (1958) described a system of pre-conceptional and prenatal care including vitamin and mineral supplements, psychotherapy, and a good doctor-patient relationship which had produced a successful outcome in 80% of 213 cases of habitual abortion. He advised a diet high in citrus fruits, providing 350 mg of vitamin C daily, as a result of supplements including vitamin C and K with bioflavonoids, as well as iron and calcium.

Ainslie (1959) reported that the administration of capsules containing vitamin C (200 mg) and citrus bioflavonoids (200 mg), four to six times daily, to patients with threatened abortion or a history of two or more previous spontaneous abortions, apparently contributed to the reduction of the abortion rate from 30 in 304 cases (or 9.9%) in the first series to 35 in 543 cases (or 6.4%) in the second series. The number of threatened abortions in each series was relatively constant. However, of those patients who developed uterine bleeding in the first series, treated with various therapies, 46.9% suffered fetal loss. In the second series, treated with vitamin C and bioflavonoids, with or without thyroid extract, only 26.7% aborted. No improvement was noted in the spontaneous abortion rate among primiparas, which approximated 4% in both series, while marked improvement in fetal salvage was noted among a small number of cases with a history of previous abortion (7 of 38 lost in 1955; 8 of 92 lost in 1957).

Baldwin and Greenblatt (1959) described the diagnostic methods used to differentiate the different causes of habitual abortion and mentioned defects in the clotting mechanism among the vascular disorders. In the experience of the present author, capillary fragility is very common in patients with recurrent abortion, and is often manifested by easy bruising, nosebleeds, and bleeding gums, but the platelet counts, the clotting times, and the partial thromboplastin times of these patients are almost always normal. They have what may be called "simple purpura", for the ordinary laboratory blood studies are normal; it is the small venules that are abnormally weak. There does not seem to be enough evidence concerning the platelet stickiness index or other platelet functions in these patients, but the platelet count is normal. In any event, their purpura slowly improves over 2 months while they are receiving vitamin C and citrus bioflavonoids, 200 mg of each, three times a day, but the treatment must be continued.

Jacobs (1965) reported on 10 years of experience with the same treatment — vitamin C and citrus bioflavonoids — which he found to be beneficial in reducing the incidence of "placental leaks", thereby reducing the risk of Rh isoimmunization. This too was no doubt due to a reduction in the tendency for decidual hemorrhage during pregnancy. Moreover, these findings concerning Rh isoimmunization were endorsed in papers presented by nine speakers at a symposium on the clinical uses of vitamin C and bioflavonoids held at Stresa, on Lago Maggiore, in the spring of 1966. Even though we now have Rhogam® to combat the major fetomaternal red cell leakage during labor and delivery, there is still an important place for vitamin C and bioflavonoids to reduce the smaller leaks that occur during pregnancy.

The report of Samborskaya and Ferdman (1966) suggests that very high doses of ascorbic acid (6 g daily for 3 d) given to women in early pregnancy (10 to 15 d overdue for menstruation) may actually have a paradoxical effect and promote abortion; for they observed vaginal bleeding in 16 out of 20 women so treated. The reason for this paradoxical effect is unknown, but it may be due to excessive levels of dehydroascorbic acid. In any event, it is reason enough to use modest doses of ascorbic acid with catechin (200 mg of each) or, better still, catechin-coated ascorbic acid instead of megadose ascorbic acid in pregnancy.

Papers by Cattaneo et al. (1967), Antuzzi (1967), and Cavalluci and Guida (1967) all described most encouraging results in the treatment of threatened and habitual abortion with vitamin C and bioflavonoids. Umanskii (1970) reported clinical signs of hypovitaminosis C, including swollen gums, looseness of teeth, and bleeding gums in 37.3% of women who did not take any ascorbic acid supplement during pregnancy, while these signs were not observed in any of the women who took 200 to 300 mg of ascorbic acid daily.

Evidence that vitamin C plays an important role in maintaining pregnancy arises from the unusual experiment reported by Paul and Duttagupta (1974). These workers studied rats, which normally manufacture enough vitamin C in the liver, but may not be able to do so on a restricted diet. Feeding pregnant rats a fixed amount of food (15 g/rat per day), which was adequate in the nonpregnant state, was found to cause abortion in 13 out of 16 rats tested. However, the addition of ascorbic acid (25 mg/rat per day) to the restricted diet allowed pregnancy to continue in all of eight rats. The fetal weight was raised by the vitamin therapy, but remained subnormal in the diet-restricted rats. The vitamin C supplement also prevented the loss of uterine weight, the loss of placental glycogen, and also prevented the fall of blood sugar in the diet-restricted animals.

Comparing the effects of three levels of dietary vitamin C intake in guinea pigs (Group 1 receiving 0.15 mg, Group 2 receiving 0.40 mg, and Group 3 receiving 10 mg/100 g body weight per day), Rivers and Devine (1975), of Cornell University, observed that "Group 3 dams required fewer matings per pregnancy, produced more litters per female, and had fewer abortions, on the basis of number of pregnancies, than did those in groups 1 and 2."

Schorah et al. (1978) found no significant difference between the leukocyte ascorbic acid levels of 67 women who aborted spontaneously and 1080 women who did not. There was, however, no study of habitual abortion and very few of the subjects were ascorbic acid deficient; only 2.5% of the women had total ascorbic acid levels below 15 µg/10^8 leukocytes.

Papers on habitual abortion by Glass and Golbus (1978) and by Tho et al. (1979) were very detailed as regards chromosome analyses and bacteriological studies, but included no clinical observations concerning purpuric phenomena, no testing for capillary fragility, no mention of ascorbic acid levels, and no mention of dietary supplements.

Habibzadeh et al. (1986), working at the University of Leeds, conducted a very pertinent study of diet in early pregnancy in Dunkin-Hartley guinea pigs. They investigated the effects of feeding deficient, intermediate, or supplemented levels of folic acid and/or ascorbic acid to pregnant guinea pigs during organogenesis, from 3 d before conception to the 19th day of gestation, and killed all the animals for examination and analysis on the 37th day of their 70-d gestational period. As a result of several experiments, it became evident that a diet which had been used as a feeding stock for nonpregnant guinea pigs, without any apparent ill effects on the health of the animals, was quite inadequate for pregnancy. Not only was there a decrease in the number of abortions or fetal resorptions when the animals received supplementary folic acid and ascorbic acid, there was also an increase in the number and the body weight of surviving fetuses (Table 4). Both ascorbic acid and folic acid supplements were found to be important in early pregnancy, but the folic acid supplement seemed to be more important in preventing abortions, and the ascorbic acid supplement seemed to have the greater effect on fetal weight. However, the authors caution that human nutritional needs may differ from those of the guinea pig. One additional finding of interest, pointed out by these authors is that, "The number of live fetuses found at termination of pregnancy in the folic acid and vitamin C-supplemented females (range 3.5 to 4.3) was greater than the published average litter size for this strain of guina pig (3.0)."

It may be many years before we can determine the individual dietary needs of women with recurrent abortion, but diet is clearly important and, lacking more precise individual data, it will be wise to give these women dietary supplements of folic acid and ascorbic acid in early pregnancy. Moreover, modest doses of ascorbic acid with catechin (200 mg of each), once, twice, or three times a day will be safer than megadose ascorbic acid alone.

Table 4
SUMMARY OF EXPERIMENTAL DATA PROVIDED BY THE WORK OF HABIBZADEH ET AL. (1986) IN GUINEA PIG FEEDING EXPERIMENTS

Folic acid (µg/d per mother)	Vitamin C (mg/d per mother)	Erythrocyte folic acid (µg/l)	Plasma ascorbic acid (µmol/l)	Leukocyte vitamin C (µmol/l blood)	Body weight of fetuses at 37 d gestation (g)	Resorptions and abortions (as % of implantations)	Mean number of living fetuses
Group A Intermediate 99	Deficient 0.35	42	0.6[a]	3.4[a]	5.0	87.1	0.6[a]
Group B Intermediate 107	Intermediate 10.7	61	5.7	9.1	4.7	68.8	1.4
Group C Intermediate 103	Supplemented 80.0	85	34.6[a]	18.2[a]	5.6[a]	68.2	1.4
Group D Supplemented 477	Supplemented 75.0	183[a]	68.1[a]	26.1[a]	6.2[a]	17.5[a]	3.5[a]

[a] Mean values were significantly different from those of Group B animals ($p < 0.01$).

REFERENCES

Ainslie, W. H. (1959), Treatment of threatened abortion, *Obstet Gynecol.*, 13, 185.

Antuzzi, P. (1967), I bioflavonoidi nelle minacce d'aborto in gravidanza iniziale, in *Bioflavonoidi*, Zambotti, V., Ed , Scuole Grafiche Artigianelli Pavoniani, Milano, 287.

Baldwin, K. R. and Greenblatt, R. B. (1959), Habitual abortion; its causes and treatment, *J Indiana State Med Assoc.*, 52, 1945

Boué, J., Boué, A., and Lazar, P. (1975), Retrospective and prospective epidemiological studies of 1500 karyotyped spontaneous human abortions, *Teratology,* 12, 11.

Cattaneo, G., Avezzu, G., and Vinci, G. W. (1967), I bioflavonoidi in ostetricia e ginecologia, in *Bioflavonoidi,* Zambotti, V , Ed., Scuole Grafiche Artigianelli Pavoniani, Milano, 139

Cavalluci, G. G. and Guida, C. (1967), I citrobioflavonoidi per endovena nella minaccia d'aborto, in *Bioflavonoidi,* Zambotti, V., Ed., Scuole Grafiche Artigianelli Pavoniani, Milano, 557.

Crampton, E. W. (1947), The growth of the odontoblasts of the incisor tooth as a criterion of the vitamin C intake of the guinea pig, *J. Nutri.*, 33, 491

Dill, L. V. (1954), Therapy of late abortion, *Med. Ann. D C.*, 23, 667.

Glass, R. H. and Golbus, M. S. (1978), Habitual abortion, *Fertil Steril ,* 29, 257.

Greenblatt, R. B. (1953), Habitual abortion. Possible role of vitamin P in therapy, *Obstet Gynecol.*, 2, 530.

Greenblatt, R. B. (1955), The management of habitual abortion, *Ann. N Y Acad. Sci.,* 61, 713.

Habibzadeh, N., Schorah, C. J., and Smithells, R. W. (1986), The effects of maternal folic acid and vitamin C nutrition in early pregnancy on reproductive performance in the guinea pig, *Br. J. Nutr.*, 55, 23.

Hertig, A. T., Rock, J., and Adams, E. C. (1956), A description of 34 human ova within 17 days of development, *Am J. Anat.*, 98, 435.

Jacobs, W. M. (1965), Citrus bioflavonoid compounds in Rh-immunized gravidas, *Obstet Gynecol.*, 25, 648.

Javert, C. T. (1954a), Repeated abortion: results of treatment in 100 patients, *Obstet Gynecol.*, 3, 420.

Javert, C. T. (1954b), Pathology of spontaneous abortion. II. Relationship of decidual hemorrhage to spontaneous abortion and vitamin C deficiency, *Tex. State J. Med.*, 50, 652.

Javert, C. T. (1955), Decidual bleeding in pregnancy, *Ann. N.Y. Acad. Sci.,* 61, 700.

Javert, C. T. (1958), Prevention of habitual spontaneous abortion with early prenatal care, *Bull. N.Y. Acad Med.,* 34, 747.

Javert, C. T. and Stander, H. J. (1943), Plasma vitamin C and prothrombin concentration in pregnancy and in threatened spontaneous and habitual abortion, *Surg. Gynecol. Obstet.*, 76, 115.

King, W. E. (1945), Vitamin studies in abortions, *Surg. Gynecol. Obstet.*, 80, 139.

Ohnell, H. (1928), Some experiences of endemic, manifest and latent scurvy in Sweden with special reference to new methods of diagnosing latent scurvy, *Acta Med. Scand ,* 68, 176.

Paul, P. K. and Duttagupta, P. N. (1974), Maintenance of pregnancy and tissue carbohydrate levels by vitamin C in rats on restricted diets, *Fertil. Steril.*, 25, 68.

Rivers, J. M. and Devine, M. M. (1975), Relationship of ascorbic acid to pregnancy and oral contraceptive steroids, *Ann. N.Y. Acad. Sci ,* 258, 465.

Samborskaya, E. P. and Ferdman, T. D. (1966), The mechanism of termination of pregnancy by ascorbic acid, *Bull. Exp. Biol. Med.*, 62, 934; as translated from *Byull. Eksp Biol. Med.*, 8, 96.

Schorah, C. J., Zemroch, P. J., Sheppard, S., and Smithells, R. W. (1978), Leucocyte ascorbic acid and pregnancy, *Br. J. Nutr.*, 39, 139.

Simpson, J. L. (1980), Genes, chromosomes and reproductive failure, *Fertil. Steril.*, 33, 107.

Taylor, F. (1956), Habitual abortion; therapeutic evaluation of citrus bioflavonoids, *West. J. Surg.*, 64, 280.

Tho, P. T., Byrd, J. R., and McDonough, P. G. (1979), Etiologies and subsequent reproductive performance of 100 couples with recurrent abortion, *Fertil. Steril.*, 32, 389.

Umanskii, S. S. (1970), On the excretion of ascorbic acid in the urine of women during pregnancy, birth and postpartum period (in Russian), *Vopr. Okhr. Materin. Det.*, 15, 90.

Warburton, D. and Fraser, F. C. (1964), Spontaneous abortion risks in man: data from reproductive histories collected in a medical genetics unit, *Am. J. Hum. Genet.*, 16, 1.

Chapter 14

ABRUPTIO PLACENTAE

Partial separation of the normally situated placenta is one of the most serious complications of pregnancy, endangering the life of the mother as well as the fetus. Less than 5% of the women report any degree of injury to the abdomen preceding the hemorrhage, so more subtle predisposing causes have been sought. Undoubtedly, women who have not received any prenatal care do have an unduly high incidence of abruptio placentae. They often present for the first time as "emergencies", already bleeding in late pregnancy due to partial separation of the placenta. Some have been tempted to interpret the lower incidence of abruptio placentae among women receiving prenatal care to the care they have received, but in all probability those who fail to obtain any medical care are the poorest of the poor, with multiple social and economic problems, so such a conclusion cannot be justified. The late arrival of many women with abruptio placentae has made studies of this problem very difficult, because they are so often seen for the first time only after the fact. Nevertheless, several factors associated with or predisposing to abruptio placentae have been identified by different writers as follows:

- Smoking
- Low socioeconomic class
- Sickle cell disease (SS or SC)
- Diphenyl hydantoin therapy
- Systemic infection
- Essential hypertension
- Preeclampsia
- Chronic nephritis
- Trauma
- Inferior vena cava pressure
- Pregnancy cycle (28-d cycle)
- Low serum ascorbic acid levels
- Disturbance of folic acid metabolism
- Megaloblastic anemia
- Histaminemia

While we are asking the question, what causes premature separation of the normally implanted placenta?, it might be more appropriate to ask, what normally holds the placenta in place? The anchoring villi are clearly not very strong, or else they are not very well anchored, for the placenta normally separates easily after birth of the child. We may conclude that the hydrostatic pressure exerted by the amniotic fluid on the fetal surface of the placenta must be higher than the pressure in the retroplacental blood to hold the placenta in place: thus, the effects of hypertension, trauma, and inferior vena cava pressure are readily understandable, for they all increase the pressure in the retroplacental space. The strange effects of the pregnancy cycle, influencing capillary fragility and having a 4-weekly influence on the timing of the onset of abruptio placentae have been discussed in Chapter 14, Volume I.

It is intended here to discuss the disturbance of folic acid metabolism, the megaloblastic anemia, the lower serum ascorbic acid levels, and the increased blood histamine levels which are associated with abruptio placentae and which may all result from a disturbance of ascorbic acid metabolism.

The Vanderbilt cooperative study of maternal and infant nutrition included an extensive study of the nutritional status of more than 2000 pregnant women. It included both dietary intake histories and chemical analyses of blood and urine for assessment of individual vitamin levels; these were correlated with pregnancy outcome and with complications of pregnancy. Reporting on the vitamin C part of this study, Martin et al. (1957) stated, "Analysis of findings relative to health of the mother and baby revealed only five categories which may possibly be associated with ascorbic acid nutriture: haematologic findings, gingivitis, premature separation of the placenta, premature birth and puerperal fever." Serum levels of vitamin C were found to vary with the recorded intake of this vitamin, but the correlation was not high. Although no correlation was found between dietary ascorbic acid intake and abruptio placentae, there was a correlation between consistently low serum ascorbic acid levels (below 0.4 mg/100 ml in both the second and third trimesters of pregnancy) and abruptio placentae. Such findings suggest an abnormality of ascorbic acid metabolism rather than a simple dietary deficiency of ascorbic acid. Indeed, considering "consistent serum vitamin C groups," 9 out of 10 patients with abruptio placentae fell into the consistently low (<0.4) and very low (<0.2 mg/100 ml) serum ascorbic acid groups.

Nevertheless, dietary ascorbic acid intake undoubtedly plays an important role, especially in northern regions of the world. Pankamaa and Räihä (1957), analyzing 116,790 deliveries at the Women's Clinic of the University of Helsinki, observed that the lowest percentage frequency of stillbirths occurred in September and the highest in January. Moreover, the State Bureau of Statistics for Finland recorded 90,553 births in 1956, of which 1657 were stillbirths, as reported by Räihä (1958) in his monograph on the placental transfer of vitamin C; the frequency of stillbirths was highest (2.03%) in December and lowest (1.50%) in September.

Studying the records of 95 women with megaloblastic anemia of pregnancy at the National Maternity Hospital in Dublin, Hourihane et al. (1960) made the original observation of an increased incidence of abruptio placentae in these women. Accidental hemorrhage (abruptio placentae) occurred in 14% of the patients with megaloglastic erythropoeisis compared with an incidence of only 2.5% for all women delivering at that hospital. Discussing the etiology of megaloblastic anemia of pregnancy, these authors suggested that there was, in every case, an absolute or relative deficiency of folic acid. Perhaps we should say — a disturbance of folic acid metabolism leading to a relative deficiency of folinic acid. This seems far more likely, for a diet which is truly deficient in folic acid must be rare. It would have to be devoid of both meat and vegetables, or else be boiled long enough to destroy all folic and ascorbic acid.

Coyle and Geoghegan (1962) observed that there was a seasonal incidence of megaloblastic anemia in Dublin, highest in the winter, and that 35 out of 77 patients with severe accidental hemorrhage had megaloblastic erythropoeisis.

Working in Liverpool, Hibbard and Hibbard (1963) utilized the formiminoglutamic acid (Figlu) excretion test to estimate the folic acid status of pregnant women. Identification of Figlu in a specimen of urine passed 5 h after a loading dose of L-histidine was considered as a positive result and indicative of folic acid deficiency. Although this is not an entirely reliable test, it is of interest that all but 1 of 73 patients with abruptio placentae, or 98.6%, had positive Figlu excretion tests, as compared with 10.7% of normal patients studied as controls. Megaloblastic erythropoeisis was detected in 46 of the 72 patients with positive Figlu tests.

In a subsequent paper, Hibbard and Hibbard (1966) reported a prospective study of pregnancies occurring in 200 women who had shown evidence of defective folate metabolism in a previous pregnancy. They found that the defect recurred in 146 instances (73%). This and the fact that 10% of their patients failed to respond satisfactorily to treatment with folic acid, led them to conclude that an intrinsic metabolic defect was more likely than a simple dietary defect.

The works of Nichol and Welch (1950), May et al. (1950a, b), and of Nichol (1953) concerning the effects of ascorbic acid deficiency on folic acid metabolism have been reviewed in Chapter 4 of this volume. These workers had already discovered that ascorbic acid is necessary for the conversion of folic to folinic acid and also to prevent losses of folinic acid *in vitro*. Moreover, studying premature monkeys and newborn human infants, they had shown that megaloblastic anemia of infancy could be cured by folic acid, by very small amounts of folinic acid, or by ascorbic acid alone. Holly (1951) had already reported a relationship between megaloblastic anemia of pregnancy and ascorbic acid deficiency, because of the effectiveness of combined ascorbic acid and vitamin B_{12} therapy in five patients, while vitamin B_{12} alone was ineffective.

Moreover, Boscott and Cooke (1954) treated five patients with megaloblastic anemia. All excreted parahydroxyphenylacetic acid in large amounts in their urine, but ceased to do so after administration of ascorbic acid. Working in San Francisco, Clemetson and Andersen (1964) observed very low plasma ascorbic acid levels and a low ratio of reduced to oxidized ascorbic acid in 4 patients following abruptio placentae. The mean values obtained 1 h after drawing blood were TAA* 0.25 and AA** 0.16 mg/100 ml following abruptio placentae, and TAA 1.02 and AA 0.90 mg/100 ml in 19 normal women in late pregnancy. There was no way of knowing whether these low ascorbate levels preceded the abruptio and might have caused it, or whether they followed abruptio and were the result of it, but they were certainly worthy of note. Nevertheless, an Editorial (1967) on folic acid and pregnancy made no mention of ascorbic acid. Many workers have since been disappointed by their inability to prevent abruptio placentae by giving dietary supplements of folic acid.

Clearly the detrimental effect of a combined deficiency of ascorbic and folic acids might account for the seasonal incidence of megaloblastic anemia of pregnancy, but ascorbate deficiency alone could precipitate decidual hemorrhage. This is probably the reason that some authors have found a definite association of folic acid deficiency and abruptio placentae in some areas of the world, while others in different regions have found no such linkage.

Just as it seems that low folate levels result from a disturbance of folate metabolism, which may be due to low ascorbate levels rather than a dietary deficiency of folic acid, so it is likely that the low ascorbate levels are due to a disturbance of ascorbic acid metabolism rather than a simple deficiency of ascorbic acid. This is exemplified by the association between cigarette smoking and abruptio placentae, for smokers have low ascorbate levels (Chapter 4, Volume I). An association between smoking and abruptio placentae was reported by Meyer et al (1976). Also, Kullander and Kallen (1971), studying women smokers in Sweden, observed among children dying before the age of 1 week a significantly increased frequency listed as following abruptio placentae.

Umanskii (1970), working with others at the Talinn Birth Clinic, observed clinical signs of hypovitaminosis C, such as swollen bleeding gums and loose teeth in 37% of the pregnant women who did not take a vitamin supplement and in none of the women who took vitamin C supplements in the amount of 200 to 300 mg/d. They reported a very high incidence of achlorhydria or hypochlorhydria (88%) in women before childbirth and in increase in gastric acidity during 1 to 3 d after delivery. It would be interesting to know the prevalence of achlorhydria during pregnancy, for this can cause losses of ascorbic acid by oxidation and hydrolysis in the stomach (Chapter 20, Volume I).

There was renewed interest in ascorbic acid levels in pregnancy when Clemetson (1980) observed that the blood histamine levels of pregnant women, and others, increase exponentially as plasma ascorbic acid levels fall. Moreover, the fact that one of the women in that study, who had a low ascorbate and a high histamine level, subsequently developed abruptio

* TAA — total ascorbic acid, reduced and oxidized forms.
** AA — ascorbic acid, reduced form.

placentae (Figure 5, Chapter 1 of this volume) was particularly notable. It reminded us of the work done more than half a century earlier, at Johns Hopkins University by Hofbauer and Geiling (1926). These workers reported that intravenous injection of histamine caused abruptio placentae in all of ten pregnant guinea pigs. This was associated with tetanic contraction of the uterus, hypotension, engorgement of the uterine blood vessels, thrombosis of decidual and hepatic veins, and degenerative changes in the convoluted tubules of the kidneys.

Thus, we were prompted to extend our study; an attempt was made to get blood samples during pregnancy, from as many women as would volunteer, at the prenatal clinic of the Methodist Hospital, Brooklyn. Three or four blood samples (20 ml each) were collected in dry heparin tubes, from nonfasting women between 9:00 and 10:30 most weekday mornings; each sample was taken directly to the department laboratory without delay. Some women gave blood twice, in midpregnancy and again in late pregnancy, while others gave blood only once. The blood samples were analyzed from whole blood histamine by the method of Shore et al. (1959), and the plasma filtrates were analyzed for ascorbic acid (AA) in triplicate by the method of Roe (1954), using all the precautions detailed by Clemetson and Cafaro (1981). Among 386 women who had blood samples analyzed during pregnancy in this way, and who had completed their pregnancies at the time of writing, Clemetson and Cafaro (1981) found 13 who were diagnosed as having had major or minor degrees of separation of a normally situated placenta during pregnancy. Most of them were minor degrees of abruptio, and at first sight, there did not appear to be any relationship between the plasma ascorbic acid levels and abruptio placentae, for abruptio had occurred in women with high, medium, and low plasma ascorbic acid (AA) levels. However, when the incidence of abruptio placentae was represented as a percentage of the total number of women in each ascorbic acid group (Table 1), it became evident that women with low plasma ascorbic acid levels did have a markedly increased incidence of premature placental separation. In fact, the difference between the incidence of abruptio placentae of 19% in women with plasma ascorbic acid levels below 0.4 mg/100 ml, and 2% in women with ascorbic acid levels above 0.4 mg/100 ml is very highly significant ($p < 0.001$), whether one considers the first blood sample (Table 1A) or the most recent blood sample (Table 1B).

A similar analysis of the whole blood histamine levels and abruptio placentae is shown in Table 2, where it is evident that women with blood histamine levels greater than 30 ng/ml had an 18.5 to 19% incidence of abruptio placentae, which was significantly greater than the 3.2 to 3.5% incidence of abruptio in women whose blood histamine levels were below 30 ng/ml. Moreover, this was found to be true, whether one considered the first or the most recent blood samples. Thus, the association between low serum vitamin C levels and abruptio placentae which was found in the Vanderbilt study has been confirmed in this study of blood plasma samples. The question as to whether low ascorbate levels predispose to abruptio placentae or whether abruptio placentae and ascorbic acid deficiency, both being associated with social deprivation, simply tend to be found in the same segment of the population, remains to be answered.

However, the fact that low ascorbate levels are associated with high histamine levels and that Hofbauer and Geiling produced abruptio placentae by injection of histamine, makes it logical to consider the chain of events shown in Figure 1, where low ascorbic acid levels play a central role. It is unlikely that increased ascorbic acid supplements in pregnancy will eliminate abruptio placentae completely, for some cases are undoubtedly due to trauma or other causes, but one might hope for a significant reduction in its incidence.

Presumably, most pregnant women can obtain enough vitamin C from orange juice and fresh fruits and vegetables, but some do not, and they need an ascorbic acid supplement. Moreover, ascorbic acid needs undoubtedly vary from person to person, being greater in the elderly, in cigarette smokers, in the presence of infection, following surgery or trauma,

Table 1
THE RESULTS OF ANALYSIS OF 417 BLOOD SAMPLES OBTAINED FROM 386 PREGNANT WOMEN ARE ENUMERATED IN 0.2 mg/100 ml ASCORBIC ACID GROUPS

A. Ascorbic Acid (First Sample)

Plasma ascorbic acid (mg/100 ml)	Pregnancy outcome		Incidence of abruptio placentae
	No abruptio placentae	Abruptio placentae	
0.0—0.199	3	3 ⎫	
0.2—0.399	23	3 ⎬	6/32 18.8%
0.4—0.599	44	1	
0.6—0.799	105	2	
0.8—0.999	117	3	
1.0—1.199	42	0	
1.2—1.399	28	1	7/354 2.0%
1.4—1.599	8	0	
1.6—1.799	2	0	
2.2—2.399	1	0	
Total	373	13	3.5%
Mean maturity at the time of sample	26.1 weeks	24.1 weeks	$\chi_c^2 = 20.48$ $p < 0.001$
Mean age of women	25.0 years	27.2 years	

B. Ascorbic Acid (Most Recent Sample)

0.0—0.199	3	1 ⎫	
0.2—0.399	18	4 ⎬	5/26 19.2%
0.4—0.599	46	2	
0.6—0.799	108	2	
0.8—0.999	118	3	
1.0—1.199	43	0	8/360 2.2%
1.2—1.399	28	1	
1 4—1.599	8	0	
1.6—1.799	1	0	
Total	373	13	3.5%
Mean maturity at the time of sample	26.5 weeks	26.8 weeks	$\chi_c^2 = 16.65$ $p < 0.001$

Note: In Table 1A, the first or only blood samples are counted, and in Table 1B, the most recent or only blood samples are counted. It may be seen that the incidence of abruptio placentae was significantly higher in women whose ascorbate levels were below 0.4 mg/100 ml, whichever way the data are analyzed.

From Clemetson, C. A. B. and Cafaro, V. (1981), *Int. J. Gynaecol. Obstet.*, 19, 453. With permission.

Table 2

**THE RESULTS OF ANALYSIS OF 416 BLOOD
SAMPLES FROM 386 PREGNANT WOMEN
ARE ENUMERATED IN 10-ng/ml WHOLE
BLOOD HISTAMINE GROUPS**

A. Histamine (First Sample)

Pregnancy outcome

Histamine (ng/ml)	No abruptio placentae	Abruptio placentae	Incidence of abruptio placentae
0—9.9	67	1 ⎫	
10—19.9	225	4 ⎬	7/359 1 9%
20—29.9	60	2 ⎭	
30—39.9	18	2 ⎫	
40—49.9	3	2 ⎬	5/26 19.2%
50—59.9	0	1 ⎭	
Total	373	12	3.2% $\chi_c^2 = 18.60$ $p < 0.001$

B. Histamine (Most Recent Sample)

0—9.9	68	1 ⎫	
10—19.9	225	6 ⎬	8/359 2.2%
20—29.9	58	1 ⎭	
30—39.9	19	2 ⎫	
40—49.9	3	2 ⎬	5/27 18.5%
50—59.9	0	1 ⎭	
Total	373	13	3.5% $\chi_c^2 = 15.78$ $p < 0.001$

Note: In Table 2A, the first or only blood samples are counted and in Table 2B, the most recent or only blood samples are counted. It may be seen that the incidence of abruptio placentae was significantly greater in women whose whole blood histamine levels were greater than 30 ng/ml, whichever way the data are analyzed.

From Clemetson, C. A. B. and Cafaro, V. (1981), *Int. J. Gynaecol. Obstet.*, 19, 453. With permission.

in people with a rheumatic diathesis, in epileptics receiving hydantoin, etc., so perhaps one needs to take blood samples from all women during pregnancy to make sure that their plasma ascorbic acid is maintained at an adequate level and does not fall as pregnancy advances. Alternatively, one could give large enough ascorbic acid supplements to ensure adequate blood levels in the vast majority of women.

Unfortunately, there is now evidence that ascorbic acid becomes mutagenic in the presence of copper, as reported by Stich et al. (1976), and there may be as much as 2 ppm of copper in the first water drawn from copper pipes in soft-water areas. Of course it is not ascorbic

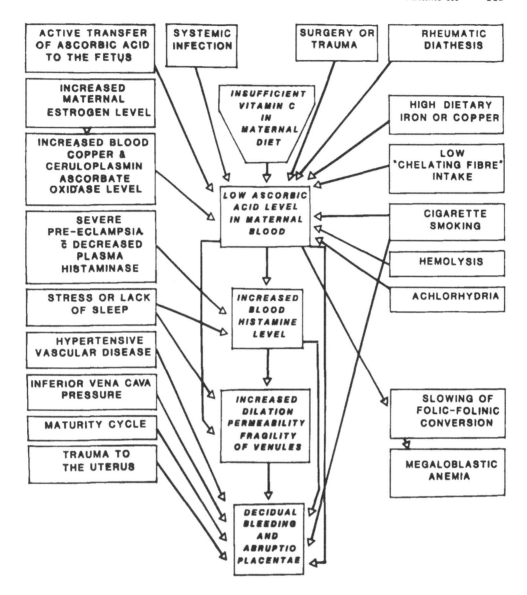

FIGURE 1.   This diagram is intended to indicate the many factors which may be responsible for low ascorbic acid and high histamine levels in maternal blood, as well as other factors which may lead directly to decidual damage and abruptio placentae. In this scheme, a disturbance of folic acid metabolism and megaloblastic anemia are considered as having a noncausal association with abruptio placentae. (From Clemetson, C. A. B. and Cafaro, V. [1981], *Int. J. Gynaecol. Obstet.*, 19, 453. With permission.)

acid itself that is toxic; it is monodehydroascorbic acid or "ascorbate free radical" that is toxic. Fortunately the copper can be inactivated by chelation, as it is in natural foods. So a plain ascorbic acid tablet should be taken with orange juice, grapefruit juice or tomato juice, rather than with tap water, until such time as catechin-coated ascorbic acid tablets are approved by the U.S. Food and Drug Administration and become generally available; there is a full discussion of this problem in Chapter 11, Volume I.

In spite of this theoretical hazard, the extensive study reported by Nelson and Forfar (1971) concerning drugs administered during pregnancy and congenital abnormalities of the fetus showed that throughout the whole of pregnancy significantly fewer mothers of infants with all major and minor abnormalities took vitamins in general, and ascorbic acid in particular. There should be no hesitation in advising ascorbic acid supplements during pregnancy.

Sharma et al. (1985) have confirmed that the blood ascorbic acid levels are lower than normal in women with abruptio placentae, but they found no difference from normal in the blood histamine levels. Sharma et al. (1986) have also observed significantly lower than normal plasma levels of vitamin A and vitamin E in women with abruptio placentae. They have therefore suggested that abruptio placentae is a condition associated with multiple vitamin deficiencies. We can readily understand that low bile acid levels due to decreased conversion of cholesterol to bile acids in vitamin C deficiency (Chapter 5 of this volume) could be responsible for decreased absorption of fat-soluble vitamins.

A recent report by Norkus et al. (1987), in New York City, has demonstrated markedly reduced ascorbic acid levels in the maternal plasma (0.40 vs. 0.89 mg/dl), the cord plasma (0.61 vs. 1.68 mg/dl), and in the placental tissue (10.1 vs. 20.9 mg/dl) in women who smoke cigarettes during pregnancy ($p < 0.01$, <0.01, and <0.01), so it is hard to escape the conclusion that smoking causes ascorbate depletion, histaminemia, and capillary fragility, which can lead to decidual hemorrhage, abruptio placentae, and even fetal death.

Dietary supplements of ascorbic acid and chelating fiber may not be enough to prevent the problem, unless we can persuade pregnant women to give up smoking.

REFERENCES

Boscott, R. J. and Cooke, W. T. (1954), Ascorbic acid requirements and urinary excretion of phpa in steatorrhoea and macrocytic anemia, *Q. J. Med.*, 23, 307.

Clemetson, C. A. B. (1980), Histamine and ascorbic acid in human blood, *J. Nutr.*, 110, 662.

Clemetson, C. A. B. and Andersen, L. (1964), Ascorbic acid metabolism and preeclampsia, *Obstet Gynecol*, 24, 774.

Clemetson, C. A. B. and Cafaro, V. (1981), Abruptio placentae, *Int. J. Gynaecol. Obstet.*, 19, 453.

Coyle, C. and Geoghegan, F. (1962), The problem of anaemia in a Dublin maternity hospital, *Proc. R. Soc. Med.*, 55, 764.

Editorial (1967), Folic acid in pregnancy, *Nutr. Rev*, 25, 325.

Hibbard, B. M. and Hibbard, E. D. (1963), Aetiological factors in abruptio placentae, *Br. Med. J.*, 2, 1430.

Hibbard, B. M. and Hibbard, E. D. (1966), Recurrence of defective folate metabolism in successive pregnancies, *J. Obstet. Gynaecol. Br. Commonw.*, 73, 428.

Hofbauer, J. and Geiling, E. M. K. (1926), Studies on the experimental production of premature separation of the placenta, *Bull. John's Hopkins Hosp.*, 38, 143.

Holly, R. B. (1951), Megaloblastic anemia in pregnancy. Remission following combined therapy with ascorbic acid and vitamin B_{12}, *Proc. Soc. Exp. Biol. Med.*, 78, 238.

Hourihane, B., Coyle, C. V., and Drury, M. I. (1960), Megaloblastic anaemia of pregnancy, *J. Ir. Med. Assoc.*, 47, 1.

Kullander, S. and Källén, B. (1971), A prospective study of smoking and pregnancy, *Acta Obstet. Gynecol. Scand.*, 50, 83.

Martin, M. P., Bridgeforth, E., McGanity, W. J., and Darby, W. J. (1957), The Vanderbilt Cooperative study of maternal and infant nutrition. X. Ascorbic acid, *J. Nutr.*, 62, 201.

May, C. D., Nelson, E. M., Lowe, C. U., and Salmon, R. J. (1950), Pathogenesis of megaloblastic anemia in infancy. An interrelationship between pteroylglutamic acid and ascorbic acid, *Am. J. Dis. Child.*, 80, 191.

May, C. D., Sundberg, R. D., and Schaar, F. (1950), Comparison of effects of folic acid and folinic acid in experimental megaloblastic anemia, *J. Lab. Clin. Med.*, 36, 963.

Meyer, M. B., Jonas, B. S., and Tonascia, J. A. (1976), Perinatal events associated with maternal smoking during pregnancy, *Am. J. Epidemiol.*, 103, 464.

Nelson, M. M. and Forfar, J. O. (1971), Associations between drugs administered during pregnancy and congenital abnormalities of the fetus, *Br. Med. J.*, March 6, 523.

Nichol, C. A. (1953), The effect of ascorbic acid on the enzymatic formation of the citrovorum factor, *J. Biol. Chem.*, 204, 469.

Nichol, C. A. and Welch, A. D. (**1950**), Synthesis of citrovorum factor from folic acid in liver slices Augmentation by ascorbic acid, *Proc Soc Exp Biol. Med.,* 74, 52.

Norkus, E. P., Hsu, H., and Cehelsky, M. R. (**1987**), Effect of cigarette smoking on the vitamin C status of women and their offspring, *Ann N.Y. Acad. Sci.,* 498, 500.

Pankamaa, P. and Räihà, N. (**1957**), Vitamin C deficiency as a factor influencing seasonal fluctuations in the frequency of stillbirths, *Etud. Neo-Natales,* 6, 145.

Räiha, N. (**1958**), On the placental transfer of vitamin C, *Acta Physiol. Scand.,* 45 (Suppl 155), 1.

Roe, J. H. (**1954**), Photometric indophenol method for determination of ascorbic acid, in *Methods of Biochemical Analysis,* Vol. 1, Glick, D., Ed., Interscience, New York, 121.

Sharma, S. C., Bonnar, J., and Dóstolová, L. (**1986**), Comparison of blood levels of vitamin A, β-carotene and vitamin E in abruptio placentae with normal pregnancy, *Int. J. Vitam. Nutr. Res.,* 56, 3.

Sharma, S. C., Walzman, M., Bonnar, J., and Molloy, A. (**1985**), Blood ascorbic acid and histamine levels in patients with placental bleeding, *Hum. Nutr. Clin. Nutr.,* 39C, 233.

Shore, P. A., Burkhalter, A., and Cohn, V. H. (**1959**), A method for the fluorometric assay of histamine in tissues, *J. Pharmacol. Exp. Ther.,* 127, 182.

Stich, H. F., Karim, J., Koropatnick, J., and Lo, L. (**1976**), Mutagenic action of ascorbic acid, *Nature (London),* 260, 722.

Umanskii, S. S. (**1970**), On the excretion of ascorbic acid in the urine of women during pregnancy, birth and post-partum period (in Russian), *Vopr Okhr. Materin. Det.,* 15, 90.

Chapter 15

PREMATURITY AND PREMATURE RUPTURE OF THE FETAL MEMBRANES

Premature birth is one of the leading causes of neonatal morbidity and mortality, and premature rupture of the membranes is a major cause of prematurity, but its cause is still unknown. When we speak of premature birth, we are referring to a child being born more than 2 weeks earlier than expected, but when we speak of premature rupture of the membranes, we refer to spontaneous rupture of the amnion and the chorion, with leakage of amniotic fluid before the onset of labor. Nevertheless, it is quite a common occurrence for premature rupture of the membranes to occur at 28 to 32 weeks, and even with conservative treatment, including bed rest in hospital, the gain of maturity is rarely more than 2 or 3 weeks, so it is a potent cause of prematurity. Not only does premature rupture of the membranes endanger the life of the fetus, it can also be a cause of morbidity in the mother if chorioamnionitis develops, and can even endanger her life if she develops septic shock.

The known causes of prematurity include multiple pregnancy, uterine myomata, essential hypertension, preeclampsia, chronic nephritis, abruptio placentae, polyhydramnios, hyperthyroidism, Rh isoimmunization, systemic maternal infections, and pyelonephritis; even asymptomatic bacteriuria in early pregnancy can predispose to premature labor, according to Kass (1959). Nevertheless, we just do not know the cause of premature labor in most instances.

Studying guinea pigs on a diet of oats and water, Ingier (1915) found that vitamin C deficiency in the earlier stages of pregnancy invariably resulted in premature birth or in dead fetuses. Older animals and those in the latter half of gestation gave birth to living and apparently mature offspring, but even those had latent scurvy, for they developed infantile scurvy or Barlow's disease within a few days when they were fed milk from a scorbutic mother. Some of the stillborn fetuses showed spontaneous fractures characteristic of Barlow's disease

One earnestly hopes that a sickly, whining, growth-retarded child with bruises and spontaneous fractures due to Barlow's disease today would be correctly diagnosed and treated and not be labeled as an "abused child" by misguided opinion. It is interesting to note that the diagnosis of Barlow's disease has become a rarity, while the diagnosis of "child-abuse" has risen enormously. Malnutrition is thought of as a thing of the past, but unfortunately, poverty, unemployment, and malnutrition are still very much a part of the modern world.

The observations of Ingier were confirmed by Reyher et al. (1928). Kramer et al. (1933) observed embryo death and abortion in vitamin C-deficient guinea pigs. Analyses of blood samples from newborn infants were reported by Fleming and Sanford (1938). They observed that the blood vitamin C (AA)* levels of seven premature infants were mostly lower than the expected average for normal infants.

In the same year, Elmby and Christensen (1928), at the University of Copenhagen, conducted a study of the vitamin C contents of the diets of 200 women during pregnancy. They classified the women into three groups: (1) those with vitamin C-poor diets, (2) those with moderate vitamin C intakes, and (3) those consuming plenty of vitamin C-rich foods. The women of Group 3 were encouraged to eat as many as six oranges or six lemons a day. Even so, these authors reported a significant fall in the serum ascorbic acid levels of women in all three groups during pregnancy, suggesting unusually high ascorbic acid needs. Table 1 shows the numbers of mature and premature infants born in each group and also the serum ascorbic acid levels of their infants. It is evident that 16 out of 74, or 22%, of the infants born to mothers in Group 1, on a vitamin C-poor diet, were premature, while 10 out of 66,

---

* AA — ascorbic acid, reduced form.

Table 1[a]

Newborn serum ascorbic acid (mg/100 ml)	Vitamin C-poor diet		Moderate vitamin C diet		Vitamin C-rich diet	
	Normal births	Premature births	Normal births	Premature births	Normal births	Premature births
3.00					1	
2.80—2.99					1	
2.60—2.79						
2.40—2.59			1		2	
2.20—2.39					2	
2.00—2.19	1		1		1	
1.80—1.99					1	
1.60—1.79					7	
1.40—1.59					6	
1.20—1 39	2		5		6	
1.00—1.19	4		9		10	
0.80—0.99	29		19	1		2
0.60—0.79	4	4	18	4		1
0.40—0.59	4	2	3	4		
0.20—0.39	14	10		1		
0.00—0.19						

Note: The serum ascorbic acid levels of newborn infants are grouped according to the calculated dietary vitamin C intakes of their mothers during pregnancy. Each vitamin C intake group is further divided into mature and premature infants. Here, 22% of the vitamin C-poor mothers, 15% of the moderate, and 8% of the high-ascorbic acid intake mothers gave birth to premature infants (see text).

[a]   Data provided by the work of Elmby and Christensen (1938).

or 15% in Group 2, and 3 out of 40, or 8% of infants in Group 3 were born prematurely. Inspection of Table 1 also shows that the premature infants had the lower ascorbic acid levels, while most of the mature infants had higher levels of vitamin C. Of course, the poor women had the poor diets, so they may have lacked other essential nutrients besides vitamin C; nevertheless, the figures are impressive.

Martin et al. (1957), reporting the serum vitamin C findings in the Vanderbilt Cooperative Study of Maternal and Infant Nutrition at Nashville, TN, observed that living premature infants were born to 22 out 442, or 5%, of women who had serum ascorbate levels consistently below 0.4 mg/100 ml in the second and third trimesters of pregnancy, while only 4 out of 198, or 2%, of premature living infants were born to women with consistently high serum ascorbic acid levels in the second and third trimesters of pregnancy. The stillbirth rates in these groups were 1.1 and 0.5% and the neonatal death rates were 2.3 and 2%, respectively.

A very interesting study was conducted by Wideman et al. (1964) on 288 women who delivered at the University Hospital in Birmingham, AL. The women all came from the lower socioeconomic group; dietary histories indicated a low ascorbic acid intake and very few were taking vitamin supplements with any regularity. The median plasma ascorbic acid (AA) level was found to be 0.35 mg/100 ml in the sixth to eighth months of pregnancy, and 0.25 mg/100 ml after delivery. The incidence of premature birth was 15.6% in 219 women with plasma AA levels below 0.6 mg/100 ml and 8.7% in 69 women with plasma AA levels above 0.6 mg/100 ml in the seventh and eighth months of pregnancy (Table 2). An analysis of the incidence of spontaneous rupture of the membranes prior to the onset of labor (Table 3) shows a very strong correlation with low prenatal ascorbic acid levels. Likewise, an analysis of spontaneous rupture of the membranes 6 and 12 h prior to delivery was suggestive of the validity of the hypothesis that adequate plasma ascorbic levels minimize the risk of early membrane rupture.

Table 2

Plasma AA level (mg/100 ml)	No. of women	Median weight of infants at birth	No. of deliveries of infants under 2500 g	Premature deliveries (%)
0 to 0.59	219	3080	34	15.6
0 60 or greater	69	3193	6	8.7

Note: The percentage of "premature" deliveries was much greater among mothers with low plasma ascorbic acid (AA) levels.

From Wideman, G L., Baird, G H., and Golding, O. T. (1964), *Am. J. Obstet. Gynecol* , 88, 592. With permission.

Table 3

Plasma AA (mg/100 ml)	No. of women	No. of premature ruptures	Premature ruptures (%)
Less than 0 20	89	13	14.6
0 20 to 0.59	130	11	8 5
0.60 or greater	69	1	1.4

Note: The relationship of plasma ascorbic acid (AA) levels to the occurrence of spontaneous rupture of the membranes prior to the onset of labor.

From Wideman, G. L., Baird, G. H., and Golding, O. T. (1964), *Am. J. Obstet. Gynecol.,* 88, 592. With permission.

The work of Levine et al. (1941a, b), who identified a defect of tyrosine and phenylalanine metabolism in premature infants, was confirmed by Light et al. (1966). The latter workers found that a marked elevation of serum tyrosine and a modest elevation of serum phenyl-alanine occurred in 25% of low-birth-weight infants, especially those under 2000 g fed on cow's milk. The abnormality, if untreated, persisted as long as 6 weeks. It could be prevented by daily administration of 100 mg of ascorbic acid, but not smaller doses (up to 15 mg daily).

The association between ascorbate deficiency and premature rupture of the membranes was very clear in the work of Wideman et al., and might be used to predict which pregnant women have the greatest risk, but these data do not allow us to deduce that ascorbic acid supplementation during pregnancy would reduce the incidence of prematurity and premature rupture of the membranes. It may simply be that ascorbate deficiency is indicative of the general nutritional state of the mother. It is also possible that premature rupture of the membranes could be due to infection and that infection lowers the ascorbate level. Another very likely situation is that low ascorbate predisposes to various infections, that infections lower the ascorbate level even more, and low ascorbate levels predispose to premature rupture of the membranes.

Certainly there is a collagen-rich stroma beneath the amnion of the guinea pig, as shown by Wynn et al. (1967), and this is needed to provide strength for the membranes. Collagen synthesis cannot occur unless there is enough ascorbic acid for the conversion of proline to hydroxyproline, as discussed in Chapter 2 of this volume.

Ganich (1973) has measured the ascorbic acid content of normal amniotic fluid, finding it to be 2.7 times higher than that of the maternal blood, but there do not seem to be any such data for the amniotic fluid from prematurely ruptured membranes.

Several workers, including Danforth et al. (1953, 1956), Embrey (1954, 1956), and Meudt and Meudt (1967), have measured the tensile strength of the membranes, bud did not find any correlation between this and the timing of their rupture. However, Artal et al. (1976) have observed that prematurely and nonprematurely ruptured membranes differ in the thickness of the membranes near the rupture site and in Young's modulus of elasticity measured near the placenta.

Koroleva (1964) reported a high copper content in the blood of women with prematurely ruptured membranes, but this was not confirmed by Artal et al. (1979) who reported lower than usual maternal and fetal blood copper levels in association with premature rupture of the membranes. Blood copper levels rise dramatically during human pregnancy as a result of the rising estrogen levels which stimulate increased synthesis of apoceruloplasmin by the liver and the release of ceruloplasmin, with eight atoms of copper per molecule, into the blood stream. The liver is the major storehouse of copper in the body and blood copper levels are not at all indicative of liver copper stores. Since blood copper levels rise progressively as pregnancy advances, as shown in Figure 6, Chapter 13, Volume I, one would expect premature delivery to be associated with somewhat less elevated maternal copper levels than full-term pregnancy, unless maternal infection, trauma, preeclampsia, or any other liver disorder caused the blood copper level to rise even higher than normal in late pregnancy. In any event, blood copper levels provide no indication of copper stores. In fact, the blood copper level is lowest in Wilson's disease, where the liver copper levels are highest.

A review of dietary histories obtained from women attending a prenatal clinic in East Harlem, New York, by Bowering et al. (1980) revealed low ascorbic acid intakes in 60% of those who delivered prematurely and in 36% of those who delivered at term. Plasma ascorbic acid levels revealed that one fifth of the women had either deficient or marginal (one being as low as 0.01 mg/100 ml); 12% were marginal, 0.3 to 0.6 mg/100 ml. Clearly, it is desirable to supplement the ascorbic acid intake of some women during pregnancy and to conduct trials of various dosage regimens, but this apparently simple matter is not without potential problems.

Cochrane (1965) suggested that an increased ascorbic acid dependency could be induced in the newborn by exposure to high levels of this vitamin *in utero*. This idea received support from Norkus and Rosso (1975) who showed that feeding high intakes of ascorbic acid during the last 30 d of pregnancy in the guinea pig caused the pups to develop signs of scurvy after 18 instead of 22 d when they were fed a deficient diet. Norkus et al. (1979) attributed this to a very efficient placental transfer of dehydroascorbic acid to the fetus, being suddenly removed at birth. However, Ginter et al. (1982) found no evidence of an increased rate of tissue ascorbate depletion in adult guinea pigs following 7 months of high-ascorbic acid intake. The half-lives of ascorbate depletion in nine organs were found to be the same in animals that had received a minimal maintenance ascorbic acid intake. Moreover, we can console ourselves that most mothers do not put their newborn infants on a diet which is totally devoid of ascorbic acid, as in the guinea pig experimental design. Nevertheless, it would seem wiser to provide pregnant women with moderate doses of ascorbic acid, say 200 mg once, twice, or three times a day, according to need, with D-catechin as a chelating antioxidant or, better still, as catechin-coated tablets, so as to minimize losses of ascorbic acid in the gastrointestinal tract and to avoid any possible adverse effects that might arise from megadose ascorbic acid treatment.

Arad et al. (1982, 1985) noted that premature infants, especially those with respiratory distress syndrome, seem to have higher ascorbic acid requirements than normal-term infants. They recommend parenteral administration of ascorbic acid, 50 mg daily, to premature infants, especially when they are breathing higher than normal concentrations of oxygen, as this vitamin has been shown to protect the lungs against oxygen toxicity (Chapter 24, Volume I).

It is particularly interesting to note that Harris et al. (1985) have observed a strong association between peripheral placental hemorrhage and premature labor. Thus, many premature deliveries may be due to minor degrees of marginal abruptio placentae, which has itself been found to be associated with low plasma ascorbic acid and high blood histamine levels (Chapter 14 of this volume).

REFERENCES

Arad, I. D., Sagi, E., and Eyal, F. G. (1982), Plasma ascorbic acid levels in premature infants, *Int. J. Vitam. Nutr. Res.,* 52, 50.

Arad, I. D., Dgani, Y., and Eyal, F. G. (1985), Vitamin E and vitamin C plasma levels in premature infants following supplementation of vitamin C, *Int. J. Vitam Nutr. Res.,* 55, 395.

Artal, R., Burgeson, R., Fernandez, F. J., and Hobel, C. J. (1979), Fetal and maternal copper levels in patients at term with and without premature rupture of membranes, *Obstet. Gynecol.,* 53, 608.

Artal, R., Sokol, R. J., Newman, M., Burstein, A. H., and Stojkov, J. (1976), The mechanical properties of prematurely and non-prematurely ruptured membranes, *Am. J. Obstet. Gynecol.,* 125, 655.

Bowering, J., Lowenberg, R. L., and Morrison, M. A. (1980), Nutritional studies of pregnant women in East Harlem, *Am. J Clin. Nutr.,* 33, 1987.

Cochrane, W. A. (1965), Over-nutrition in prenatal and neonatal life: a problem? Symposium on Nutrition, Toronto, 1964, *Can. Med. Assoc J.,* 93 (17), 893.

Danforth, D. N., McElin, T. W., and States, M. N. (1953), Studies on fetal membranes I. Bursting tension, *Am. J. Obstet. Gynecol.,* 65, 480.

Danforth, D. N., McElin, T. W., and States, M. N. (1956), On the strength of the foetal membranes, *J. Obstet. Gynaecol. Br. Emp.,* 63, 237.

Elmby, A. and Christensen, P. B. (1938), Über das Verhalten der Ascorbinsäure in der Schwangerschaft, unter der Geburt, während des Wochenbetts und in den ersten Lebenstagen des Kindes, *Klin. Wochenschr.,* 17, 1432.

Embrey, M. P. (1954), On the strength of the foetal membranes, *J. Obstet. Gynaecol. Br. Emp.,* 61, 793.

Embrey, M. P. (1956), On the strength of the membranes, *J. Obstet. Gynaecol. Br. Emp.,* 63, 757.

Fleming, A. W. and Sanford, H. N. (1938), Vitamin C content of the blood in newborn infants, *J Pediatr.,* 13, 314.

Ganich, M. M. (1973), Vitamins C, PP and B_6 in the placenta and blood during normal and complicated pregnancy and labour (in Ukrainian), *Pediatr. Akush. Ginekol.,* 2, 38.

Ginter, E., Drobna, E., and Ramacsay, L. (1982), Kinetics of ascorbate depletion in guinea pigs after long-term high vitamin C intake, *Int. J. Vitam. Nutr. Res.,* 52, 307.

Harris, B. A., Jr., Gore, H., and Flowers, C. E. (1985), Peripheral placental separation: a possible relationship to premature labor, *Obstet Gynecol.,* 66, 774

Ingier, A. (1915), A study of Barlow's disease experimentally produced in fetal and newborn guinea pigs, *J. Exp Med.,* 21, 525.

Kass, E. H. (1959), The role of asymptomatic bacteriuria in the pathogenesis of pyelonephritis, in *Biology of Pyelonephritis,* Henry Ford Hospital Int Symp., Quinn, E. L. and Kass, E. H., Eds , Little, Brown, Boston, 399.

Koroleva, M. I. (1964), Blood content of copper and vitamin C in parturient women suffering from premature loss of amniotic fluid, *Med. Zh. Uzb.,* 1, 24; as cited in *Chem. Abstr ,* 61, 2320b.

Kramer, M. M., Harman, M. T., and Brill, A. K. (1933), Disturbances of reproduction and ovarian changes in the guinea-pig in relation to vitamin C deficiency, *Am. J. Physiol.,* 106, 611.

Levine, S. Z., Gordon, H. H., and Marples, E. (1941a), A defect in the metabolism of tyrosine and phenylalanine in premature infants. II. Spontaneous occurrence and eradication by vitamin C, *J. Clin. Invest.,* 20, 209.

Levine, S. Z., Marples, E., and Gordon, H. H. (1941b), A defect in the metabolism of tyrosine and phenylalanine in premature infants. I. Identification and assay of intermediary products, *J. Clin. Invest.,* 20, 199.

Light, I. J., Berry, H. K., and Sutherland, J. M. (1966), Aminoacidemia of prematurity. Its response to ascorbic acid, *Am. J. Dis. Child.,* 112, 229.

Martin, M. P., Bridgeforth, E., McGanity, W. J., and Darby, W. J. (1957), The Vanderbilt cooperative study of maternal and infant nutrition. X. Ascorbic acid, *J. Nutr ,* 62, 201.

Meudt, R. and Meudt, E. (1967), Rupture of the fetal membranes, *Am. J. Obstet. Gynecol.,* 99, 562.

Norkus, E. P., Bassi, J., and Rosso, P. (1979), Maternal-fetal transfer of ascorbic acid in the guinea-pig, *J. Nutr.,* 109, 2205.

Norkus, E. P. and Rosso, P. (1975), Changes in ascorbic acid metabolism of the offspring following high maternal intake of the vitamin in the pregnant guinea-pig, *Ann. N.Y. Acad. Sci.,* 258, 401.

Reyher, P. E., Walkhoff, E., and Walkhoff, O. (1928), Studien über die Wirkung C-hypovitaminotischer Nahrung auf Schwangere, Feten und Neugeborene, *Muench. Med Wochenschr.*, 75, 2087.

Wideman, G. L., Baird, G. H., and Golding, O. T. (1964), Ascorbic acid deficiency and premature rupture of fetal membranes, *Am. J. Obstet Gynecol.*, 88, 592.

Wynn, R.M., Sever, P. S., and Hellman, L. M. (1967), Morphologic studies of the ruptured amnion, *Am. J. Obstet. Gynecol.*, 99, 359.

Chapter 16

MEGALOBLASTIC ANEMIA OF INFANCY, PREGNANCY, AND STEATORRHEA

I. INTRODUCTION

Proper diagnosis of megaloblastic anemia requires differentiation between (1) a deficiency of vitamin B_{12}, (2) a failure to absorb vitamin B_{12}, as in pernicious anemia due to a deficiency of intrinsic factor (Schilling, 1955), (3) a deficiency of folic acid, and (4) a disturbance of folic acid metabolism due to a deficiency of ascorbic acid (Chapter 4 of this volume). Moreover, the clinical associations and underlying causes, such as achlorhydria, total gastrectomy, blind intestinal loop, fish tapeworm, malabsorption syndrome, or pregnancy, must be sought and need to be taken into account. Fortunately pernicious anemia is mostly a disease of older people, so megaloblastic anemia of pregnancy is almost always due to a disturbance of folic acid metabolism. Young women with pernicious anemia are rare and hardly ever get pregnant or carry the pregnancy beyond the early months.

II. INFANCY

Braestrup (1936) working in Iowa City, observed that the mean plasma ascorbic acid (AA)* levels of 23 newborn infants fell very rapidly in the first 5 d of life, from 1.07 mg/100 ml in the cord blood at birth, to 0.69 in the first day, to 0.39 after 24 h, and to 0.25 mg/100 ml after 5 d. So even infants born with adequate vitamin C levels may very soon become deficient unless they are provided with an adequate ascorbic acid intake.

Further evidence of vitamin C deficiency in the newborn was provided when Levine et al. (1941a, b) observed a defect in the metabolism of tyrosine and phenylalanine in premature infants, which occurred spontaneously and could be cured by vitamin C. The urinary excretion of *p*-hydroxyphenyllacetic and *p*-hydroxyphenylpyruvic acids was completely eradicated by the administration of vitamin C.

Zuelzer and Ogden (1946) observed 25 infants under 18 months of age with severe megaloblastic anemia, mostly in association with infections under treatment with sulfonamides. Six of the infants were diagnosed as having scurvy. Nearly all of the 17 surviving infants showed a good hematopoietic response to folic acid, but many received liver extracts and ascorbic acid as well.

May et al. (1951) produced megaloblastic anemia in rhesus monkeys by feeding an ascorbic acid-deficient milk diet. Monkeys fed the same milk diet supplemented with ascorbic acid remained in good health for prolonged periods and maintained normal blood and bone marrow. The experimental megaloblastosis was virtually indistinguishable from that seen in megaloblastic anemia in human infants. The experimental megaloblastic anemia could be prevented by folic acid or by very small amounts of folinic acid, without the aid of ascorbic acid. It could also be cured by ascorbic acid alone. Vitamin B_{12} would neither cure nor prevent it. This experimental megaloblastosis was due to a disturbance in the metabolism of folic acid caused by ascorbic acid deficiency, as described in Chapter 4 of this volume.

Subsequent work by May et al. (1952) indicated that infection may cause megaloblastic anemia in newborn infants and monkeys. No doubt their milk diets were borderline as regards both ascorbic acid and folic acid. Infection decreased their ascorbate levels (Chapter 8, Volume I), which in turn decreased their folinic acid levels (Chapter 4 of this volume) and

---

* AA — ascorbic acid, reduced form.

led to megaloblastic anemia. It should be remembered that while folic acid alone will cure the anemia, it will not cure the ascorbic acid deficiency.

Light et al. (1966), at the Cincinnati General Hospital, observed elevated serum tyrosine and phenylalanine levels in premature infants on a cow milk formula. These levels fell to normal in response to ascorbic acid supplements, providing additional evidence of the persisting need for ascorbic acid supplements for premature infants.

The effects of maternal vitamin C supplements on breast milk ascorbic acid levels were studied by Bates et al. (1983) in the village of Keneba, in the Republic of The Gambia, during the rainy season when the intake of vitamin C-rich foods is very low. Plasma ascorbate increased from 0.25 to 0.72 mg/dl. Buffy coat ascorbate increased from 14.7 to 24.3 μg/ 10^8 cells, and breast milk ascorbate rose from 3.4 to 5.5 mg/dl when maternal vitamin C intake was increased from 34 to 103 mg/d.

Vobecky et al. (1985), working in eastern Canada, found that serum vitamin C levels below 0.30 mg/100 ml at 6 months of age, occurred four times more frequently in bottle-fed than in breast-fed infants. Likewise, low vitamin E levels were twice as frequent in bottle-fed babies. Many of the mothers gave vitamin supplements to their infants; nevertheless, 40% of the infants had vitamin C intakes of less than 80% of the recommended daily intake for their weight at 6 months; this percentage fell to nearly 0 at 18 months of age. The frequency of substandard serum folate levels did not differ between the breast-fed and the bottle-fed infants in this study, but no doubt megaloblastic anemia of the newborn, due to vitamin C deficiency will continue to be seen from time to time.

III. PREGNANCY

Much has been written on the successful use of folic acid in the treatment of megaloblastic anemia of pregnancy by Goodall (1961), Hibbard and Hibbard (1966), and many others, but there does not seem to have been enough work on the possible role of ascorbic acid deficiency in this condition. However, Holly (1951), writing from the University of Minnesota, reported five patients with megaloblastic anemia in pregnancy. A relationship of this anemia to a deficiency of ascorbic acid was suggested by the effectiveness of combined ascorbic acid and vitamin B_{12} treatment where B_{12} alone was ineffective. Two patients were refractory to adequate trial with vitamin B_{12}; one of these patients later had a complete remission when ascorbic acid was added to the treatment; the other responded adequately to folic acid therapy and blood transfusions. Two patients were treated for short periods of time with ascorbic acid alone; a minimal response in reticulocytes was observed in one; subsequent addition of B_{12} to the therapy, after saturation with ascorbic acid, produced a complete hematological response; one patient had a complete remission after delivery. In three patients, combined therapy with ascorbic acid and vitamin B_{12} produced a complete remission. Holly (1958) reported that the blood ascorbic acid level in patients with megaloblastic anemia of pregnancy is zero.

Pritchard et al. (1962) stated that while all three vitamins — folic acid, B_{12}, and ascorbic acid — have been implicated as causes of megaloblastic anemia of pregnancy, their own studies led them to conclude that folic acid is the most likely major deficiency. Nevertheless, the writer believes that ascorbic acid supplements should always be given, as well as folic acid, in the treatment of megaloblastic anemia of pregnancy, for the reasons given in Chapter 14 of this volume.

IV. STEATORRHEA, MALABSORPTION SYNDROME

Working at the Queen Elizabeth Hospital in Birmingham, England, Boscott and Cooke (1954) observed excessive quantities of *p*-hydroxyphenylacetic acid (*p*HPA) in the urines of 20 patients with steatorrhea and megaloblastic anemia. Moreover, the response of these patients to ascorbic acid therapy, as judged by the disapperance of *p*HPA, was very much slower than that seen in a few normal subjects excreting *p*HPA. Thus, greatly increased doses of ascorbic acid and longer periods of treatment were required before *p*HPA excretion was significantly reduced. They also found that five patients with megaloblastic anemia of pregnancy also excreted *p*HPA in amounts comparable to those found in idiopathic steatorrhea. On the other hand, in five patients with pernicious anemia, *p*HPA was present in small amounts only. Unfortunately, these workers did not have an opportunity to test the response of the pregnant patient to ascorbic acid. They did, however, carry out ascorbic acid saturation tests on those with steatorrhea and megaloblastic anemia.

Ascorbic acid saturation tests — "Six healthy normal subjects, on a daily dosage of 600 mg of ascorbic acid, excreted one-third of the test dose (600 mg) in the first 6 hours after 3 to 7 days treatment. In six patients with steatorrhoea and megaloblastic anaemia, doses totalling 13.5 to 20 gm. were administered over periods varying from 23 to 40 days before significant amounts were excreted in the urine. The maximum that was excreted in any one patient was 176 mg in the six hours following the administration of the test dose, after 40 days of therapy. The administration of large amounts of ascorbic acid intramuscularly in cases 1, 2 and 3 did not result in any more rapid saturation."

These patients had macrocytic megaloblastic anemia of nutritional origin and needed much more ascorbic acid than would a person with classical scurvy, but they showed no clinical signs of scurvy. Their anemia was very refractory to treatment. The authors concluded that neither folic acid, nor vitamin B_{12}, nor ascorbic acid, when administered alone, could be guaranteed to restore the blood picture to normal, but that intensive ascorbic acid therapy appears to synergize the effects of folic acid and vitamin B_{12} in some patients with megaloblastic anemia.

Gross et al. (1975) have reported depressed cell-mediated immunity in megaloblastic anemia due to folic acid deficiency both in pregnant and in nonpregnant individuals. So it is easy to see how a vicious cycle of infection, ascorbate deficiency, folinic acid deficiency, and megaloblastic anemia can lead to further infection, unless the cycle is broken by proper nutrition. Moreover, the role that ascorbic acid may play in the proper nutrition of a patient with megaloblastic anemia is well illustrated by the following clinical report published by Stokes et al. (1975) from the Nutritional and Intestinal Unit of the General Hospital, Birmingham, England.

AC, now aged 64 years, is a bachelor living by himself who was crippled in childhood and consequently has been unable to work for most of his life. He first presented at another hospital in 1963 with a severe pancytopenia and megaloblastic anemia with hemoglobin of 5.1 g/100 ml. There was free gastric acid present and he responded well to a combination of ward diet, vitamin B_{12}, and ascorbic acid. In 1968 he again presented with pancytopenia and hemoglobin of 4.1 g/100 ml and on this occasion responded well to folic acid alone, with a good reticulocyte response and subsequent rise in hemoglobin to normal. In 1971, he again developed a megaloblastic anemia with hemoglobin of 3 g/100 ml, serum B_{12} 120 pg/ml *(Lactobacillus leichmanii)* and serum folate *(L. casei)* 0.9 ng/ml, and was transferred to the Nutritional and Intestinal Unit. Clinical scurvy was evident with the presence of follicular hyperkeratosis, curled body hairs, petechiae and characteristic peridontal gum changes. The plasma ascorbic acid was zero (control normal subject 0.64 mg/100 ml), serum folate *(L. casei)* 1.4 ng/ml, hemoglobin 5.6 g/100 ml, mean corpuscular volume 92 cubic microns, serum iron 52 mcg/ml. Gastric pH was 2.6 and hepatic uptake of ^{58}Co-vitamin B_{12} was normal. Barium studies of the small intestine, fecal fat excretion and jejunal biopsy were all normal.

Review of his diet indicated a maximum intake of 100 μg of folic acid and 6 mg ascorbic acid per day. He showed an excellent response to combined treatment with folinic acid and ascorbic acid, and detailed studies showed that his folic acid metabolism was restored to normal following the ascorbic acid therapy.

REFERENCES

Bates, C. J., Prentice, A. M., Prentice, A., Lamb, W. H., and Whitehead, R. G. (1983), The effect of vitamin C supplementation on lactating women in Keneba, a West African rural community, *Int. J Vitam. Nutr Res.*, 53, 68

Boscott, R. J. and Cooke, W. T. (1954), Ascorbic acid requirements and urinary excretion of P-hydroxyphenylacetic acid in steatorrhoea and macrocytic anaemia, *Q. J Med.*, 91, 307.

Braestrup, P. W. (1936), The content of reduced ascorbic in blood plasma in infants, especially at birth and in the first days of life, *J. Nutr*, 16, 363.

Goodall, H. B. (1961), Megaloblastic anaemia of pregnancy and the puerperium, *Pathol. Microbiol*, 24, 682.

Gross, R. L., Reid, J. V. O., Newberne, P. M., Burgess, B., Marston, R., and Hift, W. (1975), Depressed cell-mediated immunity in megaloblastic anemia due to folic acid deficiency, *Am. J. Clin. Nutr.*, 28, 225

Hibbard, B. M. and Hibbard, E. D. (1966), Recurrence of defective folate metabolism in successive pregnancies, *J. Obstet. Gynaecol Br. Commonw.*, 73, 428.

Holly, R. G. (1951), Megaloblastic anemia in pregnancy Remission following combined therapy with ascorbic acid and vitamin B_{12}, *Proc Soc. Exp. Biol Med.*, 78, 238

Holly, R. G. (1958), Anemia in pregnancy, *Clin Obstet Gynecol.*, March, 15.

Levine, S. Z., Gordon, H. H., and Marples, E. (1941a), A defect in the metabolism of tyrosine and phenylalanine in premature infants. II. Spontaneous occurrence and eradication by vitamin C, *J Clin. Invest.*, 20, 209.

Levine, S. Z., Marples, E., and Gordon, H. H. (1941b), A defect in the metabolism of tyrosine and phenylalanine in premature infants I Identification and assay of intermediary products, *J Clin. Invest.*, 20, 199.

May, C. D., Stewart, C. T., Hamilton, A., and Salmon, R. J. (1952), Infection as a cause of folic acid deficiency and megaloblastic anemia, *Am J. Dis. Child.*, 84, 718.

May, C. D., Sundberg, R. D., Schaar, F., Lowe, C. U., and Salmon, R. J. (1951), Experimental nutritional megaloblastic anemia: relation of ascorbic acid and pteroylglutamic acid, *Am J Dis. Child.*, 82, 282.

Pritchard, J. A., Mason, R. A., and Wright, M. R. (1962), Megaloblastic anemia during pregnancy and the puerperium, *Am. J. Obstet. Gynecol.*, 83, 1004.

Schilling, R. F. (1955), The absorption and utilization of vitamin B_{12}, *Am. J. Clin. Nutr.*, 3, 45.

Stokes, P. L., Melikian, V., Leeming, R. L., Porter-Graham, H., Blair, J. A., and Cooke, W. T. (1975), Folate metabolism in scurvy, *Am. J. Clin. Nutr.*, 28, 126.

Vobecky, J. S., Vobecky, J., Shapcott, D., Demers, P. L., Blanchard, R., and Fisch, C. (1985), The vitamin status of infants in a free living population, *Int. J Vitam. Nutr. Res.*, 55, 205.

Zuelzer, W. W. and Ogden, F. N. (1946), Megaloblastic anaemia in infancy. A common syndrome responding specifically to folic acid, *Am. J. Dis. Child.*, 71, 211.

Chapter 17

GASTROINTESTINAL ULCERS AND HEMORRHAGE

I. HUMAN OBSERVATIONS

Foote (1926) reported the case history of a $2^1/_2$-year-old child who had symptoms of abdominal spasms and passed blood-streaked stools. The gums were normal and although the clinical picture closely simulated intussusception, exploration was delayed for a few hours until purpuric spots developed, when a diagnosis of scurvy was made. Complete and rapid recovery followed the usual anti-scorbutic treatment.

Hutter (1928) observed that the peak incidence of peptic ulcer occurs during late winter and early spring, which are the times of the year when vitamin C intake is least and body stores are lowest (Chapter 19, Volume I).

Davidson (1928), Wood (1935), and Platt (1936) all recorded their observations of scurvy in individual patients with peptic ulcers. The vitamin deficiency was attributed to the low ascorbic acid intake of the usual ulcer diet and to the extensive use of alkalis by these patients, but, as we shall see from guinea pig studies, it is also possible that the ascorbate deficiency may have contributed to the development of peptic ulceration. So a vicious cycle may develop. Indeed, Davidson suggested that, "peptic ulcer itself may possibly be looked on as an example of tissue destruction, like scurvy, secondary to vitamin deficiency." Troutt (1932) pointed out that the diets of Sippy, Lenhartz, von Leube, and Alvarez, used for controlling the acute symptoms of peptic ulcer, make no provision for vitamin C. He therefore recommended the addition of about half a glass (3 oz) of orange juice after meals and reported that this caused no discomfort in the majority of patients. However, some ulcer patients have an instinctive aversion to orange juice; tomato juice may be more acceptable than orange juice for some ulcer patients.

The causes of death of 51 patients following partial gastrectomy at St. Bartholemew's Hospital in London, were analyzed by Payne (1936). He found that general or local peritonitis was the cause of death in 16 cases. In 12 of these, "the peritoneal infection appeared to have arisen as a result of bacterial leakage at the site of anastomosis following almost complete absence of any fibrinous response along the suture line." He stated that in the majority of peritonitis cases the infection appeared to be due to failure of response on the part of the patient rather than to any technical defect in the actual method of suturing. Archer and Graham (1936), observing one of these post-mortem examinations, reported that, "the stitches stood out as though the operation had just been completed and there was no sign of fibrin or healing." This prompted them to study the vitamin C status of patients with peptic ulcer disease. Using ascorbic acid saturation tests, they found that six out of nine men with duodenal ulcers were in a subscorbutic state and conjectured that vitamin C deficiency might have been responsible for the development of their ulcers. They also noted that the ulcer diets consisting mainly of milk, eggs, bread and butter might have contributed to their vitamin C deficiency. It was difficult to know whether the ulcers healed more quickly after treatment with ascorbic acid, but there was no doubt that these patients needed an extra supply of vitamin C and they seemed to do well with it.

Harris et al. (1936) estimated the vitamin C in the urine of voluntary hospital patients and found that among those with the lowest excretions, "a quite disproportionate number were patients with gastric or duodenal ulcer."

Studies by Rivers and Carlson (1937) at the Mayo Clinic revealed low blood and urine vitamin C levels and a capillary fragility state in several patients who suffered repeated gastrointestinal hemorrhages, with or without demonstrable peptic ulceration. Lazarus (1937) reported the state of vitamin C nutriture of 15 patients admitted to Bispebjerg Hospital in Copenhagen during the preceding year. Ascorbic acid saturation tests demonstrated that 13 of the 15 were in a condition of vitamin C subnutriture or "subclinical scurvy", and this was of severe degree in 7 of them. Three patients with peptic ulcers uncomplicated by hemorrhage were also found to be in a state of vitamin C subnutrition. It was suggested that the existence of fever or the consumption of alkalis might have contributed to the ascorbic acid deficiency in some cases, but it was thought that the subscurvy state was in the main due to an inadequate dietary ascorbic acid intake.

Witts (1937) recommended medical rather than surgical management of patients with hematemesis and melena and advocated early feeding, starting on the day of admission to hospital, with a diet including orange juice, marmite, and cod liver oil, as well as the usual bland purée, etc. He stated that, "they look, feel and do much better than any previous series of cases of gastro-duodenal haemorrhage I have seen."

Peters and Martin (1937) analyzed samples of human gastric juice for ascorbic acid (AA)* and found a range from 0.046 to 1.04 mg/100 g, with a mean value of 0.397 mg/100 ml. The gastric juice of dogs showed a range from 0.33 to 1.51 mg/100 ml, with a mean of 0.69 mg/100 ml. These workers also measured the ascorbic acid content of the mucosa of the fundus and pylorus of the stomach, as well as the duodenal, ileal, and colic mucosa in two dogs. The ascorbic acid concentration was found to be highest in the duodenal mucosa (11 and 22 mg/100 g), lower in the mucosa of the gastric fundus (5 and 6 mg/100 g), and lowest in the mucosa of the pyloric portion of the stomach (0.6 and 4.0 mg/100 g).

Ingalls and Warren (1937), working at the Peter Bent Brigham Hospital in Boston, using an indophenol titration method, studied the plasma ascorbic acid (AA) levels of 19 patients with gastric or duodenal ulcers and found values of 0.2 mg/100 ml or less in 13 of them, 0.4 mg/100 ml or less in 3 others, and 0.6 mg/100 ml in one other. They gave values of 0.65 to 2.0 mg/100 ml as the range found in 150 normal individuals by this method. So it would appear that 13 out of 19, or 68%, had very low, and 17 out of 19, or 89%, had low plasma ascorbic acid levels at the time of testing. None of them had frank scurvy. Most of the patients had been taking an inadequate amount of vitamin C in the diet, and their blood plasma levels rose to normal when an adequate ascorbic acid intake was provided. These patients had been treated for their ulcer disease with one form or another of bland diet. Some had been following the "first-week Sippy diet", which consists of hourly feedings of milk and cream and six small feedings of either cereal or eggs. Others had been following the "fourth-week Sippy diet", which, in addition to the hourly feedings of cream and milk, allows three small meals of meat, potatoes, puréed vegetables, and stewed fruit. All the patients on the latter diet were allowed tomatoes and were advised to take orange juice once daily.

Nielsen (1938) recorded a mean serum ascorbic acid level of 0.23 mg/100 ml in 20 patients with gastric or duodenal ulcers or pylorogastritis in the month of April and a mean of 0.34 mg/100 ml in 84 control subjects studied in the same month.

Roux (1938), working in Paris, and Chamberlin and Perkin (1938), in Boston, also found chemical evidence of subclinical ascorbic acid deficiency. Rao (1938), in south India, reported normal blood ascorbic acid levels in 15 patients with peptic ulcer disease, but showed no control values from his own laboratory for comparison.

Portnoy and Wilkinson (1938) compared the ascorbate status of 25 patients with peptic ulceration, 31 patients with hematemesis and 51 control subjects by several methods including (1) urinary ascorbic acid excretion, (2) ascorbic acid saturation tests, (3) plasma ascorbic acid (AA) determinations, (4) oral ascorbic acid tolerance tests, (5) intravenous ascorbic acid tolerance tests, and (6) by an intradermal indophenol test. They found the ulcer patients and the hematemesis patients to be markedly deficient in ascorbic acid by all these tests.

---

* AA — ascorbic acid, reduced form.

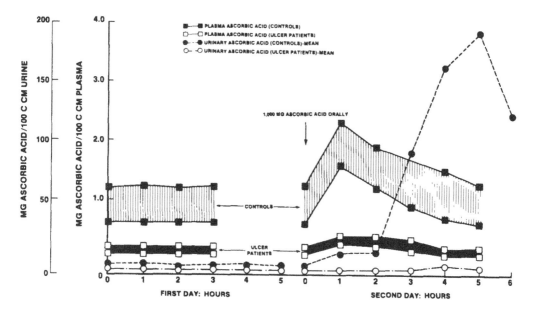

FIGURE 1. Plasma ascorbic acid levels of peptic ulcer patients and controls, before and after oral administration of 1000 mg of ascorbic acid. The corresponding mean urinary values are also shown. (From Portnoy, B. and Wilkinson, J. F. [1938], *Br. Med. J.*, 1, 554. With permission.)

The results of the oral and intravenous ascorbic acid tolerance tests conducted by these workers are illustrated in Figures 1 and 2 which are taken from their work.

Bourne (1938) studied the dietary histories and also the capillary fragility of the skin of patients with peptic ulcer disease. Using Göthlin's standards of 1 to 4 petechiae as normal, 5 to 8 as a transitional stage of vitamin C deficiency, and more than 8 petechiae as indicating a definite deficiency, 20 out of 22 normal control subjects gave normal test results. However, only 6 out of 28 patients with gastric ulcer disease and 8 out of 14 patients with duodenal ulcers gave normal test results. The mean petechial counts were 1.82 for normal controls, 7.10 for gastric ulcer patients, 7.36 for those with duodenal ulcers, and 2.86 for a variety of patients with other disease states. Most of the patients had inadequate diets, but there was no definite evidence that lack of antiscorbutic vitamins (vitamin C and bioflavonoids) was directly responsible for the occurrence of ulceration; the possibility that it might play a part could not be excluded. It seemed to Bourne more likely that the deficiency of anti-scorbutic vitamins in the therapeutic diet might be a factor influencing the transition from the acute to the chronic condition, predisposing to delay in healing, relapses, and possibly hematemesis.

Studies conducted by Warren and Pijoan (1939) on five duodenal ulcer patients after they had been saturated with ascorbic acid, suggested that such patients utilize 20% more ascorbic acid per kilogram of body weight per day than do normal subjects. Nevertheless, they stated that this could be provided by including one or two good-sized fresh oranges in their diet every day.

Croft and Snorf (1939) found among their first 100 determinations of ascorbic acid (AA) in private patients, 38 were below 0.40 mg/100 ml. Of these, 15 (or 40%) had peptic ulcers, and 13 of them had experienced recent episodes of gastrointestinal bleeding. Issler and Demole (1939) conducted gastric analyses and studied the hydrochloric acid content, the ascorbic acid concentration, and the volume of gastric juice secreted in 1 h following a standard dose of histamine, before and again after oral administration of ascorbic acid for several days in eight subjects. There was a definite increase in the ascorbic acid content of the gastric juice (from 0.5 to 0.8 mg/100 ml) after saturation with vitamin C, but no change

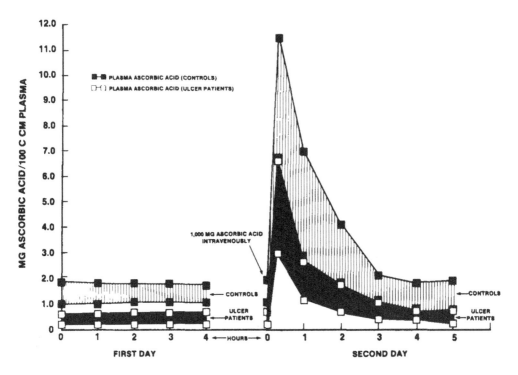

FIGURE 2. Plasma ascorbic acid levels of peptic ulcer patients and controls, before and after intravenous administration of 1000 mg of ascorbic acid. (From Portnoy, B. and Wilkinson, J. F [1938], *Br. Med. J.,* 1, 554. With permission.)

in the titratable acid nor in the volume of gastric juice secreted. Since ascorbic acid is now known to decrease the blood histamine level in ascorbic acid deficient subjects (Chapter 1 of this volume), it would be more interesting to study the acidity of the gastric contents following a standard test meal, before and after ascorbate saturation in patients with duodenal ulcer disease.

Bartlett et al. (1940) found a mean plasma ascorbic acid (AA) level of 0.34 mg/100 ml in 34 patients with gastric and duodenal ulcers compared with a mean of 1.24 mg/100 ml in 13 normal control subjects. However, these authors also found low levels of ascorbic acid in patients with many other disease states. Field et al. (1940), working at Ann Arbor, MI, also found low plasma ascorbic acid levels in the majority (39 of 58) of patients with peptic ulcer and reached the conclusion that hemorrhage is associated with the lowest ascorbic acid levels. Moreover, they considered that patients with hematemesis but no ulcers may have had a scorbutic bleeding.

Stewart et al. (1941) reported that intravenous injection of ascorbic acid (1 g) prolonged the life of cats bled to 50% of their original blood volume, so the use of large doses of ascorbic acid in patients with hematemesis would seem to be indicated from several points of view. Pelner (1943) pointed out the need for supplements of all vitamins in patients with gastrointestinal disease. Lund (1944) studied 18 patients before and during healing in non-radical operations for gastric or duodenal ulcer disease at the Boston City Hospital and observed that nearly all of them had severe ascorbic acid deficiency, as judged by dietary vitamin C intake histories (mean 14.7 mg/d), their plasma ascorbic acid levels (mean 0.15 mg/100 ml), their leukocyte ascorbic acid levels (mean 4.3 mg/100 g), and also as judged by ascorbic acid saturation tests. A few of them had such low ascorbic acid reserves that they must have been very close to scurvy. Postoperative complications, especially wound disruption, were more common in those who had the lowest ascorbic acid reserves.

Studies by Ebbesen and Rasmussen (1944) of 52 untreated patients with gastric or duodenal ulcers, on the other hand, showed the serum ascorbic acid levels of these patients to be normal for the season of the year. These workers did, however, observe a very rapid depletion of the ascorbic acid level of the patients as soon as they started receiving treatment with the usual milk, egg, and gruel ulcer diet in hospital. In fact, a 75% fall in the serum ascorbic acid level during the first 12 to 16 d of ulcer treatment was stated to be typical of the group. Two of the patients developed clinical signs of scurvy after 17 and 20 d, respectively, of the dietary ulcer treatment.

Hoffman and Dyniewicz (1946) studied the effect of aluminium hydroxide gel (Amphojel® Wyeth) on the absorption of ascorbic acid by human volunteers. This "antacid" caused a slight lowering of the average maximal rise of the plasma ascorbic acid level following the oral administration of ascorbic acid (500 mg), but this did not reach statistical significance. It is, however, possible that antacids might be found to cause an appreciable loss of ascorbic acid if smaller amounts of ascorbic acid, equivalent to a normal meal, were studied. Indeed, the frequent consumption of antacids may contribute to, or compound, the low ascorbic acid levels of peptic ulcer patients.

Crescenzo and Cayer (1947), writing from the Bowman Gray School of Medicine of Wake Forest College in Winston-Salem, NC, reported on the relation between ascorbic acid status and the development or recurrence of peptic ulcer and of hemorrhage as a complication of the disease in 55 individuals. All the ulcer patients showed both a low initial level of the vitamin and a somewhat less steep saturation curve following the load test (Figure 3). Those with hemorrhage had the lowest initial value and the poorest response to the load test. Patients who were asymptomatic and were taking more adequate diets had the highest levels, and these compared favorably with levels in the control group. There was no evidence indicating that these low vitamin C values were the result of increased metabolism, destruction, or poor absorption.

Two patients with very severe but unsuspected ascorbic acid deficiency were reported by Zerbini (1947); both were men undergoing gastric resection for duodenal ulcer. One had no detectable ascorbic acid in the leukocytes or plasma of a blood sample drawn before surgery. He had no clinical manifestations and the laboratory results were not available before surgery. He developed profound shock with evidence of peripheral circulatory collapse during surgery. The extremities were cold and cyanotic and he seemed almost moribund, but when the results of the ascorbic acid determinations became known, he was given ascorbic acid (2 g i.v.); his blood pressure rose and the color of his arms and legs improved within a few minutes and he made an uneventful recovery. The other patient, with a known absence of ascorbic acid in plasma and white cells, completely corrected before operation, exhibited after operation a fall to zero in the plasma level, with no corresponding fall in the white cell level, in spite of the daily injection of 200 mg of the vitamin (Figure 4). On removal of the stitches on the seventh postoperative day, the wound showed no evidence of healing.

Hatherley (1947) recorded one instance and an Editorial in the *American Journal of Digestive Diseases* (1948) recorded two instances where ascorbic acid was life saving in patients with previously uncontrollable gastrointestinal hemorrhage. Hatherley's patient, who was a 38-year-old woman on a restricted diet because of a peptic ulcer, had required transfusion of 20 pt of blood for gastrointestinal bleeding and also hemoptysis. Within half an hour of an intravenous injection of ascorbic acid, 1000 mg, there was a dramatic improvement in her general condition. Instead of being weak, apathetic, and apparently dying, she became alert, bright, and cheerful. However, ascorbic acid saturation was markedly delayed; after receiving the 1 g intravenously and another 7.2 g orally, during 3 d, a 24-h urine specimen was found to contain only 11.25 mg of ascorbic acid. Only after 3 weeks of treatment with ascorbic acid, 200 mg daily, did her urine indicate that her tissues were saturated.

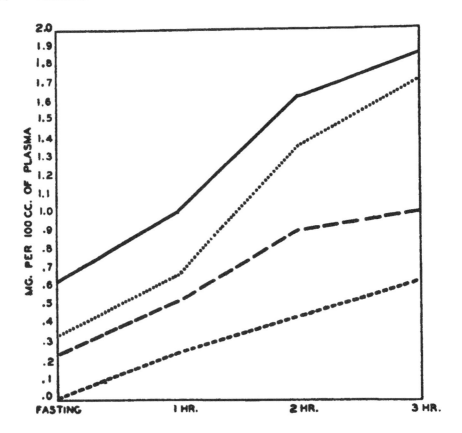

FIGURE 3.   Mean plasma ascorbic acid levels following oral administration of ascorbic acid (1 g). (----) Peptic ulcer with hemorrhage (7 patients); (– – – –) active peptic ulcer (13 patients); (• • • • • •) convalescent ulcer group (10 patients); (———) control group (25 patients). (From Crescenzo, V. M. and Cayer, D. [1947], *Gastroenterology*, 8, 754. ©1947 Williams & Wilkins Co., Baltimore With permission.)

Doll and Pygott (1952) conducted a trial of admission to hospital, phenobarbitone, and ascorbic acid as independent variables in the treatment of patients with gastric ulcers, all of whom received a peptic ulcer diet and alkalis. Admission to hospital was found to be beneficial, but ascorbic acid (50 mg three times a day by mouth) was not found to affect the rate of healing of the ulcers. A study of higher doses of ascorbic acid (with D-catechin) in the treatment of duodenal ulcers would be more valuable; it is duodenal ulcers that are so closely dependent on histamine levels and gastric hyperacidity.

Nash (1952) conducted a comparative study of antacids, with an ascorbic acid supplement in the treatment of 8 patients and without any ascorbic acid supplement in 11 patients with peptic ulcer disease, over periods ranging from 12 to 30 months. The addition of ascorbic acid (900 mg daily in the beginning, later reduced to about 400 mg) was followed by longer remissions and fewer relapses than the control group. No differences in the ulcer healing time were noted between the two treatments. Adams (1954) reported the occurrence of frank scurvy in a 31-year-old man, 7 months after surgery for a perforated duodenal ulcer due to rigid adherence to an ulcer diet. Freeman and Hafkesbring (1954) reported an average blood ascorbic acid level of 0.43 mg/100 ml in patients with peptic ulcer disease and stated that this was 56% lower than normal. The gastric juice ascorbic acid level was 0.39 mg/100 ml and this was 59% below normal.

Weiss et al. (1955) reported excellent results from the use of citrus bioflavonoids and ascorbic acid in the treatment of patients with bleeding duodenal and gastric ulcers. They

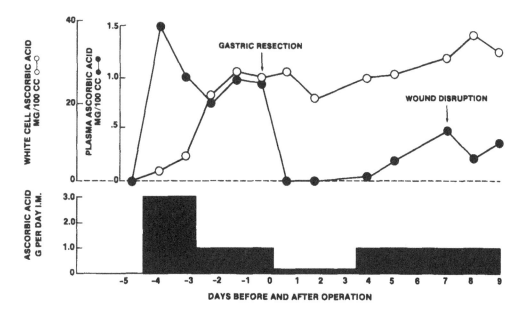

FIGURE 4   A 42-year-old man with persistent vomiting and loss of weight was admitted to hospital with duodenal ulcer, complete duodenal stenosis, and melena He was found to have no detectable ascorbic acid in his blood plasma or leukocytes. In spite of saturation with ascorbic acid before surgery and continuation of ascorbic acid 200 mg daily (i.m.) after surgery, his plasma ascorbic acid level fell to zero in association with a postoperative pneumonia, and he developed a wound disruption on the seventh postoperative day (From Zerbini, E. de J. [1947], *Arch. Surg* , 54, 117 ©1947 American Medical Association. With permission)

obtained good results in 14, satisfactory in 4, and a failure in only 1 patient who had a cicatrized duodenal ulcer with pyloric obstruction. Their results with this treatment in hemorrhagic ulcerative colitis were not as good, but they did record five good results, five satisfactory, and four failures in such patients.

Drummond (1955) reported excellent results in 54 patients with peptic ulcer by combined treatment with high-dose ascorbic acid, corticotropin, and atropine. This regimen afforded rapid relief of pain. In four patients with active hemorrhage, bleeding was rapidly controlled. Over a period of 5 years only three patients required surgery. It was felt that the corticotropin reduced the inflammatory reaction at the ulcer site, counteracted histamine release, and produced local conditions favorable to the healing activity of vitamin C.

Further studies by Freeman and Hafkesbring (1957), at the Woman's Medical College of Pennsylvania, included 79 patients with peptic ulcers who had an average fasting blood ascorbic acid level of 0.43 mg/100 ml and a fasting gastric juice ascorbic acid level of 0.39 mg/100 ml, both of which showed highly significant differences ($p < 0.001$) from the blood and gastric juice levels of 113 patients with nongastrointestinal diseases which averaged 0.88 and 0.62 mg/100 ml, as shown in Table 1 and Figure 5.

Among 11 cases of adult scurvy reported by Cutforth (1958), 2 were men on diets for duodenal ulcer; another had painless hematuria for several days and had passed a black, tarry stool, but investigation of the gastrointestinal and urinary tracts revealed no local lesion. Another patient presented with a large tender abdominal mass due to a retroperitoneal hematoma, which gradually disappeared on treatment of the scurvy.

Bodi and Weiss (1960) reported encouraging results from the use of ascorbic acid and hesperidin in the treatment of five patients with bleeding peptic ulcers. Morris (1960) reported two cases of hemorrhagic gastritis due to scurvy which were successfully treated by the administration of large doses of vitamin C and citrus fruit juice. He stated that, "Scurvy occurs in over 5 per cent of patients seen in normal practice, but it is often not properly diagnosed."

Table 1
RESULTS OF TESTS, SUMMARIZED IN AVERAGES

Category	No. of cases	Blood ascorbic acid (mg%)		Gastric ascorbic acid (mg%)		Free HCl (clinical units)		Total acid (clinical units)	
		Average	Range	Average	Range	Average	Range	Average	Range
Unselected controls	113	0.88	0—1.83	0.62	0.12—1.98	15	0—70	35	4—109
Pathologic cases	108	0.44	0—1.20	0.36	0—1.33	22	0—69	45	5—142
Peptic ulcer	79	0.43	0—1.70	0.39	0—1.33	27	0—69	52	5—142
Pernicious anemia	6	0.47	0.24—0.80	0.18	0.02—0.38	0	—	9	5—16
Gastritis	17	0.42	0.17—1.20	0.29	0.13—0.64	15	0—69	35	7—99
Gastric malignancy	5	0.62	0.35—0.80	0.21	0.04—0.49	0	—	11	7—15
Benign gastric tumor	1	0.56	—	0.79	—	0	—	11	—

Note: The unselected control group consisted of 113 routine hospital, clinic, and private patients with some illness other than known gastrointestinal disease. There was a highly significant difference ($p < 0.001$) between both the blood and gastric juice ascorbic acid levels of the peptic ulcer patients and those of the control patients.

From Freeman, J. T. and Hafkesbring, R. (1957), *Gastroenterology*, 32, 878. ©1957 Williams & Wilkins Co., Baltimore. With permission.

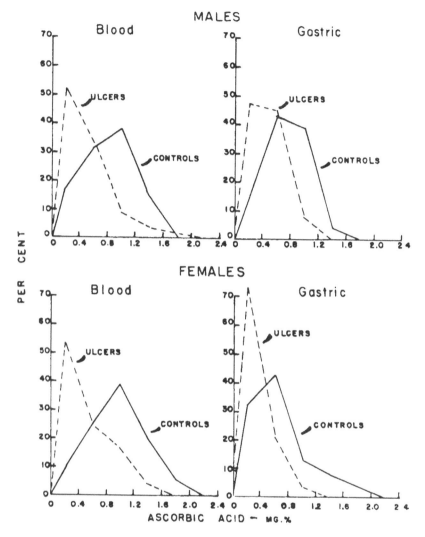

FIGURE 5.   Percentage distribution of fasting blood and fasting gastric juice ascorbic acid levels in control and peptic ulcer patients. (From Freeman, J. T. and Hafkesbring, R. [1957], *Gastroenterology*, 32, 878. ©1947 Williams & Wilkins Co., Baltimore. With permission.)

Shafar (1965) presented three patients with adult scurvy arising as a consequence of dietary restrictions adopted on medical advice as treatment for disorders of the alimentary tract. Two were men with duodenal ulcers; the other was a woman with ulcerative colitis. Shafar said these were instances of iatrogenic scurvy.

Working at the Glasgow Royal Infirmary, Cohen and Duncan (1967) reported a study of the leukocyte ascorbic acid levels of 14 patients with gastroduodenal disorders and of 14 control surgical patients, matched for age and sex, who had no known alimentary disease. The mean leukocyte ascorbic acid (TAA)* level of the patients with gastroduodenal disease was 11.0 μg/10^8 cells, and in the control patients was 22.9 μg/10^8 cells. Eight patients with proven duodenal ulcer disease had a reduced dietary intake of ascorbic acid (27 mg/d) and a mean leukocyte ascorbic acid level of only 8.7 μg/10^8 cells when tested before surgery. It was suggested that ascorbic acid supplements should be mandatory in such patients before surgery.

---

* TAA — total ascorbic acid, reduced and oxidized forms.

Table 2
BUFFY LAYER ASCORBIC ACID LEVELS IN
GASTRODUODENAL DISORDERS

Group	No. studied	Buffy layer ascorbic acid (μg/10^8 WBC)	
		Range	Mean $\pm$ SD
Control	25	18.7—53.5	28.71 $\pm$ 8.08
Duodenal ulcer	34	4.5—25 1	11.66 $\pm$ 4.49
Duodenal ulcer with stenosis	9	5.8—18.4	10.12 $\pm$ 4.49
Previous gastric surgery — in good health	39	7.3—39.0	20.44 $\pm$ 8.37
Previous gastric surgery with symptoms	27	0.9—26 3	12.96 $\pm$ 6.77

From Dymock, I W., Turck, W. P. G., Brown, P. W., Sircus, W., Small, W. P., and Thomson, C. (1968), *Br. Med. J.*, 1, 179. With permission.

Cox et al. (1967) investigated five patients with scurvy and anemia. One was an alcoholic, one a recluse, one was unemployed and two were taking scorbutogenic diets to avoid diarrhea associated with jejunal diverticulosis and diverticulosis coli, respectively; but occult blood tests of their feces were consistently negative in all five of these patients.

Studies by Esposito and Valentini (1968) confirmed the existence of low leukocyte ascorbic acid levels in patients with duodenal ulcers, irrespective of whether they showed high or normal gastric acid secretion in response to histamine. These workers thought that inadequate absorption of ascorbic acid is probably the most important factor in causing the low ascorbic acid levels in these patients, but also suggested that inadequate intake or utilization of the vitamin may also play a part.

Dymock et al. (1968) studied the leukocyte ascorbic acid levels of 109 patients with duodenal ulceration before and after surgery and concluded that ascorbic acid depletion is commonly found in patients with duodenal ulcer, duodenal stenosis, and also in patients with post gastrectomy syndromes (Table 2).

Another extensive study was reported by Russel et al. (1968) of Glasgow. They studied the leukocyte ascorbic acid levels of 60 patients admitted to hospital with gastrointestinal hemorrhage. These were matched for age and sex with a group of patients with uncomplicated peptic ulcer, and with a group of volunteers (healthy shipyard workers). The mean leukocyte ascorbic acid (TAA) level in the patients with gastrointestinal hemorrhage was 14.2 μg/10^8 cells, compared with 17.6 in the peptic ulcer group without hemorrhage and 23.7 in the healthy control group (Table 3). Only six of the patients with gastrointestinal hemorrhage showed evidence of skin purpura, and sublingual hemorrhages were present in two of these, but subclinical vitamin C deficiency was felt to have been a major factor contributing to the hemorrhage in the majority of them. Of the 60 patients with gastrointestinal hemorrhage, 23 (38%) were found to have a peptic ulcer (duodenal in 21 and gastric in 2); 40 of the 60 patients with gastrointestinal hemorrhage (67%) had taken aspirin, alcohol, or both in the week before admission to hospital. The mean leukocyte ascorbic-acid level was significantly lower in patients who had taken aspirin or alcohol than in those with no precipitating factor: this was especially notable in patients over 45 years of age (Table 4). The authors concluded that subclinical scurvy is a factor, which combines with aspirin-induced gastric erosion, in precipitating gastrointestinal hemorrhage, especially in patients over the age of 45.

Croft (1968) suggested that both ascorbate and folate may be important for replacement of the epithelial cells of the gastric mucosa when they have been damaged and exfoliated as a result of contact with aspirin or other irritants. It was suggested that lack of these essential nutrients leads to hemorrhage and to gastric erosion.

Table 3
MEAN (± SEM) LEUKOCYTE LEVELS (μg/10⁸ WBC) IN PATIENTS AND HEALTHY VOLUNTEERS

Age	Gastrointestinal hemorrhage (A)	Peptic ulcer controls (B)	Healthy controls (C)	Significance (p) of differences		
				A vs. B	A vs. C	B vs. C
Whole series	14.2 (± 0.9)	17.6 (± 0.9)	23 7 (± 1.3)	<0.005	<0 001	<0.005
Under 45	19.3 (± 1.7)	20.6 (± 1.7)	23.7 (± 1.2)	>0.2	<0.05	>0.2
Over 45	11.3 (± 0.8)	16.4 (± 1.0)	23 7 (± 2.3)	<0.005	<0.001	<0.001

Note: Comparison of the mean leukocyte ascorbic acid (TAA) levels of (A) 60 patients admitted to hospital with gastrointestinal hemorrhage, (B) an age- and sex-matched group of peptic ulcer patients without hemorrhage, and (C) a group of healthy shipyard workers. There was a highly significant correlation between low leukocyte ascorbic acid levels and gastrointestinal hemorrhage.

From Russel, R. I , Williamson, J. M., Goldberg, A., and Wares, E. (1968), *Lancet*, 2, 603. With permission

Table 4
MEAN (± SEM) LEUKOCYTE AA LEVELS (μg/10⁸ WBC) IN RELATION TO HISTORY OF ALCOHOL OR ASPIRIN INGESTION

Age	Patients with aspirin or alcohol ingestion	Patients with no aspirin or alcohol ingestion	p
Whole series	12.6 (± 0.9)	18.2 (± 0.8)	<0.005
Under 45	16.4 (± 1.1)	23.1 (± 0.9)	<0.05
Over 45	10.9 (± 0.8)	14.9 (± 1.2)	<0.02

Note: Of the 60 patients with gastrointestinal hemorrhage, 40 had taken aspirin, alcohol, or both in the week before admission to hospital. The mean leukocyte ascorbic acid level was found to be significantly lower in patients who had taken aspirin or alcohol than in those with no precipitating factor.

From Russel, R. I., Williamson, J. M., Goldberg, A., and Wares, E. (1968), *Lancet*, 2, 603. With permission.

Hansky and Allmand (1969) found that the dietary intake of vitamin C is significantly lower in patients admitted to hospital with gastrointestinal hemorrhage than in those of an age- and sex-matched control group of hospital patients. Scobie (1969) observed pyloric stenosis in association with frank scurvy in one patient and reported that he became fit to undergo partial gastrectomy after 2 weeks of ascorbic acid therapy.

A study of the dietary ascorbic acid intake and the leukocyte ascorbic acid levels of 33 patients, before and after vagotomy for peptic ulcer, was conducted by MacDonald and Cohen (1972). Although the daily ascorbic acid intake rose from 26 mg before surgery to 53 mg after surgery, there was little change in the leukocyte ascorbic acid (TAA) levels which rose from 18.3 to 19.7 μg/10⁸ cells. It was postulated that the hypochlorhydria resulting from vagotomy impaired the absorption of ascorbic acid.

The relationship between hypovitaminosis C and ulceration of the gastric and duodenal mucosa was summarized and tabulated by Booth and Todd (1972) and also by Wilson (1974). Stewart (1974, 1975) expressed his belief that subclinical scurvy could be responsible for hematemesis after aspirin. Coffey and Wilson (1975) suggested that age-induced physiological desaturation of tissue ascorbate stores could certainly predispose to hematemesis associated with aspirin consumption in older patients.

Studies by Gerson (1975) have shown that patients with regional ileitis have low plasma and leukocyte ascorbic acid levels, which may be due to their restricted diets. He found that patients with fistula formation, a common complication of ileitis, had ileal tissue and blood ascorbate levels lower than those in patients without fistulas. It was postulated that fistulas might more easily form in tissue with depressed vitamin C content because of defective collagen formation.

The incidental discovery of clinical scurvy in a patient with hiatus hernia led Hiebert (1977) to diagnose ascorbic acid deficiency in nine other individuals with hiatus hernia, esophagitis, and/or gastroesophageal reflux. Most of them had an aversion to acid foods. They all had easy bruising and increased capillary fragility and four were noted as being depressed. The plasma ascorbic acid levels were unreliable, possibly because of failure to dessicate the ascorbic acid used as a standard, but all of the symptoms improved when vitamin C was administered, so there can be no doubt of the diagnosis.

Bacelar et al. (1982) reported lower plasma ascorbic acid levels in patients with schistosomiasis following gastrointestinal hemorrhage (0.25 mg/100 ml) than in similar patients without hemorrhage (0.44 mg/100 ml), but the lower levels were probably a result of the hemorrhage rather than a secondary or precipitating cause, as the leukocyte ascorbate levels of the two groups did not differ significantly.

II. STUDIES OF MONKEYS

Talbot et al. (1913) observed extensive ulcerative colitis in a rhesus monkey *(Macaca mulata)* which died with classical scurvy after 3 months on a diet of condensed milk. Mackie and Chitre, cited by McCarrison (1931), found that the colon of the monkey is very sensitive to deficiency of vitamins, especially vitamin C, which causes a condition varying from local congestion and thickening of the mucous membrane to one indistinguishable from ulcerating and sloughing dysentery. Miura and Okabe (1933) kept a female Formosan long-tail macaque monkey on a vitamin C-deficient diet. They reported an episode of bloody diarrhea after 4 months of ascorbic acid deficency. Machlin et al. (1976) reported finding an ulcer approximately 4 mm in diameter, with a surrounding scirrhous reaction, in the gastric pylorus of a rhesus monkey with scurvy produced by feeding ascorbic acid 2-sulfate instead of ascorbic acid. An abcess was also present in the mesentery at the level of the ileocecal valve.

III. GUINEA PIG STUDIES

Holst and Frölich (1907), in their studies of experimental scruvy, noted the occurrence of peptic ulcer with occasional perforation in some of the guinea pigs. The diet on which the animals were fed consisted of bread and grains of different types. Talbot et al. (1913) induced scurvy in 19 guinea pigs by feeding them on a diet of bread and water. Hemorrhagic spots were noted in the mucous membranes of the stomach and small intestine of one animal and in the intestine of another, but there was no record of any ulceration. Peptic ulcer was reported only once, and gastric hemorrhages once in all of the scorbutic guinea pigs observed by Jackson and Moore (1916). Gastrointestinal bleeding was noted in some of the scorbutic guinea pigs fed a soybean cracker diet by Cohen and Mendel (1918), but these workers did not describe the stomach or duodenum. They paid particular attention to the caecum, which is the equivalent of the appendix in the human, and made the following statement: "Summarizing our experience with nearly one hundred scorbutic animals, we conclude that actual impaction of the feces in the cecum occurred in about one-quarter of the cases; and visible damage to wall, i.e., congestion or hemorrhage, or impaction or both was found in perhaps half of the cases." Hess (1920) mentioned the occurrence of duodenal ulcers in scorbutic guinea-pigs, but gave no data as to their frequency.

McCarrison (1921) fed nine guinea pigs on a diet of crushed oats and autoclaved milk; they were in the "prescorbutic" state when they were killed. The duodenum in every case was congested and a hemorrhagic infiltration extended downwards from the pylorus for a distance of 1 to 2 in. on examination with a hand lens, the duodenal mucosa showed ecchymoses and occasional punched-out areas were seen extending to the peritoneal coat. Three of the animals also had punched-out necrotic areas in the stomach. In 1, as many as 15 were counted in the base of the ulcer extending down to the peritoneal coat, the mucous membrane being studded with brownish debris of hemorrhages into the interior of the organ. These changes were also present in the stomach and duodenum of guinea-pigs dying as a result of this diet, although they showed none of the classical appearances of scurvy. In a clinical sense, they were prescorbutic. Indeed, it seems that peptic ulcers are much more frequently seen in animals maintained for some time on a low ascorbic acid intake than they are in acute scurvy brought on by a completely deficient diet.

Mme Randoin (1923) in her description of scurvy in guinea pigs noted that the feces became very fluid and finally became hemorrhagic. Höjer (1924) recorded the presence of ulcers in the ventricle or the duodenum of guinea pigs with scurvy. He stated that these ulcers in the duodenum are placed above the papilla of Vater and that a strong congestive hyperemia is common in that place. Meyer and McCormick (1928) stated that duodenal and cecal hemorrhages were common in their guinea pigs with scurvy.

Magee et al. (1929) kept guinea pigs on diets of barley and grass or barley and turnip. These diets were adequate, as the animals all grew normally, remained healthy, and showed no organic pathology. However, guinea pigs fed the same diets with restricted intakes of grass or turnip showed definite gastrointestinal pathology. Among 32 animals on diets low in or completely lacking turnips or grass, 81% showed congestion and hemorrhage in the mucosa of the pylorus, 22% showed ulcer or erosion in the mucosa of the pyloric region, and 37% showed congestion, hemorrhages, or ulcers of the cecum (appendix) or colon. There may have been factors other than vitamin C missing from the barley diet, but vitamin C deficiency was clearly a major defect.

McCarrison (1931) reported that he had repeated his earlier experiment, feeding guinea pigs on a scorbutic diet of crushed oats and autoclaved milk. He obtained the same results, observing duodenal ulceration at all stages of the process up to perforation (in one case), the resulting abscess being localized.

While studying the relationship of vitamin C deficiency to intestinal tuberculosis in the guinea pig, McConkey and Smith (1933) observed 6 nontuberculous duodenal ulcers, 1 pyloric ulcer, and 1 cecum with several small necrotic areas among 52 animals maintained in a prescorbutic state (no ascorbic acid for 3 weeks and then 2 or 3 cc of tomato juice daily), but no such lesions among 46 animals on full diets or diets lacking only vitamins A and D. There were also 26 tuberculous ulcers of the ileum among 37 animals fed tubercle bacilli by mouth while on prescorbutic diets. Only 2 tuberculous ulcers of the ileum were found among 35 animals fed tubercle bacilli while on diets supplemented by an adequate amount of vitamin C. Describing the origin of the ileal ulcers, these authors stated, "We found that in the early scorbutic lesions submucosal hemorrhage was the outspoken feature; the mucosa overlying the area of hemorrhage becomes necrotic, sloughs, and forms an ulcer. The initial tuberculous lesion also occurs in the submucosa and apparently goes on to ulceration when there is not sufficient vitamin C in the diet. Apparently, an adequate supply of vitamin C in the diet protects the guinea pig against ulcerative intestinal tuberculosis, even in the presence of submucous tubercle."

These findings prompted further studies by Smith and McConkey (1933) who confirmed that guinea pigs kept in the subscurvy state were liable to develop duodenal ulcers. When a sufficiency of vitamins A and D was given to correct any deficiency in the diet the ulcers still occurred. In one experiment, 15 guinea pigs were divided into 3 groups: those in Group I were given a good diet; those in Groups II and III a scorbutogenic diet of rolled oats and autoclaved milk, with the addition of 0.3 cc of cod-liver oil to correct the deficiency of A and D. At the end of 15 d, there was no gross evidence of scurvy, but they knew from experience that the signs would appear in 7 to 10 d. All the animals were then subjected to surgery under aseptic conditions. The abdominal cavity was opened, a hole bored in the stomach with a cautery, and the wall of the duodenum scarified with a dental burr. The hole in the stomach was sealed with a purse-string suture, and the abdomen was closed. The animals in Group I were killed at the end of 18 d; the wounds were healed and the stomach and duodenum appeared normal. The animals in Groups II and III did not show any gross signs of scurvy at the time of operation, but the incisions bled more easily and the tissues were slightly edematous. The animals in Group II were given 10 cc of canned tomato juice each day, and killed after 21d. The wounds had healed completely and showed no abnormality except a little edema of the subcutaneous tissues. The stomach and duodenum were normal. The animals of Group III died in 1 to 2 weeks after the operation. The skin incision showed no signs of healing. In two animals an ulcer had developed at the site of puncture into the stomach, and four out of five had developed large ulcers of the duodenum. Summarizing their experiments, they recorded 20 peptic ulcers in 75 guinea pigs which were in a state of partial but chronic vitamin C deficiency, and only 1 peptic ulcer in 80 guinea pigs on diets providing adequate supplies of vitamin C.

Spellberg and Keeton (1940), at the University of Illinois, studied guinea pigs on low-ascorbic acid diets and reported that some of the animals developed multiple gastric ulcers, most commonly located in the greater curvature of the stomach.

Ungar (1942, 1943) showed that standardized lethal trauma to the thigh muscles of guinea pigs caused hemorrhagic lesions in the stomachs of 75% of the animals, and in some of them perforation of the organ. He also reported that subcutaneous injection of ascorbic acid (not less than 100 mg/kg), given within 15 min of the trauma, protected all of the animals from a degree of trauma that was otherwise 100% fatal. In subsequent studies, Ungar (1944, 1945) observed that similar protective effects against the lethal effects of trauma could be obtained by prior sublethal stress and by adrenocorticotropic hormone or adrenal extracts and could be blocked by hypophysectomy or adrenalectomy. However, the protective effect of ascorbic acid was not blocked by hypophysectomy or adrenalectomy. He also observed that all substances providing protection against death from traumatic shock were associated with inhibition of the normal release of histamine from the blood cells and that D-catechin was one such substance. The effects of stress on ascorbic acid metabolism are discussed in Chapter 16, Volume I, of D-catechin on ascorbic acid metabolism in Chapter 11, Volume I, and of ascorbic acid on histamine metabolism in Chapter 1 of this volume.

Autopsy studies of eight scorbutic guinea pigs, reported by Fabianek (1956), revealed hemorrhages in the duodenum in two, in the small intestines of four, and in the cecum and ileocecal valve in two animals.

Russel and Goldberg (1968) allocated 16 guinea pigs to each of four groups: (1) normal diet without aspirin; (2) scorbutogenic diet without aspirin, (3) normal diet with added aspirin. and (4) scorbutogenic diet with added aspirin. The animals were killed after 2 weeks and before any of them had developed scurvy. Bleeding points were found significantly more often in the gastric mucosae of the animals on the scorbutogenic diet than in those on the normal diet (Table 5); when aspirin had been administered, the difference was even more striking.

Studying gallstone formation in guinea pigs, Pavel et al. (1969) fed the animals a vitamin C-free mixed diet, with liver as the principal protein source. Not only did the animals all develop gallstones on this vitamin C-deficient diet (Chapter 13, Volume II), most of them also developed ulcers of the stomach and duodenum and hemorrhagic lesions of the colon.

Table 5
NUMBER OF GUINEA PIGS WITH
OBVIOUS BLEEDING POINTS ON
GASTRIC MUCOSA

	No. of animals with bleeding points on	
	Scorbutogenic diet	Normal diet
With aspirin	9/16	1/16
Without aspirin	5/16	0/16
Both groups	14/32	1/32

Note: After 2 weeks on a scorbutogenic diet, guinea pigs were in a prescorbutic state and bled from the gastric mucosa significantly more often than those on a normal diet. The addition of aspirin to the scorbutogenic diet further increased the likelihood of gastric mucosal bleeding.

From Russel, R. I. and Goldberg, A. (1968), *Lancet*, September 14, 606. With permission.

IV. RAT STUDIES

Cheney and Rudrud (1974) reported that only one fifth of the rats given ascorbic acid in their drinking water, prior to and during three consecutive 47-h food-deprivation periods, developed stress ulcers of the stomach, while all of those given water and the majority of those given deactivated L-ascorbic acid did. These findings were unexpected, as the rat can synthesize ascorbic acid in its liver and is not usually believed to need any more in the diet.

Glavin et al. (1978) failed to confirm these findings; in fact, they found some evidence that the addition of ascorbic acid to the drinking water may have exacerbated the development of stomach ulcers. Close inspection of their data suggests that ascorbic acid was toxic only when it was allowed to stand in the drinking bowls for up to 24 h and may have been protective when administered by oral injection in distilled water (Table 6). If we add the rumenal and the glandular ulcers of Table 6 together, we find that the worst results occurred in the rats which had been pretreated for 8 d with ascorbic acid in their drinking water (nine ulcers in ten rats), while the best results (two ulcers in nine rats) occurred in those which received ascorbic acid by daily oral injection in distilled water before being stressed by three consecutive 47-h periods of starvation. The toxic effect may have been due to cupric ions in the drinking water, for ascorbic acid is known to be toxic in the presence of copper and oxygen, as shown by Stich et al. (1976), almost certainly because of the release of ascorbate-free-radical during the catalytic oxidation of the vitamin. It would be most interesting to repeat this experiment using ascorbic acid with a chelating flavonoid or catechin to inactivate heavy metal catalysts.

V. DOG STUDIES

Nasio (1946) observed that administration of ascorbic acid in doses of 500 mg daily exerts, in 60% of cases, a marked protective effect against cinchophen ulcer. He also observed in several instances a healing effect on the ulcerative lesions and a reduction in the hemorrhagic tendency following cinchophen.

Table 6
SUMMARY OF STOMACH PATHOLOGY FOR THE ASCORBIC ACID PRETREATMENT AND STARVATION STUDY

Pretreatment group	Rumenal ulcer incidence	Mean length of ulcers (mm)	Glandular ulcer incidence	Mean length of glandular ulcers (mm)
Ascorbic acid, 750 mg in 5 ml of distilled water by oral injection daily for 8 d	1/9[a]	0.66	1/9	3.30
Ascorbic acid, 30 g/l in drinking water changed daily for 8 d	5/10	10 20	4/10	7.33
Water, by oral injection	4/10	19.50	0/10	
Drinking water, *ad libitum*	3/10	6.90	0/10	

Note: Rats given ascorbic acid by oral injection had less gastric ulcers than control animals given water *ad libitum* or water by oral injection in this study of starvation stress effects. However, ascorbic acid allowed to stand in the drinking water for up to 24 h seems to have had deleterious effects (see text).

[a]   One rat in this group was killed because of respiratory illness.

From Glavin, G. B., Paré, W. P., and Vincent, G. P., Jr. (1978), *J Nutr.*, 108, 1969. © American Institute of Nutrition. With permission.

VI. OBSERVATIONS IN COWS AND SHEEP

Rosenow (1923) found peptic ulcers and areas of confluent hemorrhagic infiltration in the stomachs of sheep, cows, and calves at the Armour abattoir in Chicago, and he cultured streptococci from them. He learned that, ''the incidence of ulcer in the cow and the sheep was highest each year during the latter part of the Winter and early Spring months, and that in their opinion the ulceration was the result of eating dry, coarse foods during these months, since the condition disappeared entirely each year during the Summer and Fall months when the animals were in pasture.'' Although cows and sheep, like most animals, can synthesize ascorbic acid in the liver, it is entirely possible that they can benefit from the additional ascorbic acid they obtain from fresh grass. Dogs and rabbits likewise synthesize ascorbic acid in the liver, but unpublished observations by the writer have shown that their plasma ascorbic acid levels can be raised and their blood histamine levels are reduced by administration of ascorbic acid and green vegetables, respectively.

VII. DISCUSSION

The etiology of peptic ulceration in man is unknown. Heredity, personality type, male sex, occupation, infection, tobacco, alcohol, aspirin, spicy foods, lack of sleep, worry, irregular meals, frustration of appetite, and trauma to the gastic or duodenal mucosa have all been suggested as causative factors, but each on its own seems to be an inadequate explanation.

Certainly duodenal ulcers are associated with gastric hyperacidity. Moreover, histamine is known to cause the secretion of hydrochloric acid by the stomach and even hemorrhages and ulceration if pushed to excess. Indeed, McIlroy (1928) produced peptic ulcers in cats by injection of histamine, and also demonstrated that histamine causes extension instead of healing of experimental wounds of the pyloric antrum of the stomach. Moreover, histamine receptor-blocking agents (H_2 blockers) are widely and successfully used in the treatment of peptic ulcers.

It is therefore interesting to note that ascorbic acid administration decreases the whole blood histamine levels, when they are elevated due to ascorbate insufficiency, both in guinea pigs and in human subjects (Chapter 1 of this volume). This vitamin is, in fact, the best antihistamine available, for if consumed as it occurs in fresh fruits and vegetables or in tablet form with a chelating antioxidant such as D-catechin, it is totally nontoxic.

Clearly, duodenal ulcer is a multifactorial disease. Stresses, such as long hours and lack of sleep, can elevate the blood histamine level (Chapter 17, Volume I). Ascorbic acid deficiency prevents the detoxification of histamine by preventing its conversion to hydantoin-5-acetic acid, and causes histamine to accumulate in the blood. This presumably stimulates excessive acid production by the stomach.

Hafkesbring et al. (1952) demonstrated that both the free and total gastric acid levels of ten healthy women volunteers, aged 20 to 32 years, were reduced when their blood ascorbic acid levels were raised from a mean of 1.07 to a mean of 1.88 mg/100 ml. The free acid fell by an average of 16% and the total acid fell 22%. These are very impressive results when one considers that none of the subjects had a low initial ascorbic acid level. How much more change might one expect in ascorbic acid-deficient patients with duodenal ulcer disease?

A hemorrhagic or purpuric area in the mucosa and submucosa of the duodenum, resulting from ascorbic acid deficiency, may easily be converted into an ulcer by acid from the stomach, and the erosion will likely continue until enough ascorbic acid is provided to reduce the blood histamine and gastric acid levels and to promote healing of the ulcer.

However, a paradox arises from the finding by May et al. (1949) that the feeding of an iron-supplemented, ascorbic acid-deficient diet to monkeys caused scurvy (with megaloblastic anemia) after 3 or 4 months, and that 10 out of 11 monkeys developed histamine refractory gastric achlorhydria. If ascorbic acid deficiency could be relied upon always to cause the oxyntic cells of the stomach to become refractory to histamine, then it would seem that the problem would be self-limiting. Unfortunately, it is not, and gastric hyperacidity is often associated with very low blood ascorbic acid levels, as shown by Freeman and Hafkesbring (1957) and recorded in Table 1. So it would seem that either histamine refractory achlorhydria is an inconstant result of ascorbic acid deficiency, or else it occurs only in the late stages of scurvy.

REFERENCES

Adams, J. F. (1954), Scurvy occurring in a patient on an ulcer diet: report of a case, *Glasgow Med. J.*, 35, 64.

Archer, H. E. and Graham, G. (1936), The subscurvy state in relation to gastric and duodenal ulcer, *Lancet*, 2, 364.

Bacelar, T. S., Ferraz, E. M., Aguiar, J. L. de A., Filho, H. A. F., and Kelner, S. (1982), Ascorbic acid in schistosomiasis mansoni with or without digestive haemorrhage, *Int. J. Vitam. Nutr. Res.*, 52, 442.

Bartlett, M. K., Jones, C. M., and Ryan, A. E. (1940), Vitamin C studies on surgical patients, *Ann. Surg.*, 111, 1.

Bodi, T. and Weiss, B. (1960), Experimental and clinical considerations on hesperidin-ascorbic acid in upper gastrointestinal bleeding, *Am. J. Gastroenterol.*, 34, 402.

Booth, J. B. and Todd, G. B. (1972), Subclinical scurvy — hypovitaminosis C, *Geriatrics*, 27, 130.

Bourne, G. (1938), Vitamin C deficiency in peptic ulceration estimated by the capillary resistance test, *Br. Med. J.*, 1, 560.

Chamberlin, D. T. and Perkin, H. J. (1938), The level of ascorbic acid in the blood and urine of patients with peptic ulcer, *Am. J. Digest. Dis.*, 5, 493.

Cheney, C. D. and Rudrud, E. (1974), Prophylaxis by vitamin C in starvation induced rat stomach ulcer, *Life Sci.*, 14, 2209.

Coffey, G. and Wilson, C. W. M. (1975), Ascorbic acid deficiency and aspirin-induced haematemesis, *Br. Med. J.*, January 25, 308.

Cohen, B. and Mendel, L. B. (1918), Experimental scurvy of the guinea pig in relation to the diet, *J. Biol. Chem.*, 35, 425

Cohen, M. M. and Duncan, A. M. (1967), Ascorbic acid nutrition in gastroduodenal disorders, *Br. Med. J.*, 4, 516.

Cox, E. V., Meynell, M. J., Northam, B. E., and Cooke, W. T. (1967), The anaemia of scurvy, *Am J Med*, 42, 220.

Crescenzo, V. M. and Cayer, D. (1947), Plasma vitamin C levels in patients with peptic ulcer. Response to oral load test of ascorbic acid, *Gastroenterology*, 8, 754.

Croft, D. N. (1968), Aspirin, vitamin-C-deficiency, and gastric haemorrhage, *Lancet*, October 12, 831.

Croft, J. D. and Snorf, L. D. (1939), Cevitamic acid deficiency. frequency in group of 100 unselected patients, *Am. J. Med. Sci.*, 198, 403.

Cutforth, R. H. (1958), Adult scurvy, *Lancet*, 1, 454, 456

Davidson, P. B. (1928), The development of deficiency disease during therapeutic diets, *JAMA*, 90, 1014

Doll, R. and Pygott, F. (1952), Factors influencing the rate of healing of gastric ulcers; admission to hospital, phenobarbitone and ascorbic acid, *Lancet*, 1, 171

Drummond, J. (1955), Peptic ulcer, *S Afr. Med. J*, 29, 581.

Dymock, I. W., Turck, W. P. G., Brown, P. W., Sircus, W., Small, W. P., and Thomson, C. (1968), Vitamin C and gastroduodenal disorders, *Br Med. J.*, 1, 179.

Ebbesen, I. and Rasmussen, M. (1944), Some investigations on the ascorbic acid content in serum from patients suffering from peptic ulcer II, *Acta Med. Scand.*, 117, 507.

Editorial (1948), Ascorbic acid in the treatment of hemorrhage in peptic ulcer, *Am. J. Digest. Dis.*, 15, 354.

Esposito, R. and Valentini, R. (1968), Vitamin C and gastroduodenal ulcers, *Br. Med. J.*, 2, 118.

Fabianek, J. (1956) Sur le scorbut réalisé chez le cobaye au moyen de régimes artificiels I, *C. R. Soc. Biol.*, 150, 274.

Field, H., Robinson, W. D., and Melnick, D. (1940), Vitamins in peptic ulcer, *Ann. Intern. Med.*, 14, 588.

Foote, R. R. (1926), Scurvy simulating acute intussusception, *Br. Med. J.*, 1, 1035.

Freeman, J. T. and Hafkesbring, R. (1954), Ascorbic acid levels in blood and gastric secretions. III. Gastrointestinal diseases, *Fed. Proc. Fed. Am. Soc. Exp. Biol.*, 13, 48.

Freeman, J. T. and Hafkesbring, R. (1957), Comparative studies of ascorbic acid levels in gastric secretion and blood. III. Gastrointestinal diseases, *Gastroenterology*, 32, 878.

Gerson, C.D. (1975), Ascorbic acid deficiency in clinical disease including regional ileitis, *Ann. N.Y Acad. Sci.*, 258, 483.

Glavin, G. B., Paré, W. P., and Vincent, G. P., Jr. (1978), Ascorbic acid and stress ulcer in the rat, *J. Nutr.*, 108, 1969.

Hafkesbring, R., Freeman, J. T., and Caldwell, E. K. (1952), Comparative study of ascorbic acid levels in gastric secretion, blood, urine and saliva. II. Saturation studies, *Am. J. Med. Sci.*, 224, 324.

Hansky, J. and Allmand, F. (1969), Gastro-intestinal bleeding: the role of vitamin C, *Australas. Ann. Med.*, 18, 248

Harris, L. J., Abbassy, M. A., and Yudkin, J. (1936), Vitamins in human nutrition. Vitamin-C reserves of subjects of the voluntary hospital class, *Lancet*, 1, 1488.

Hatherley, L. I. (1947), A case of vitamin C deficiency, *Br Med. J.*, 1, 679.

Hess, A. E. (1920), *Scurvy, Past and Present*, J. B. Lippincott, Philadelphia.

Hiebert, C. A. (1977), Gastroesophageal reflux and ascorbic acid deficiency, *Ann. Thorac. Surg.*, 24, 108.

Hoffman, W. S. and Dyniewicz, H. A. (1946), The effect of alumina gel upon the absorption of amino-acids, ascorbic acid, glucose and neutral fat from the intestinal tract, *Gastroenterology*, 6, 50.

Höjer, J. A. (1924), Studies in scurvy, *Acta Paediatr. Scand.*, 3, (Suppl. 1), 8.

Holst, A. and Frölich, T. (1907), Experimental studies relating to ship beri-beri and scurvy II. On the etiology of scurvy, *J. Hyg.*, 7, 634.

Hutter, K. (1928), Seasonal variations in occurrence of gastric and duodenal ulcers, *JAMA*, 91, 2030.

Ingalls, T. H. and Warren, H. A. (1937), Asymptomatic scurvy: its relation to wound healing and its incidence in patients with peptic ulcer, *N. Engl. J. Med.*, 217, 443.

Issler, A. and Demole, M. (1939), Influence de la saturation en vitamine C sur le suc gastrique, *C. R. Soc. Biol.*, 130, 1227.

Jackson, L. and Moore, J. J. (1916), Studies on experimental scurvy in guinea pigs, *J. Infect. Dis.*, 19, 478.

Lazarus, S. (1937), Vitamin C nutrition in cases of haematemesis and melaena, *Br. Med. J.*, 2, 1011.

Lund, C. C. (1942), Ascorbic acid deficiency associated with gastric lesions, *N. Engl. J. Med.*, 227, 247.

MacDonald, J. A. E. and Cohen, M. M. (1972), Effect of vagotomy on ascorbic acid nutrition in patients with peptic ulcer, *Br. Med. J.*, 2, 738.

Machlin, L. J., Garcia, F., Kuenzig, W., Richter, C. B., Spiegel, H. E., and Brin, M. (1976), Lack of antiscorbutic activity of ascorbate 2-sulfate in the rhesus monkey, *Am. J. Clin. Nutr.*, 29, 825.

Magee, H. E., Anderson, W., and McCallum, J. (1929), Diet and peptic ulcers in cavies, *Lancet*, 1, 12.

May, C. D., Nelson, E. N., and Salmon, R. J. (1949), Experimental production of megaloblastic anemia; an interrelationship between ascorbic acid and pteroylglutamic acid, *J. Lab. Clin. Med.*, 34, 1724.

McCarrison, R. (1931), Some surgical aspects of faulty nutrition, *Br. Med. J.*, 1, 966.

McCarrison, R. (1921), *Studies in Deficiency Disease*, Oxford Medical, London, 95.

McConkey, M. and Smith, D. T. (1933), The relation of vitamin C deficiency to intestinal tuberculosis in the guinea pig, *J. Exp Med.*, 58, 503.

McIlroy, P. T. (1928), Experimental production of gastric ulcer, *Proc. Soc. Exp. Biol. Med.*, 25, 268.

Meyer, A. W. and McCormick, L. M. (1928), Studies on scurvy, *Stanford Univ. Publ. Univ. Ser Med. Sci.*, II (2), 133.

Miura, M. and Okabe, N. (1933), On the antiscorbutic factor in commercially sterilized milk and Japanese green tea, *Sci. Pap. Inst. Phys. Chem. Res.*, 20, 145.

Morris, G. E. (1960), Hemorrhagic gastritis due to avitaminosis C, *Postgrad. Med.*, 27, 207.

Nash, E. C. (1952), A comparative study of an antacid with and without vitamin C in the treatment of peptic ulcer, *Am. Pract.*, 3, 117.

Nasio, J. (1946), Influence of some vitamins and hormones in the prevention of experimental cinchophen peptic ulcer, *Rev. Gastroenterol.*, 13, 195.

Nielsen, H. E. (1938), Serumascorbinsyren under normale og forskellige pathologiske forhold, *Bibl. Laeger*, 130, 20.

Pavel, I., Chisiu, N., and Sdroboci, D. (1969), La lithiase biliaire chez le cobaye avec dysnutrition scorbutique, *Nutr. Dieta*, 11, 60.

Payne, R. T. (1936), The post-mortem findings after partial gastrectomy, *St. Bartholemew's Hosp. Rep.*, 191.

Pelner, L. (1943), Vitamins in gastro-intestinal disease, *Am. J. Digest. Dis.*, 10, 414.

Peters, G. A. and Martin, H. E. (1937), Ascorbic acid in gastric juice, *Proc. Soc. Exp Biol Med.*, 36, 76.

Platt, R. (1936), Scurvy as the result of dietetic treatment, *Lancet*, 2, 366.

Portnoy, B. and Wilkinson, J. F. (1938), Vitamin C deficiency in peptic ulceration and haematemesis, *Br. Med. J.*, 1, 554.

Randoin, L. (1923), La question des vitamines. II. Le facteur antiscorbutique, *Bull. Soc. Chim. Biol.*, 5, 806.

Rao, N. R. (1938), Vitamin C and peptic ulcer, *Indian J. Med. Res.*, 26, 171.

Rivers, A. B. and Carlson, L. A. (1937), Vitamin C as a supplement in the therapy of peptic ulcer: Preliminary report, *Proc Staff Meet. Mayo Clin.*, 12, 383.

Rosenow, E. C. (1923), Etiology of spontaneous ulcer of stomach in domestic animals, *J. Infect. Dis.*, 32, 384.

Roux, J.-C. (1938), Hypovitaminose C et ulcère gastroduodénal, *Arch. Mal Appar. Dig.*, 28, 835.

Russel, R. I. and Goldberg, A. (1968), Effect of aspirin on the gastric mucosa of guinea pigs on a scorbutogenic diet, *Lancet*, September 14, 606

Russel, R. I., Williamson, J. M., Goldberg, A., and Wares, E. (1968), Ascorbic-acid levels in leucocytes of patients with gastrointestinal haemorrhage, *Lancet*, 2, 603.

Scobie, B. A. (1969), Scurvy in the adult, *N. Z. Med. J.*, 70, 398.

Shafar, J. (1965), Iatrogenic scurvy, *Practitioner*, 194, 374.

Smith, D. T. and McConkey, M. (1933), Peptic ulcers (gastric, pyloric and duodenal) Occurrence in guinea pigs fed on a diet deficient in vitamin C, *Arch. Intern. Med*, 51, 413.

Spellberg, M. A. and Keeton, R. W. (1940), The production of fatty and fibrotic livers in guinea pigs and rabbits by seemingly adequate diets, *Am. J. Med. Sci.*, 200, 688.

Stewart, C. P., Learmonth, J. R., and Pollock, G. A. (1941), Intravenous ascorbic acid in experimental acute haemorrhage, *Lancet*, 1, 818.

Stewart, H. C. (1974), Aspirin and what else?, *Br. Med. J.*, 3, 525.

Stich, H. F., Karim, J., Koropatnick, J., and Lo, L. (1976), Mutagenic action of ascorbic acid, *Nature (London)*, 260, 722.

Talbot, F. B., Dodd, W. J., and Peterson, H. O. (1913), Experimental scorbutus and the Roentgen ray diagnosis of scorbutus, *Boston, Med. Surg. J.*, 169, 232.

Troutt, L. (1932), Quality studies of therapeutic diets. I. The ulcer diet, *J. Am. Diet. Assoc.*, 8, 25.

Ungar, G. (1942), Effect of ascorbic acid on the survival of traumatized animals, *Nature (London)*, 149, 637.

Ungar, G. (1943), Experimental traumatic "shock"; factors affecting mortality and effect of therapeutic agents (ascorbic acid and nupercaine), *Lancet*, 1, 421.

Ungar, G. (1944), Endocrine reaction to tissue injury, *Nature (London)*, 154, 736.

Ungar, G. (1945), Endocrine function of the spleen and its participation in the pituitary-adrenal response to stress, *Endocrinology*, 37, 329.

Warren, H. A., Pijoan, M., and Emery, E. S., Jr. (1939), Ascorbic acid requirements in patients with peptic ulcer, *N. Engl. J. Med.*, 220, 1061.

Weiss, S., Weiss, J., and Weiss, B. (1955), Gastrointestinal haemorrhage: therapeutic evaluation of bioflavonoids. Report on 55 cases, *Am. J. Gastroenterol.*, 24, 523.

Wilson, C. W. M. (1974), Vitamin C: tissue saturation, metabolism and desaturation, *Practitioner*, 212, 481.

Witts, L. J. (1937), Haematemesis and melaena, *Br Med. J.,* 1, 837.

Wood, P. (1935), A case of adult scurvy, *Lancet,* 2, 1405

Zerbini, E. de J. (1947), Vitamin C in gastric resection for peptic ulcer, *Arch. Surg.,* 54, 117.

Chapter 18

OCULAR LESIONS

I. CLASSICAL SCURVY

Acute scurvy seldom produces any obvious ocular lesions; indeed, Lind did not record any such problems among the sailors under his care. However, proptosis of the eyeball due to postorbital hemorrhage was described by Kato (1932) in a 1-year-old child with frank scurvy admitted to the Children's Hospital in Los Angeles. The child presented with pain and swelling of the lower extremeties, loss of weight, pallor, bulging of the right eye, and petechial hemorrhages. This seems particularly anachronous in the land of "sun-kissed oranges" and other citrus fruits, but it is recorded that, "orange juice was not taken very well."

Hess (1935) pointed out the need to keep scurvy in mind as a cause of proptosis, as it may appear before other signs and symptoms have rendered the diagnosis obvious. Moreover, Hood and Hodges (1969), in a review of the literature on ocular lesions in scurvy, cited three references to the subject even before the publication of Lind's classical monograph in 1753. "In 1609, Mathew Martinus recorded transient and recurring swelling of the eyes and decreased visual acuity and in 1624 Daniel Sennertus reported two cases of loss of vision in scorbutic patients. Nitzch in 1747 mentioned periobital and conjunctival hemorrhages." Many writers have since recorded subperiosteal or periorbital hemorrhages causing proptosis, but more frequently hemorrhages are seen in the bulbar and palpebral conjunctivae, in the episcleral and subconjunctival tissues, and in the skin of the eyelids. Papilledema, retinal, and macular exudates, a macular star, congestion of the retinal vessels, and thrombosis of the central retinal vein have all been observed in patients with scurvy.

II. EXPERIMENTAL HUMAN SCURVY

Bartley et al. (1953) observed no ocular lesions in ten volunteers deprived of ascorbic acid for periods ranging from 186 to 269 d, even though they conducted ophthalmoscopic and slit lamp studies on these subjects every 6 weeks. Hood and Hodges (1969) observed conjunctival lesions in five out of nine volunteers in a study of experimental scurvy after 74 to 95 d of total ascorbic acid deprivation. These lesions varied from minute bulbar conjunctival hemorrhages and varicosities to a large subconjunctival hemorrhage accompanied by palpebral petechial hemorrhages and conjunctival congestion. No retinal hemorrhages or exudates were seen at any time during the study. During repletion with ascorbic acid, the conjunctival lesions disappeared rapidly in one subject who received 32 mg daily and faded more slowly in a man who was given 8 mg daily. One subject who received 4 mg daily had red spots on the conjunctiva which persisted for 3 months.

In a subsequent report, Hood et al. (1970) described the development of keratoconjunctivitis sicca and other signs of Sjögren's syndrome in three men who lived on a vitamin C-free diet for 84 d and then an inadequate ascorbic acid intake (6.5 mg daily) until 213 d. The lesions occurred in spite of recommended allowances of all other vitamins and healed when vitamin C was restored. During deficiency, dryness of the mouth, tender swelling of the parotid glands, dry, scaly skin, and conjunctivitis were present, in addition to the usual signs of scurvy, such as perifollicular hemorrhages, bleeding gums, loose teeth, and joint effusions. The men complained of dryness, redness, irritation, stickiness and mattering of the eyes, intolerance to light, and transient blurring of vision relieved by blinking.

III. HYPOVITAMINOSIS C IN MONKEYS AND GUINEA PIGS

Talbot et al. (1913) produced scurvy in a 6-month-old male Java monkey by keeping it on a diet of condensed milk. The gums became swollen, discolored, and ulcerated after 13 weeks; when the animal died 1 week later, its sclera were reported to be discolored and totally obscured by subconjunctival hemorrhages. It also showed numerous hemorrhages in the small intestine and well-developed ulcerative colitis. A control female monkey of the same age, fed on cow milk, carrots, raw apples, bread, and bananas remained in good health.

Friedenwald et al. (1943) found no significant difference in intraocular pressure between normal and vitamin C-deficient guinea pigs, but they did report that the rate of refilling of the anterior chamber of the eye, after withdrawal of aqueous fluid, was markedly slower than normal in ascorbic acid deficiency. They concluded that the rate of secretion of aqueous humour by the ciliary body was impaired in scurvy and proposed a secretory mechanism in which ascorbic acid was important as a mediator in a "redox pump" system. They proposed that the movement of water across the membrane was accomplished by an electric current generated by a difference in redox potential between the ciliary stroma and the epithelium. However, Bárány (1951), using a steady-state *p*-aminohippuric acid distribution method to study the rate of flow of aqueous humor in the intact eye of the guinea pig, found it to be normal, or even above normal, in scurvy.

Christiansson (1957) found the viscosity of the vitreous humor of guinea pigs with scurvy to be markedly reduced. This substance passed through a 12G5 filter with a 1-μ pore size in 30 instead of 330 min. This was interpreted as being most probably due to depolymerization of the hyaluronic acid in the vitreous body. Other findings included a decrease in the glucosamine concentration of an ultrafiltrate of the vitreous humor and an increase in the glucosamine concentration of the aqueous humor. There was also an increase in the hexosamine level of the serum from 103.0 to 133.6 mg/100 ml.

Kahan (1958) also concluded that there is depolymerization of the mucopolysaccharides in the vitreous humor of scorbutic guinea pigs. Moreover, hyaluronidase caused the same changes when injected into the vitreous humour of rabbits. Boyd, in 1961, studying guinea pigs on a plentiful ascorbic acid intake (10 mg/d), found a lower than normal total ascorbic acid (TAA)* level in the vitreous humor and liquefaction of that substance following removal of the lens, but he did not think that the liquefaction was a direct consequence of the decreased ascorbic acid level.

Experiments on guinea pigs by Greco et al. (1980) showed that an induced hypovitaminosis C (2 weeks of scorbutogenic diet followed by a maintenance dose of 0.5 mg of ascorbic acid) caused the appearance of widespread retinal hemorrhages and a significant decrease in the blood ascorbic acid from 0.36 mg/100 ml in controls to 0.14 mg/100 ml in the test animals. Ophthalmoscopic examinations carried out on days 20 and 40 on the diet showed that hypovitaminosis C was accompanied by retinal hemorrhages and that prolonged hypovitaminosis caused the hemorrhagic spots to extend throughout the retina. No hemorrhages were seen in the control animals.

IV. NORMAL ASCORBATE LEVELS

Harris (1933) discovered a high concentration of ascorbic acid in the aqueous humor of the eye, and this has since been confirmed by many investigators. There are variations between individuals and between different species of animal, and the level is affected by age, being low prenatally, rising after birth, and falling again in old age.

Müller and Buschke (1934) and Goldman and Buschke (1935) reported that the ascorbic

---

* TAA — total ascorbic acid, reduced and oxidized forms.

Table 1
LEVELS OF OCULAR ASCORBIC ACID AFTER HEALING

	Aphakic eyes		Control eyes		
Substance	Ascorbic acid (mg/100 g)	No. of eyes	Ascorbic acid (mg/100 g)	No. of eyes	Probability of chance occurrence
Cornea	32.09 (±5.62)[a]	18	47 29 (±14 03)	18	$p < 0.01$
Aqueous	9.70 (±0 62)	17	12.45 (±0.79)	17	$p < 0.02$
Vitreous	12 10 (±0 94)	17	16.69 (±1 03)	16	$p < 0.1$

Note: Guinea pigs were maintained on a diet including 20 mg of ascorbic acid a day. Their eyes were analyzed 24 d after removal of the lens from the left eye and control surgery on the right eye. The mean ascorbic acid (AA) concentration in the aqueous humor of the aphakic left eye was found to be significantly lower than in the intact right eye, suggesting that the lens in some way aids in storage of ascorbic acid in the aqueous.

[a]   Figures in parentheses indicate SEM.

From Boyd, T. A S. (1955), *Br. J. Ophthalmol*, 39, 204. With permission

acid (AA)* content of the aqueous humor is markedly decreased after removal of the lens, and only a small quantity of dehydroascorbic acid (DHAA) remains. Moreover, Müller and Buschke showed that *in vitro* a lens in aqueous exerts a preserving action on the vitamin C content of the aqueous. Fischer (1934) suggested the possibility that the lens might be able to synthesize ascorbic acid from simple sugars, but Bellows (1936 b) provided evidence that the lens acts as a preservative for ascorbic acid, "by an antioxidative action."

Friedenwald et al. (1943) demonstrated that ascorbic acid injected intravenously into rabbits reaches the ciliary body within 5 min and before there has been any increase of this reducing substance in the aqueous humour, thus confirming the existing belief that ascorbic acid travels from the blood to the ciliary body and is then secreted with the aqueous humor into the posterior chamber. They gave their reasons for believing that ascorbic acid occupies the role of a mediator in the redox chain connecting the oxidase activity of the ciliary epithelium with the dehydrogenase activity of the ciliary stroma. They suggested that oxidation of ascorbic acid by the ciliary epithelium is probably a necessary step in the intraocular secretion of this substance. Presumably, it must be rapidly reduced again in the aqueous humour.

Kronfeld (1952), Purcell et al. (1954), and Chatterjee and Ghosh (1956) concluded that there was no change in the level of ascorbic acid after removal of the lens, but the work of Boyd (1955) provided new evidence that the lens may play a role in maintaining a normal level of ascorbic acid in the aqueous humor and the cornea of guinea pigs. Statistically significant decreases in the ascorbate (AA) concentrations in the aqueous humor and in the cornea were found in complete aphakia (Table 1). However, Boyd could not rule out the possibility that this might have been due to slight residual inflammation in the eyes following surgery.

Subsequently, Heath et al. (1961) studying partially aphakic and completely aphakic eyes of guinea pigs showed that there were significant decreases in the ascorbic acid (TAA) concentrations in the aqueous and in the vitreous in complete aphakia, but that normal levels were gradually restored when traces of lens fibers remained (Tables 2 and 3). The fact that the level in the partially aphakic eyes was almost normal indicates that the lowered level in complete aphakia was not caused either by surgical trauma or by inflammation. The high ascorbic acid content of the retina was not altered by the absence of the lens. "The lower

* AA — ascorbic acid, reduced form.

Table 2

THE MEAN ASCORBIC ACID CONTENT OF TISSUES FROM APHAKIC EYES AND NORMAL EYES FROM THE SAME ANIMALS

Tissue	Number of eyes	Aphakic eyes ascorbic acid (mg/100 g)	Control eyes ascorbic acid (mg/100 g)	Blood ascorbic acid (mg/100 ml)
Aqueous	3	10 41 ± 1 83	15 30 ± 1 62	0 92 ± 0 12
Vitreous	8	11.05 ± 1 61	17.78 ± 0 77	0 75 ± 0.09
Sclera	8	14 93 ± 1 69	20.13 ± 0 87	0.75 ± 0.09
Retina	6	23.60 ± 1 46	25.70 ± 1.33	0 74 ± 0 12
Choroid	6	26.55 ± 2 79	28.40 ± 1.64	0.86 ± 0 07
Ciliary body and iris	5	20.86 ± 1.65	34.26 ± 5.18	0 81 ± 0 10
Cornea	8	25 51 ± 2.42	39 49 ± 2.51	0.75 ± 0 09

Note: The figures given are the arithmetic means and the standard errors The ascorbic acid (TAA) concentrations of the aqueous and the vitreous humors of guinea pigs were found to be significantly reduced following complete removal of the lens, but in the presence of residual lens fibres, the ascorbate level was restored to normal (see Table 3)

From Heath, H , Beck, T. C., Rutter, A C., and Greaves, D. P. (1961), *Vision Res.*, 1, 274. ©1961 Pergamon Press, Ltd With permission.

Table 3

THE MEAN ASCORBIC ACID CONTENT OF TISSUES FROM PARTIALLY APHAKIC EYES AND NORMAL EYES FROM THE SAME ANIMALS

Tissue	Number of eyes	Partially aphakic eyes ascorbic acid (mg/100 g)	Control eyes ascorbic acid (mg/100 g)	Blood ascorbic acid (mg/100 ml)
Lens	10	25.96 ± 2.76	10.34 ± 0.46	0 85 ± 0.08
Aqueous	7	17.71 ± 4.25	16.39 ± 1.69	0.85 ± 0.11
Vitreous	10	17.01 ± 2.30	17.95 ± 1.83	0.85 ± 0.08
Sclera	10	19.48 ± 2.25	19.53 ± 1.83	0.85 ± 0.08
Retina	8	28.55 ± 1.53	30.31 ± 2.61	0.87 ± 0.10
Choroid	9	37.89 ± 5.51	36.71 ± 3.09	0.85 ± 0 09
Ciliary body and iris	9	34.04 ± 3.45	41.03 ± 3.56	0.85 ± 0 09
Cornea	10	33.29 ± 3.22	45.14 ± 2.67	0.85 ± 0.08

Note: The figures given are the arithmetic means and the standard errors. The ascorbic acid (TAA) concentrations of the aqueous and the vitreous humors of guinea pigs were found to be significantly reduced following complete removal of the lens (see Table 2), but in the presence of residual lens fibers, the ascorbate level was restored to normal.

From Heath, H., Beck, T. C., Rutter, A. C., and Greaves, D P. (1961), *Vision Res.*, 1, 274. ©1961 Pergamon Press, Ltd. With permission.

levels of ascorbic acid in the aqueous of aphakic eyes was attributed not to any lack of ability of ascorbic acid to enter the aphakic eye, but to the failure to maintain ascorbic acid in the more stable reduced form.''

Davson (1956, 1962) noted that the plasma ascorbic acid level is about 1 mg/100 ml in most animals, but the ascorbic acid level in the aqueous humor is between 15 and 30 mg/

Table 4

**THE DISTRIBUTION OF ASCORBIC
ACID IN THE TISSUES OF GUINEA
PIG, RABBIT, AND COW EYES**

	Ascorbic acid (mg/100 g)		
Tissue	Guinea pig	Rabbit	Cow
Lens	13.4	14.7	33.7
Aqueous	16.9	12.8	18.9
Vitreous	17.8	8.9	14.7
Sclera	19.5	11.1	5.1
Retina	28.2	19.8	21.7
Choroid	31.2	36.4	15.1[a]
Ciliary body	33.0	33.6	29.8
Iris		32.7	24.4
Cornea	43.7	34.4	30.3

Note: There are considerable differences in the way
ascorbic acid is distributed within the eye in
different species.

[a] This value is for choroid and tapetum combined

From Heath, H., Beck, T. C., Rutter, A. C., and
Greaves, D. P. (1961), *Vision Res.*, 1, 274. ©1961
Pergamon Press, Ltd. With permission.

100 ml in the rabbit, ox, horse, guinea pig, sheep, monkey, and man, between 5 and 15 mg/100 ml in the pig and the dog, between 1 and 5 mg/100 ml in the cat and the frog, less than 1 mg/100 ml in the galago, and close to 0 in the rat, so the aqueous plasma ascorbate ratio (R·aq/pl) varies widely. High levels of ascorbic acid have also been found in other ocular tissues, notably the ciliary body, the lens, the cornea, the vitreous body and the retina, but there are wide differences in the distribution of ascorbic acid within the eye in different species (Table 4).

Mitsukawa (1963) expressed the belief that under normal conditions the vitamin C of the aqueous humor comes mostly from the ciliary body, but under loaded conditions it comes also from the iris. The same worker (Mitsukawa, 1964) gave reasons for believing that there is an outflow route for ascorbic acid from the aqueous humor to the vortex vein through the vitreous.

Shimizu (1964) investigated the uptake of ^{14}C-labeled ascorbic acid by the ocular tissues of rabbits and guinea pigs; 15 to 30 min after intravenous administration of the tracer, absorption of ascorbic acid into the ciliary body reached a maximum which exceeded that of the liver. Ascorbic acid spread gradually from the ciliary body to the aqueous, the vitreous, and the lens.

Many workers have favored the concept of an active transfer of ascorbic acid from the plasma to the aqueous humor by the epithelial cells of the ciliary body. This may indeed be so, but another possibility is that ascorbic acid may be stored within the eye as a result of a low oxidation-reduction potential provided by reduced glutathione, especially in the lens. Ascorbic acid crosses cell membranes mainly in the oxidized form, as DHAA, which is lipid soluble, but it is stored mainly as ascorbic acid, which is less soluble in the lipoproteins of the cell membranes. Heath (1962) therefore suggested that lenticular glutathione may be responsible for the reduction of DHAA to ascorbic acid. Such a concept was supported by the work of Hughes and Maton (1968), studying vitamin C passage across the erythrocyte membrane. Ascorbic acid storage would then depend upon the ability of the lens to convert glutathione to its reduced form; this will depend on the presence of the enzyme glutathione reductase, the activity of which is dependent on the presence of FAD derived from dietary riboflavin and NADP derived from dietary nicotinic acid.

Davson (1956) cited evidence that ACTH causes a 15% increase in the ascorbic acid concentration in the aqueous humor. This suggests to the writer that cortisone may aid in the conversion of DHAA to ascorbic acid within the eye; indeed, this may be one of its most beneficial actions.

Studying ascorbic acid depletion, Hughes et al. (1971) found that after 14 d without ascorbic acid, the guinea pig lens retained 27% of its ascorbic acid, while the aqueous humor retained only 3% of its original level. This certainly does seem to suggest that there may be a special need to retain a high ascorbate level in the lens. However, according to the data of Hughes and Jones (1971), the ascorbic acid level in the lens of the guinea pig decreases with aging ($p < 0.01$), and this phenomenon may be due to a decreased efficiency of gastrointestinal absorption.

Buck and Zadunaisky (1975), as a result of their studies of the corneas of rabbits and frogs, concluded that ascorbic acid produces an increase in the cyclic AMP content of the corneal epithelium and other tissues by inhibition of $3':5'$-cyclic-AMP phosphodiesterase. They believed that this occurrence may be one of the main functions of ascorbic acid in ocular tissues.

Varma et al. (1979) chose to study the lens of the eye of the rat, which has a very low ascorbate content. Studying light-induced damage *in vitro*, they found that riboflavin alone did not provide protection, but riboflavin and ascorbic acid together did protect against light-induced superoxide damage to the lens. They suggested that the low ascorbate level of the eyes of the galago, the rat, and the cat may be related to their nocturnal habitat and that, likewise, the low ascorbate levels of other species before birth may also be related to decreased need, because of decreased exposure to light. They suggested that essential nutrient deficiencies in the lens may accelerate the development of cataracts.

Killey et al. (1980), studying rabbits, have reported data showing an aqueous/plasma ascorbate ratio of 40:1 and a vitreous/plasma ratio of 20:1.

V. VALUE OF ASCORBIC ACID AND BIOFLAVONOIDS* IN VARIOUS EYE DISEASES

A. Hemorrhagic Lesions

There is plenty of evidence that retinal hemorrhage is associated with skin "capillary fragility" in hypertensives, in diabetics, in nondiabetic nonhypertensives, and also in Eales' disease (recurrent hemorrhages into the retina and vitreous humor). Moreover, there is evidence that treatment with the bioflavonoid rutin, or with rutin and ascorbic acid corrects or improves the "capillary fragility" in the majority of such patients, as seen in Figure 1, which is based on work by Donegan and Thomas (1948). Objective evidence of improvement in the retinal picture is less frequent, especially in diabetics. Few investigators have felt justified in observing an untreated control group of people who are gradually losing their vision. However, Beardwood et al. (1948) did report an excellent controlled study. Among 321 consecutive clinic patients receiving treatment for diabetes mellitus, they found that 46% showed increased capillary fragility, and that 24% of the total number (77 subjects) showed an associated retinopathy. Another 46 patients with capillary fragility and retinitis were added from the private practice of the authors. Successive patients were given rutin, hesperidin, or placebo. The initial dose of rutin was 20 mg t.i.d., increased to 50 mg t.i.d. if the capillary fragility did not return to normal in 4 or 5 weeks. The dose of hesperidin

* The relationship between bioflavonoids and ascorbic acid metabolism is discussed in Chapter 11, Volume I.

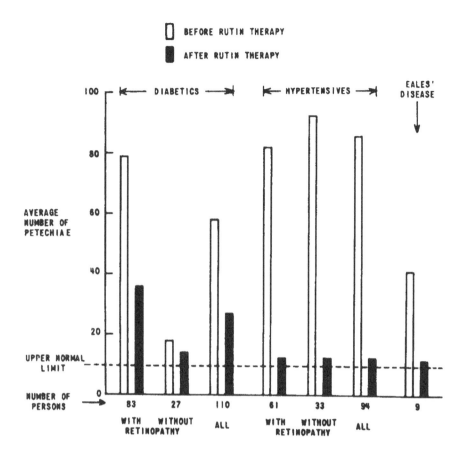

FIGURE 1.   Effect of rutin treatment on capillary fragility in diabetics and in hypertensives with and without retinopathy and in patients with Eales' disease Based on the work of Donegan and Thomas. (From Griffith, J. Q., Krewson, C. F., and Naghski, J. [1955], *Rutin and Related Flavonoids. Chemistry, Pharmacology, Clinical Applications,* Mack, Easton, PA. With permission.)

was 50 mg twice a day. After 6 months of treatment, a definite improvement in the retinal picture was observed in 25% of each of the two test groups and in only 11% of the control group receiving the placebo. The capillary strength returned to normal in 80% of the flavonoidtreated patients. In the other 20%, the capillary fragility was more refractory to treatment and, in some cases, remained abnormal at the end of 6 months of study (Table 5).

Beardwood et al. observed that Kimmelsteil-Wilson's disease, or intercapillary glomerular nephrosclerosis, characterized by diabetes mellitus, hypertension, and albuminuria, showed the most marked degree of capillary fragility of any of the cases studied.

In one case with death from uremia, the capillary fragility was so great that it was practically impossible to read the Göthlin's index.* The others had indices from 50 to 100. One of these patients is of particular significance . . . D.T., a girl of 23 years who had been a diabetic from the age of 12. Her diabetes had gone along under fairly satisfactory control. Two years before admission she rather suddenly developed many capillary haemorrhages in the retina and impaired vision. Her Göthlin index at that time was 50. She was placed on rutin and within a month her vision had improved; the hemorrhages had disappeared and there was no fresh bleeding. She left Philadelphia for a period of six months and stopped taking rutin. She was again seen after her return, at which time she had multiple extensive hemorrhages of the retina and her blood pressure was 160/90. The urine showed a cloud of albumin, and there was evidence of peripheral edema. The blood urea nitrogen was 35 mgms and her total proteins were 5.2 g. We felt that undoubtedly she was developing intercapillary glomerular nephrosclerosis.

* Göthlin's method of measuring the "capillary fragility" of the skin is described in Chapter 1, Volume II.

Table 5
ANALYSIS OF CASES STUDIED

	Rutin	Hesperidin	Placebo
Göthlin improved, no change in retina	20	22	0
Göthlin improved, improvement in retina	10	10	0
Göthlin improved, retina worse	5	5	0
Göthlin same, retinopathy same	5	3	32
Göthlin same, retinopathy improved	0	0	4
	40	40	36

Note: Diabetic patients with capillary fragility and retinopathy were treated with rutin, hesperidin, or placebo. The results of 6 months of treatment are summarized for comparison.

From Beardwood, J. T., Roberts, E., and Trueman, R. (1948), *Proc. Am. Diabetes Assoc.*, 8, 241. © American Diabetes Association, Inc. With permission.

She was given 50 mgms of rutin 4 times a day and after a period of two months her eye grounds had entirely returned to normal, her blood pressure was now 135/80, urinalysis showed only a faint trace of albumin and her blood chemistry has returned to normal.

Yudkin (1951) reported that:

Vitamin C is valuable in many cases of ocular disturbance where the vascular system has not been functioning normally because of some alteration in the blood stream or changes in the blood vessels. If the hemorrhagic retinopathy is associated with diabetes, the latter condition must be treated properly before favourable results can be expected.

Many of the hemorrhages appearing in the vitreous because of local arteriosclerosis and general hypertension are improved if large amounts of cevitamic acid* (300 mg to 500 mg) and rutin are given daily. Edema of the macular region produced by vascular decompensation often responds more rapidly when this amount of cevitamic acid and 10 to 32 ounces of orange or grapefruit juice are given daily for several days.

Griffith et al. (1955) in their book entitled "Rutin and related flavonoids" reviewed the extensive literature on this subject and their own experience with more than 3000 hypertensive patients. They reported that capillary fragility could usually be corrected and retinal hemorrhages could often be arrested in nondiabetics, but they found the retinal hemorrhage of diabetics to be much more resistant to treatment, perhaps because of the numerous microaneurysms which are so widespread in their retinal capillaries. Nevertheless, they felt rutin treatment to be worthwhile in diabetic retinopathy.

Loewe (1955) used a combination of vitamin C and citrus bioflavonoids in addition to "lipotropic factors" consisting of choline, inositol, methionine, vitamin B_{12}, and liver extract in the treatment of 50 patients with hemorrhagic retinopathy and 25 with vitreous opacities. He obtained excellent results, with rapid absorption of hemorrhages and prevention of new ones. Surprisingly, the best results were reported in diabetic retinitis, but it is impossible to know which of the many dietary supplements was most effective in achieving these results.

Hamilton (1958), of Hobart, Tasmania, reported four cases of intraocular hemorrhage which were diagnosed as being due to scurvy and which responded well to ascorbic acid therapy.

A male, aged 41 years, a returned serviceman, who never ate fresh fruit or uncooked vegetables owing to dyspepsia, developed right peripheral retinal haemorrhages and left gross vitreous haemorrhage. He was referred back to his country general practitioner with a diagnosis of scurvy.

* An old name for vitamin C.

Table 6
BLOOD ASCORBIC ACID LEVELS IN PATIENTS SUFFERING FROM HEMORRHAGIC EYE DISEASE AND IN HEALTHY SUBJECTS

No. of pts.	Diagnosis	Blood ascorbic acid (μg/100 ml; mean $\pm$ SEM)
30	Hemorrhagic clouding of the vitreous	560 $\pm$ 55[a]
15	Central-vein thrombosis	460 $\pm$ 40[a]
10	Apoplectic glaucoma	435 $\pm$ 25[a]
45	Healthy	980 $\pm$ 70

[a] $p < 0.001$ with respect to control group.

From Greco, A. M., Fioretti, F., and Rimo, A. (1980), *Acta Vitaminol. Enzymol.*, 2, 21. With permission

A male, aged 61 years, a widower and batching, had right retinitis and left gross vitreous haemorrhage. He was returned to his country general practitioner with a provisional diagnosis of scurvy. He was treated with 200 mg of ascorbic acid daily, with marked improvement in right and left vision to 6/9 and 6/24 with correction.

A female, aged 57 years, with dyspepsia, had loss of vision in the left eye for 10 months, treated with Nyal's eye drops. She had complete left vitreous haemorrhage, with vision reduced to hand movements. The physician's report indicated long abstinence from essential food factors owing to dyspepsia. Ascorbic acid, 300 mg, was given by mouth, and four hours later she passed urine containing less than 5 mg of ascorbic acid. Nine years later the left vision stood at 6/6 with correction.

A female, aged 54 years, had almost complete right vitreous haemorrhage with less than 6/60 vision with correction, and she had one patch of retinitis proliferans. Six years later she had right retinal detachment. A request was made to her physician regarding vitamin C deficiency. He reported a mild degree of vitamin C deficiency. Vitamin C therapy was then commenced, but was obviously too late.

Mann (1974) has suggested that DHAA may be transported into tissues by an insulin-dependent mechanism and that the high glucose levels of diabetes mellitus may impede ascorbic acid transport by competitive inhibition. Impairment of insulin function, whether by its absence, as in juvenile diabetes, or by its inhibition, as in adult onset diabetes, would thus lead to impaired transport of vitamin C. He pointed out that the hallmark of chronic diabetes is vascular disease and that this disease might be slowed by administration of both insulin and ascorbic acid. The decreased ascorbic acid levels found in the platelets of insulin-dependent diabetics by Sarji et al. (1979) may possibly play a role in the tendency of diabetics to develop venous thrombosis.

Greco et al. (1980) reported a study of the blood ascorbic acid levels in 55 patients with hemorrhagic ocular diseases and in 45 healthy control subjects between the ages of 20 and 60 years; 30 patients had hemorrhagic clouding of the vitreous, 15 had central vein thrombosis, and 10 had glaucoma. Table 6 shows that in all three of these conditions the mean blood ascorbic acid levels were significantly lower than normal ($p < 0.001$). In patients with low levels of ascorbic acid, daily vitamin C supplements caused a progressive amelioration of the condition and a clear regression of the hemorrhagic spots. Most of the patients affected by these ocular pathologies said that their diet was poor in vegetables and fruits. Some of them stated that they never ate these fresh foods.

The work of Verlangieri and Sestito (1981) is certainly pertinent to the development of diabetic microangiopathy, for these workers demonstrated that insulin promotes the uptake of ascorbic acid by endothelial cells and that hyperglycemia impairs the uptake of ascorbic acid by these cells (Chapter 2, Volume II).

Studies of experimental subretinal hemorrhage in rabbits by Glatt and Machemer (1982) have led them to, "speculate that subretinal hemorrhage leads to iron toxicity of the retina which leads to the histologic pattern of senile macular degeneration." Clearly the accumulation of hemosiderin beneath the retina following a hemorrhage could cause a local loss of ascorbic acid, just as hemosiderosis causes scurvy in the Bantu (Chapter 10, Volume I). The absence of lymphatics from the retina undoubtedly impairs the rate of clearance of hemosiderin from this tissue. Moreover, oxidative local loss of ascorbic acid due to the hemosiderin may well be a factor in the often-made observation that when a retinal hemorrhage has once occurred, it tends to recur, and often in the same general retinal area. No wonder the response to rutin treatment is slow, if this heavy metal chelating agent has to remove the iron from the retinal tissues before a normal ascorbate level and a normal AA/DHAA ratio can be restored.

B. Eye Infections

Lyle and McLean (1941) found the intravenous administration of ascorbic acid to be of definite value in the treatment of inflammatory conditions of the cornea. Summers (1946), writing on the treatment of hypopyon ulcers or corneal ulcers associated with pus in the anterior chamber, concluded that penicillin heals the ulcers, but does not much affect the hypopyon. He observed that penicillin and vitamin C together produced a much quicker resolution than penicillin alone or vitamin C alone. Yudkin (1951) observed that many cases of chronic infection of the cornea improved when large doses of vitamin C were administered. Ogielska (1963) found that bacterial iritis produced experimentally in rabbits was associated with a decrease in the ascorbate content of the vitreous humor. It was thought that this was due to damage of the ciliary body by the inflammation.

Foster and Goetzl (1978) provided successful treatment for a *Candida albicans* corneal ulcer in a 24-year-old man by the use of topical amphotericin B and 12 g of oral 5-fluorocytosine daily for 30 d. However, he warranted further investigation because of recurrent severe infections. He was found to have a phagocytic dysfunction known as the Hill-Quie syndrome, involving impaired neutrophil and monocyte chemotaxis; the latter responded well to ascorbic acid supplementation.

The syndrome consists of atopy, hyperimmunoglobulinemia E, variable blood eosinophilia, and defective polymorphonuclear neutrophil chemotaxis, resulting in recurrent severe bacterial infections in different locations. This man's skin was markedly lichenified over the face, neck and flexural surfaces and there were multiple pustules on the face and neck as well as a superficial punctate keratitis. His past history included eczema and mild asthma since childhood, as well as the following infections: persistent staphylococcal pyoderma and carbuncles, presumed scrofula, *Staphylococcus aureus* pericarditis; *S. aureus* perinephric abscess with retroperotineal psoas and hip abscess, mixed bacterial pneumonia, *S. aureus* subcutaneous and joint abscess, and a *Proteus mirabilis* urinary tract infection. He did indeed have hyperimmunoglobulinemia and eosinophilia.

Leukocyte migration studies, performed after the patient had been on a regimen of oral ascorbic acid, 3 g daily for 30 d (which raised the serum ascorbate level more than four fold), revealed considerable improvement in spontaneous and chemotactic migration of monocytes. The polymorphonuclear leukocyte migration values, however, were unchanged after the course of ascorbic acid. Phagocytosis by monocytes and by polymorphs was normal in this patient before and after ascorbate therapy.

Foster and Goetzl (1978) pointed out that, "chemotactic peptides and histamine, liberated by immediate hypersensitivity reactions, are capable of suppressing leucocyte chemotaxis at very low concentrations, as are other products of inflammatory response."

It may have been the ability of ascorbic acid to decrease blood histamine levels (Chapter 1 of this volume) that improved this man's monocyte chemotaxis. He was reported to have remained well for a year without further evidence of infection at the time of publication.

C. Wounds of the Eye

The cornea is composed largely of collagen, and it is normally rich in ascorbic acid, so one would expect this vitamin to be important in the healing of corneal wounds. Galloway et al. (1948) studied the rate of healing of very superficial injuries made with a dental burr in the corneas of guinea pigs under local anesthesia and found no delay of healing when the animals were maintained on a subscorbutic diet. However, these were superficial wounds involving only the epithelium, which seems to heal by proliferation and spilling over of epithelial cells into the denuded area; no new collagen was required. Deeper wounds penetrating Bowman's membrane and involving the collagen fibers of the cornea do require adequate supplies of ascorbic acid for normal healing.

Studying experimental nonperforating incised wounds of the corean in guinea pigs after 10 d on a scorbutogenic diet, Barber and Nothacker (1952) observed normal regeneration of the corneal epithelium, but 18 d later when the animals had scurvy, the ground substance was found to be scanty and the collagen fibers of the corneal lamellae were immature and poorly oriented in relation to the surface of the cornea.

Geever and Levenson (1960) showed that vitamin C deficiency caused a greater impairment of collagen deposition in polyvinyl sponge implants in the abdominal wall than in the anterior chamber of the eye, presumably because the eye holds on to ascorbic acid longer. Nevertheless, Sabatine et al. (1961), in a similar experiment, observed that hydroxyproline deposition in polyvinyl sponges in the anterior chamber of the guinea pig eye was reduced to 43% of normal after 16 d on an ascorbic acid-deficient diet; collagen deposition was also markedly impaired.

D. Corneal Burns

Campbell et al. (1950) studied standard superficial and deep heat injuries of the cornea in guinea pigs on high and low ascorbic acid intakes, (20 mg daily and 0.5 mg on alternate days). Using a standard sodium fluorescein solution to stain the lesions, it was found that the healing of superficial burns, confined to the corneal epithelium, was not impaired by ascorbic acid deficiency. On the other hand, deeper lesions involving the substantia propria of the cornea healed more slowly. Moreover, the healing lesions were structurally weaker in the ascorbic acid-deficient animals. Epithelial covering of the deeper wounds was delayed in the prescorbutic animals for the want of a fibrous tissue substratum.

Further studies of the healing of deep corneal heat injuries in guinea pigs on high and low ascorbic acid intakes were reported by Campbell and Ferguson (1950). They used a 2% pontocaine hydrochloride solution to anesthetize the eye, and inflicted standard deep burns 1 mm wide by means of an electrocautery needle. There was a highly significant increase in the time taken for complete healing in the guinea pigs with protracted moderate scurvy ($p < 0.01$), as shown in Table 7.

Boyd (1955) studied the rate of healing of small standardized corneal burn injuries inflicted on the normal right and aphakic left eyes of guinea pigs by touching with a platinum-wire electrocautery needle for 1 s under cocaine anesthesia. The animals were receiving a known high intake of ascorbic acid, 20 mg daily, but the mean ascorbic acid level in the cornea of the aphakic left eye was significantly lower than that in the cornea of the intact right eye (Table 1). The time taken for complete healing was significantly less in the normal eye with the higher ascorbate level (Table 8).

E. Alkali Burns

Severe alkali burns of the cornea frequently result in corneal ulceration or even perforation, but much evidence has now accumulated showing that systemic or topical ascorbic acid treatment can significantly reduce the incidence of these serious complications.

Levinson et al. (1976), studying rabbits, observed depression of the ascorbate content of

Table 7
TIME FOR EPITHELIAL HEALING OF DEEP CORNEAL HEAT INJURIES IN GUINEA PIGS

No. of wounds	Mean time of healing (h)	Differences between controls and deficient animals (h)
32 control	94.0 ($\pm$ 6.02)[a]	31.8
32 scorbutic	125 8 ($\pm$ 7.02)	(highly significant)

Note. $t = 3.43$; $p < 0.01$. The healing of deep corneal burn injuries was significantly retarded in guinea pigs with protracted moderate scurvy.

[a]   SEM

From Campbell, F. W. and Ferguson, I. D. (1950), *Br. J. Ophthalmol.*, 34, 329 With permission.

Table 8
COMPARISON OF HEALING IN APHAKIC AND CONTROL CORNEAE

Assessment of healing process	Aphakic eyes		Control eyes		Probability of chance occurrence
	Mean	No. of eyes	Mean	No. of eyes	
Time (days)	17.3 ($\pm$0 84)	20	10.8 ($\pm$2.2)	19	$p < 0.001$
Relapses	6.1 ($\pm$1.94)	19	2.8 ($\pm$0.53)	19	$p < 0.001$
Relapse frequency	3.9 ($\pm$0.39)	19	2.7 ($\pm$0.44)	19	$p < 0.05$
No. of vascularized sectors	11.3 ($\pm$2.15)	20	6.7 ($\pm$1.56)	19	$p < 0.1$

Studies of the rate of healing of small electrocautery burns in the corneas of guinea pigs revealed that healing was significantly slower in aphakic eyes which had lower ascorbic acid levels (see Table 1).

[a]   Figures in parentheses indicate SEM.

From Boyd, T. A. S (1955), *Br. J. Ophthalmol.*, 39, 204. With permission.

the aqueous humor following alkali burns. After a 20-s, 12-mm-diameter, 1 *N* sodium hydroxide burn, aqueous humor glucose levels were depressed, but soon returned to normal; ascorbic acid levels, however, remained significantly depressed for up to 30 d. These corneas became markedly ulcerated in about 60% of the animals and frequently perforated. However, similar alkali burns in rabbits treated with ascorbic acid, 1.5 g daily, by subcutaneous injection, rarely developed corneal ulcerations and the corneas did not perforate.

Pfister et al. (1980) in their first experiment found no significant difference in the number of perforations between rabbits receiving ascorbic acid subcutaneously and untreated controls, but in a second experiment they did observe a reduction of ulcerations and perforations from 80 to 30% in rabbits receiving 10% ascorbic acid topically.

Nirankari et al. (1981), also studying alkali burns of rabbit corneas, found that treatment with ascorbic acid or with the enzyme superoxide dismutase by subconjunctival injection provided significant protection against serious complications, as shown in Table 9. These authors concluded that both ascorbic acid and superoxide dismutase were beneficial because they acted as scavengers of the superoxide radical (O_2^-). However, subconjunctival injections of glutathione did not have the beneficial effect that was anticipated. This would be understandable if glutathione tended to draw ascorbic acid out of the cornea into the subconjunctival space at the site of its injection.

Table 9
CLINICAL OBSERVATIONS OF EXPERIMENT 1 (DAY 30)

	Normal saline (control) (10 eyes)	% Incidence	Ascorbic acid treatment (12 eyes)	% Incidence	Glutathione treatment (10 eyes)	% Incidence	Superoxide dismutase treatment (9 eyes)	% Incidence
Ulcers	6		3		4		2	
Descemetoceles	0	90	0	33	0	80	1	33
Perforations	3		1		4		0	
No Ulcers	1	10	8	67	2	20	6	67

Note: Difference in number of ulcers, descemetoceles, and perforations (combined) between control (saline-treated) group and ascorbic acid- and superoxide dismutase-treated group is statistically significant ($p < .05$). The corneas of anesthetized rabbits were exposed to 1 N sodium hydroxide over a circular area with a diameter of 11 mm for 20 s. The eyes were then washed and treated by daily subconjunctival injection of either saline or ascorbic acid or glutathione or superoxide dismutase. Both ascorbic acid and superoxide dismutase treatment reduced the incidence of complications such as ulceration and perforation of the cornea.

From Nirankari, V. S., Varma, S. D., Lakhanpal, V., and Richards, R. D. (1981), *Arch. Ophthalmol.*, 99, 886. ©1981 American Medical Association. With permission.

Table 10
RELATIONSHIP OF MEAN HEALING TIME TO DEPTH OF CORNEAL ULCERS

Mean healing time (d)

Depth of infiltration	Principals (receiving 1.5 g additional ascorbic acid daily)	No. of cases	Controls (no additional ascorbic acid)	No. of cases
Superficial	A 3.63 ($\pm$0.54)[a]	11	B 3.80 ($\pm$0.50)	15
Deep	C 4.36 ($\pm$0 40)	11	D 6 15 ($\pm$0.50)	13

Note: Deep corneal ulcers healed significantly more quickly in people receiving an ascorbic acid supplement of 500 mg three times a day by mouth than in control patients receiving placebo tablets.

[a] SEM

From Boyd, T. A. S. and Campbell, F. W. (1950), *Br. Med. J.*, 2, 1145. With permission.

F. Corneal Ulcers

Boyd and Campbell (1950) studied the rate of healing of small acute corneal ulcers in 51 patients presenting at the Glasgow Eye Infirmary. The patients received either penicillin or sulfacetamide eye drops and either ascorbic acid, 500 mg, or placebo tablets, three times a day by mouth. There was no significant difference in the rates of healing between the penicillin- and the sulfacetamide-treated patients. The administration of large dosee of ascorbic acid had no significant effect upon the healing time of superficial ulcers, but significantly accelerated healing of deep ulcers (Table 10). Pyorrhea was found significantly more frequently in patients with corneal ulcers than in others with noninfective eye complaints, suggesting possible transmission of infection from the mouth to the eye and/or decreased resistance to infection due to ascorbate need.

G. Cataracts

The normal lens has a plentiful supply of ascorbic acid within its fibers, but studies by von Euler and Martius (1933) and others revealed that both the ascorbic acid and glutathione levels are markedly diminished in cataractous lenses. Moreover, Josephson (1935) reported that the administration of vitamin C and glutathione to patients with cataracts due to dinitrophenol toxicity resulted in a definite improvement of visual acuity. Monjukowa and Fradkin (1935) observed that opacity of the lens developed in a small percentage of guinea pigs fed a vitamin C-deficient diet. Bellows (1936a) reported lower mean plasma ascorbic acid levels in 20 patients with cataracts (0.61 mg/100 ml) than in 20 normal subjects (1.02 mg/100 ml) and also observed that larger quantities of ascorbic acid were required to raise the plasma ascorbate levels of the cataractous patients. These figures are not impressive, and the difference may have been due to the somewhat older age of the patients with cataracts (68 years) than the normal controls (52 years), but they were not inconsistent with the concept of an abnormality of ascorbate metabolism in patients with cataracts.

Hawley and Pearson (1938) also found the cataractous lens to have a low ascorbic acid content, but they concluded that this was most probably the result rather than the underlying cause of the cataractous changes. Bouton (1939) studied patients and staff at the Hastings State Hospital in Detroit and found 12 patients with severely impaired vision to be among

Table 11
MANIFEST READINGS OF VISUAL ACUITY FOR PATIENTS BEFORE, DURING, AND AFTER TREATMENT

Case no.	Patient	Sex	Age (years)	March 15	April 19 or later	May 31
4	M Ha.	F	56	O D 20/70	O.D. 20/40	O D 20/40 + 2
				O.S 20/70	O.S. 20/40	O.S 20/40
6	G.B.	M	54	O D. 20/00	O.D 20/00	O.D. 20/200
				O.S. 20/70	O.S. 20/50	O.S. 20/50
7	E.Bo.	F	41	O.D. 20/70	O.D. 20/40 − 1	O.D. 20/30 − 2
				O.S. 20/50 − 1	O.S. 20/40 − 2	O.S 20/30 − 2
10	S.B	F	56	O D 20/70 + 2	O.D. 20/40 − 2	O.D. 20/50 + 1
				O S. 20/00	O.S. 20/200 + 1	O.S 20/100 + 1
11	H.J.	M	64	O.D. 20/200	O.D. 20/100 + 1	O.D. 20/70
				O.S. 20/200	O.S. 20/200 + 2	O.S. 20/70 − 1
12	J.C.	M	55	O.D. ?	O.D. 20/40 + 2	O D. 20/30 − 1
				O.S. ?	P.S. 20/30 − 2	O.S. 20/30

Note: Showing the improvement in visual acuity that occurred in 6 out of 12 elderly patients with defective vision at the Hastings State Hospital in Detroit following dietary supplementation with ascorbic acid, 350 mg daily Those whose main problem was clouding of the vitreous showed the greatest improvement, and those whose main problem was cataract showed no improvement of vision.

From Bouton, S. M. (1939), *Arch Intern. Med.*, 63, 930 ©1939 American Medical Association. With permission.

the most deficient, as regards ascorbic acid saturation. Six of them had early, moderately advanced, or mature cataracts and six had clouding of the vitreous. All were treated with ascorbic acid, 350 mg daily, for 2 to 4 weeks. There was a marked improvement in the visual acuity of those with vitreous clouding (Table 11), but those with mature cataracts showed no improvement as a result of this treatment. The benefits were noted as being associated with clearing of opacities of the vitreous, possibly as a result of decreased capillary fragility. Associated with these objectively determinable changes was subjective improvement of vision, expressed by such statements as these "My eyes are stronger"; "I can do my needle work without my glasses now"; "I can read much easier"; "Things seem so much clearer than they did." Moreover, both objective and subjective improvement set in rapidly when it occurred at all, and in most cases did not progress much after 2 weeks of daily treatment.

Gapeev (1941) presented the vitamin C content of the aqueous and the lens of the normal eye as 7.7 to 15.8 mg/100 ml and those of cataractous eyes as 2.8 to 7.4 mg/100 ml. Likewise, the normal blood ascorbic acid was 0.3 to 1.2 mg/100 ml and only 0.08 to 0.46 in cataractous patients. Similarly, Lindahl (1941) wrote that senile cataract was associated with deficiencies of vitamin C and glutathione.

Atkinson (1952) mainly treating share croppers in the corn belt of Texas, reported his exprience with 450 patients who had incipient cataract; he expressed his belief that these cataracts were preventable with proper diet. He advised an increased intake of the green tops of vegetables as well as milk, eggs, and a high fluid intake; most of the incipient cataracts were arrested. Ogino et al. (1954) reported finding a disturbance of ascorbic acid metabolism in cataractous patients. Purcell et al. (1954) found the average ascorbic acid (AA) concentration in the aqueous humor of seven patients with cataract (12.1 mg/100 ml) to be less than that in three patients with normal lenses (18.0 mg/100 ml). However, because the concentration of the vitamin in the serum was also less in the group with cataracts, the ratio of the concentration in the aqueous to that in the serum (R·aq/ser) was almost the same in the two groups — mean 19.2 and 21.2, respectively. They concluded that the aqueous (and the lens) had not lost the ability to concentrate ascorbic acid and stated that, "the therapeutic administration of vitamin C to patients with cataracts would appear to be irrational." Unfortunately, these workers did not include analyses of any lenses in their study.

Maurice (1962) expressed the opinion that, "scurvy has never been shown to cause cataract," but Mann (1974) was still of the opinion that ascorbic acid deficiency plays an important role in cataract formation, for he wrote, "All these relationships suggest that human cataracts may be a late consequence of a marginal intake of ascorbic acid." Lee et al. (1977) reported the mean value of the ascorbate concentration in the aqueous humor from the eyes of patients with senile cataract to be 11.55 ± 3.01 mg/100 ml, ranging from 9.1 to 15.4 mg/100 ml.

It is hard to know whether the low ascorbic acid levels in cataractous lenses are a potential cause, a result, or simply an accompaniment of the lenticular disease.

Studies of rat eye lenses in organ culture, by Varma et al. (1984) have lent support to the hypothesis that cataract development is promoted by the photocatalytic production of superoxide $(O_2^-\cdot)$ and other oxygen radicals. These workers also provided evidence that this oxidative damage can be retarded by ascorbic acid and also by vitamin E.

Chandra et al (1986) measured the blood, aqueous humor, and lens ascorbic acid (TAA) levels in both cortical and nuclear cataracts at the time of surgical removal. The ascorbic acid level of the aqueous was found to be significantly lower in association with cortical than with nuclear cataracts (14.0 vs. 28.3 mg/100 ml; $p < 0.001$), but there was no significant difference between the blood ascorbate or the lens ascorbate levels of the two varieties of cataract. These workers suggest that a sluggish transport of ascorbic acid from the blood to the aqueous humor may contribute to the development of cortical cataracts.

Varma (1987), Lohmann (1987), and Blondin (1987) all agree that antioxidants, such as ascorbic acid and α-tocopherol, protect lens structural proteins against photooxidation and may therefore be useful in delaying the development of cateracts. However, it may well be that the intraocular concentrations of reduced glutathione and other sulfhydrils may be more important than the plasma ascorbic acid level in controlling the redox potential and the ascorbic acid content of the lens.

Clearly, cataracts develop slowly over many years, so long-term studies will be needed to elucidate the question of their relationship to ascorbic acid levels and ascorbic acid metabolism.

H. Glaucoma

Kronfeld (1942) studied the ascorbic acid concentration in the vitreous humor from eyes with wide-angle glaucoma and found it to be very variable, ranging from normal to unusually low, but without any evident relationship to the intraocular pressure or the stage of the disease.

It was the works of Virno et al. (1966) in Rome that first showed the value of intravenous sodium ascorbate in the emergency treatment of acute glaucoma; they found it to be even more effective as an osmotic agent than mannitol. In a dosage of 0.4 to 1.0 g/kg body weight intravenously, administered as a 20% solution at pH 7.2 to 7.4, vitamin C (sodium salt) induced marked ocular hypotony in approximately 60 to 90 min. Pain and decreased visual acuity were associated with intraocular hypertension in one man, and both symptoms were relieved in a few minutes; his ocular pressure returned to normal within 1 h.

Subsequently, Virno et al. (1967) discovered that, both in man and in experimental animals, the oral administration of 20% ascorbic acid produces a significant reduction of the intraocular pressure in normal and glaucomatous eyes. The usual dose was 0.5 g/kg of body weight.

Table 12
CORNEA AND INTRAOCULAR PRESSURE IN NORMAL RABBITS AND IN RABBITS WITH HEREDITARY BUPHTHALMIA

Group	Strain (genotype)	Phenotype	Cornea Diameter (mm)	Cornea Clouding[a]	Intraocular pressure (mmHg)
1	AX/J (+/+)	Normal	13.8 ± 0.2[b]	0.0 ± 0.0	21.5 ± 2.8
2	AXBU/J (bu/bu)	Subclinical glaucoma	14.8 ± 0.4	0.1 ± 0.4	22.4 ± 2.7
3	AXBU/J (bu/bu)	Mild glaucoma	16 7 ± 0.6	0 7 ± 0.2	23.2 ± 2.0
4	AXBU/J (bu/bu)	Severe glaucoma	18.6 ± 0.6	2 8 ± 0 3	17 0 ± 4 0

[a] Clouding based on a scale of 0 (clear) to 4 (opaque).
[b] Mean ± SE.

From Lam, K -W., Lee, P.-F., and Fox, R. (1976), *Arch. Ophthalmol.*, 94, 1565. ©1976 American Medical Association. With permission

The hypotensive action was most marked with chronic simple glaucoma. The authors were able to normalize the ocular pressure in some patients who could not be controlled with miotics and carbonic anhydrase inhibitors.

Studies on 25 patients by Linner (1969) in Sweden confirmed that ascorbic acid administered orally or topically produced a fall in intraocular pressure, but it amounted to only 2 mm Hg about 2 d after administration. The fall was attributed to a reduction in the rate of aqueous flow, although bulk drainage by posterior uveoscleral routes could not be excluded. Hood et al. (1970) found no change in the intraocular pressure in any of five volunteers subjected to complete deprivation of ascorbic acid for 84 to 97 d, nor in three of them who continued on an inadequate diet of 2.5 to 4 mg/d until they developed Sjögren's syndrome and obvious signs of scurvy after 213 d.

Lam et al. (1976) used liquid chromatography for estimation of ascorbic acid in very small samples of fluid aspirated from the anterior chamber of the eyes of rabbits. Studying normal rabbits, they found the ascorbate level in the aqueous humor to be 22.5 ± 2.3 mg/100 ml, but in fluid from a mutant strain of buphthalmic rabbits, the ascorbate level was lower (11.7 ± 1.7 mg/100 ml), even before they developed any meaningful clinical signs of glaucoma, and much lower ascorbate values were obtained in the aqueous humor of those with mild (8.9 ± 3.8 mg/100 ml) and severe glaucoma (1.9 ± 0.4 mg/100 ml), as shown in Tables 12 and 13. Moreover, Fox et al. (1977) observed a rapid drop ($p < 0.01$) in the intraocular pressure of both normal and buphthalmic rabbits following the intravenous administration of a 20% solution of ascorbic acid (1.5 g/kg), buffered to pH 7.2 to 7.4. In normal rabbits the intraocular pressure fell from 23.8 to 13.1 mmHg in 10 min and rose again to 18.3 mmHg after 60 min. The response in buphthalmic rabbits was slower, but more prolonged, falling from 28.4 to 16.3 mmHg at 60 min and being maintained at 19.8 mmHg after 240 min. Serum and aqueous ascorbate analyses clearly demonstrated that the slower response of the buphthalmic rabbits was associated with a much slower loss of ascorbic acid from the serum and a much slower transfer of ascorbic acid to the aqueous humor.

Further work on these buphthalmic rabbits by Lee et al. (1978) demonstrated that the low ascorbic acid content of the aqueous is associated with a reduction of secretion by the ciliary processes and decreased outflow facility. Fox et al. (1982) reported that the low ascorbate concentration of the aqueous in buphthalmic rabbits is also found in the lens and the cornea, which derive their ascorbate from the aqueous, but the ascorbate concentrations in other tissues of the body, including the adrenals, pituitary, spleen, ciliary body, iris, and retina are normal. It appears that the ciliary body-iris has not lost the abilty to concentrate ascorbic

Table 13
ASCORBATE CONCENTRATION IN AQUEOUS HUMOR AND SERUM

		Ascorbate concentration (mg/100 ml)			
		Aqueous humor		Serum	
Group	Phenotype	Average ± SE	Range	Average ± SE	Range
1	Normal	25 5 ± 2.3	18.3—28.3	0.45 ± 0.80	0.20—0.58
2	Subclinical glaucoma	11.7 ± 1.7	7.8—13.3	0.30 ± 0.08	0.20—0.50
3	Mild glaucoma	8.9 ± 1.7	3.2—19.4	0.55 ± 0.90	0.30—0.70
4	Severe glaucoma	1.9 ± 0.4	1.1—2.7	0.34 ± 0.90	0.10—0.50

Note: Mutant rabbits with hereditary buphthalmia have been found to have lower than normal ascorbic acid levels in the aqueous humor, which fall lower and lower as the animals develop glaucoma, even though the serum ascorbic acid level remains relatively unchanged.

From Lam, K.-W., Lee, P.-F., and Fox, R. (1976), *Arch. Ophthalmol.*, 94, 1565. ©1976 American Medical Association. With permission.

acid within itself, but has an impairment of its ability to secrete ascorbic acid. The authors suggest that a defect in the ability of the ciliary body to oxidize ascorbic acid may be the primary fault preventing the secretion of ascorbic acid.

Lee et al. (1977) studied the ascorbic acid content of the aqueous humor aspirated from the anterior chambers of the eyes of 35 patients at the time of surgery for open-angle glaucoma and found them to be normal (22.4 ± 12.9 mg/100 ml) with a broad range varying from 1.5 to 62.0 mg/100 ml. They concluded that there is no evidence of ascorbic acid deficiency and no indication for ascorbic acid treatment in open-angle glaucoma.

However, when Lin et al. (1982) studied 14 patients with neovascular glaucoma, 14 with primary open-angle glaucoma, and 14 with angle closure glaucoma, they found lower than normal aqueous ascorbate levels in those with neovascular glaucoma. Further subdividing the neovascular glaucoma into eight patients with diabetic retinopathy and six with central retinal vein occlusion, it became evident that the neovascular glaucoma of diabetes mellitus was associated with a significantly lower than normal aqueous ascorbate level of 10.6 mg/100 ml vs. 20.9 in open-angle and 26.9 in angle closure glaucoma. There was a higher than normal aqueous protein level in the diabetics, so it was suggested that protein leakage from the blood vessels into the aqueous and ascorbate leakage out of the aqueous into the blood might account for the high protein and low ascorbate levels in the aqueous humor of diabetics. However, one must also consider the fact that ascorbate uptake by leukocytes has been shown to be impaired in diabetics (Chapter 2, Volume II), so the ascorbate levels may be lower than normal in many tissues of diabetics.

I. Toxic Retinopathy

Young et al. (1968) have studied chloroquine retinopathy, which is usually considered to be irreversible; "the scotomas and retinal lesions usually progress for several months after diagnosis and chloroquine withdrawal." But, in the exprience of these workers, careful routine testing, prompt withdrawal of chloroquine, and the administration of ascorbic acid and pyridoxine will usually protect the vision of these patients. They gave ascorbic acid, 4000 mg, and pyridoxine, 800 mg, daily for years and reported very encouraging results.

J. Retinal Detachment

Discussing the possible role of ascorbic acid in retinal detachment, Heath (1962) made the following observations. "The structure of the vitreous body depends upon both the integrity of the meshwork of collagen fibrils and on the maintenance of the polymeric state of the mucopolysaccharide, hyaluronic acid. Disorganization of the vitreous body with the formation of fibrous bands is often associated with retinal detachment. To what extent this is related to abnormal ascorbic acid metabolism is not known. Christiansson (1957) has shown that depolymerization of vitreal hyaluronic acid occurs in scorbutic guinea-pigs and the viscosity of the hyaluronic acid returns to normal on treating the animals with ascorbic acid.''

As mentioned earlier, Hamilton (1958) reported the successful treatment of vitreous hemorrhage in a 54-year-old woman by the use of ascorbic acid. When he saw her again 6 years later with retinal detachment in the same eye, he wondered whether expulsive hemorrhages from scurvy could be the cause of some retinal detachments.

However, Wichard et al. (1982) have reported very high ascorbate levels (27.4 ± 2.1 mg/dl) in the subretinal fluid of people with rhegmatogenous retinal detachment. They gave reasons for the belief that the aqueous humor flows backwards through the retinal hole into the subretinal space, causing a loss of both pressure and ascorbic acid concentration in the anterior chamber of the eye.

K. Osteogenesis Imperfecta

It has already been mentioned in Chapter 5, Volume II, that high dose ascorbic acid supplements have been found useful in the treatment of osteogenesis imperfecta. Kurz and Eyring (1974) observed that ascorbic acid treatment (25 to 50 mg/kg/d) caused a decreased tendency to bone fracture and, hence, allowed much greater physical activity by these children. Pale blue sclera are typical of this disease, probably as a result of defective collagen formation, but it was observed by these authors and by the parents of some of the children that the sclera seemed to have whitened following high-dose ascorbic acid treatment. Of course, the increased bone strength is of paramount importance, but the apparent whitening of the sclera is certainly of interest as regards the nature of the defect and can perhaps be quantified in future studies.

L. Ocular Palsies

Sir Stewart Duke-Elder, in his seven-volume *Textbook of Ophthalmology* (1932 to 1954), reported that oculomotor and unilateral and bilateral abducens paresis have been reported in scurvy. This finding was confirmed by Hamilton (1958) who reported seeing in consultation at Hobart-on-the-Derwent in 1954, a lady who subsisted on tea, toast, biscuits, and pies. She had developed a sixth nerve palsy, and tests showed her to be vitamin C deficient; this ocular palsy was completely cured by 2 weeks of treatment with ascorbic acid.

VI. SUMMARY

Ascorbic acid is usually plentiful in the lens, the cornea, the sclera, the aqueous, and the vitreous humors. Along with other vitamins, it is undoubtedly important in maintaining the integrity of these tissues.

Proptosis of the eyeball due to a postorbital hemorrhage, ocular palsies due to subperiosteal hemorrhage, and even blindness due to retinal hemorrhage have all been described in scurvy, but lesser degrees of ascorbic acid deficiency are much more common. Ascorbic acid supplements accelerate the healing of deep corneal wounds, ulcers, and thermal or caustic burns and also increase resistance to infection.

Ascorbic acid and bioflavonoids have been found to be beneficial in the treatment of retinal hemorrhages associated with capillary fragility. The role of prolonged subclinical ascorbic acid deficiency or abnormal ascorbate metabolism in the etiology of cataracts remains unknown. Likewise, the roles, if any, of ascorbic acid deficiency in the development of glaucoma or retinal detachment are unknown.

Whitening of the sclera in osteogenesis imperfecta and improved spontaneous and chemotactic migration of leukocytes in Hill-Quie syndrome throw interesting side lights on the activity of this vitamin.

VII. CONCLUSIONS

Improved methods for the cultivation, preservation, and transportation of fruits and vegetables have made scurvy rare today, but very low plasma ascorbate levels approaching scurvy are often found in elderly, sick, or institutionalized people. Moreover, local or systemic abnormalities of ascorbate metabolism are much more common than a simple ascorbic acid deficiency. Further studies are needed to increase our understanding of the relationships of subclinical ascorbic acid deficiency to retinal hemorrhage and other slowly progressive pathological changes in the eye.

REFERENCES

Atkinson, D.T. (1952), Malnutrition as an etiological factor in senile cataract, *Eye Ear Nose and Throat Mon.,* 31, 79.

Bárány, E. H. (1951), Rate of flow of aqueous humor in normal and scorbutic guinea pigs, *AMA Arch. Ophthalmol.,* 46, 326.

Barber, A. and Nothacker, W. G. (1952), Effects of cortisone on nonperforating wounds of cornea in normal and scorbutic guinea pigs, *Arch. Pathol.,* 54, 334.

Bartley, W., Krebs, H. A., and O'Brien, J. R. P. (1953), Vitamin C requirements of human adults, *Med. Res. Counc (G. B.) Spec. Rep. Ser.,* 280.

Beardwood, J. T., Roberts, E., and Trueman, R. (1948), Observations on the effect of rutin and hesperidin in diabetic retinitis, *Proc. Am. Diabetes Assoc.,* 8, 241.

Bellows, J. (1936a), Biochemistry of the lens. V. Cevitamic acid content of the blood and urine of subjects with senile cataract, *Arch. Ophthalmol.,* 15, 78.

Bellows, J. (1936b), Biochemistry of the lens. VII. Some studies on vitamin C and the lens, *Arch. Ophthalmol.,* 16, 58.

Blondin, J., Baragi, V., Schwartz, E. R., Sadowski, J. A., and Taylor A. (1987), Dietary vitimin C delays UV-induced eye lens protein damage, *Ann. N.Y. Acad. Sci.,* 498, 460.

Bouton, S. M. (1939), Vitamin C and the ageing eye; an experimental clinical study, *Arch. Intern. Med.,* 63, 930.

Boyd, T. A. S. (1955), Influence of local ascorbic acid concentration on collagenous tissue healing in the cornea, *Br. J. Ophthalmol.,* 39, 204.

Boyd, T. A. S. and Campbell, F. W. (1950), Influence of ascorbic acid on the healing of corneal ulcers in man, *Br. Med. J.,* 2, 1145.

Buck, M. G. and Zadunaisky, J. A. (1975), Stimulation of ion transport by ascorbic acid through inhibition of 3':5'-cyclic-AMP phosphodiesterase in the corneal epithelium and other tissues, *Biochem. Biophys. Acta,* 389, 251.

Campbell, F. W. and Ferguson, I. D. (1950), The role of ascorbic acid in corneal vascularization, *Br. J. Ophthalmol.,* 34, 329.

Campbell, F. W., Ferguson, I. D., and Garry, R. C. (1950), Ascorbic acid and heat injuries in the guinea pig cornea, *Br. J. Nutr.,* 4, 32.

Chandra, D. B., Varma, R., Ahmad, S., and Varma, S. D. (1986), Vitamin C in the human aqueous humor and cataracts, *Int J. Vitamin. Nutr. Res.,* 56, 165.

Chatterjee, B. M. and Ghosh, B. P. (1956), Total ascorbic acid in the aqueous humor and serum in Indian patients with and without cataract, *AMA Arch. Ophthalmol.,* 56, 756.

Christiansson, J. (1957), Changes in the vitreous body in scurvy, *Acta Ophthalmol.,* 35, 336.

Davson, H. (1956), *Physiology of Ocular and Cerebrospinal Fluids,* Little, Brown, Boston, 259.

Davson, H. (1962), *The Eye,* Vol. 1, Academic Press, New York, 90.

Donegan, J. M. and Thomas, W. A. (1948), Capillary fragility and cutaneous lymphatic flow in relation to systemic and retinal vascular manifestations: rutin therapy, *Am. J. Ophthalmol.,* 31, 671.

Duke-Elder, S. (1932 to 1954), *Textbook of Ophthalmology*, Vol. 4 and 7, London, 4113 and 6966.

Duke-Elder, S. and Gloster, J. (1968), *System of Ophthalmology*, Vol. 4, Duke-Elder, S , Ed., Henry Kimpton, London.

Fischer, F. P. (1934), Über das C-Vitamin der Linse, *Klin. Wochenschr.*, 13, 596

Foster, S. and Goetzl, E. J. (1978), Ascorbate therapy in impaired neutrophil and monocyte chemotaxis, *Arch Ophthalmol.*, 96, 2069

Fox, R. R., Lam, K.-W., and Coco, J. F. (1977), Effect of ascorbic acid on intraocular pressure of normal and buphthalmic rabbits, *J Hered.*, 68, 179.

Fox, R. R., Lam, K.-W., Lewen, R., and Lee, P.-F. (1982), Ascorbate concentration in tissues from normal and buphthalmic rabbits, *J. Hered.*, 73, 109.

Friedenwald, J. S., Buschke, W., and Michel, H. O. (1943), Role of ascorbic acid (vitamin C) in secretion of intraocular fluid, *AMA Arch. Ophthalmol.*, 29, 535.

Galloway, N. M., Garry, R. C., and Hitchin, A. D. (1948), Ascorbic acid and epithelial regeneration, *Br. J Nutr.*, 2, 228

Gapeev, P. (1941), Content of ascorbic acid (vitamin C) in aqueous, lens and blood of cataractous and non-cataractous persons, *Vestn. Oftalmol.*, 18, 154; as cited in *Arch. Ophthalmol.*, 27, 1219, 1942.

Geever, E. F. and Levenson, S. M. (1960), Pathogenesis of the collagen defect in experimental scurvy, *AMA Arch Ophthalmol.*, 63, 812.

Glatt, H. and Machemer, R. (1982), Experimental subretinal hemorrhage in rabbits, *Am J Ophthalmol.*, 94, 762.

Goldmann, H. and Buschke, W. (1935), Ascorbinsäure (C-vitamin) und Blutkammerwasserschranke, *Klin. Wochenschr.*, 14, 239.

Greco, A. M., Fioretti, F., and Rimo, A. (1980), Relationship between hemorrhagic ocular diseases and vitamin C deficiency: clinical and experimental data, *Acta Vitaminol. Enzymol.*, 2, 21.

Griffith, J. Q., Krewson, C. F., and Naghski, J. (1955), *Rutin and Related Flavonoids. Chemistry, Pharmacology, Clinical Applications*, Mack Easton, PA.

Hamilton, J. B. (1958), Eyes and scurvy, *Trans. Ophthalmol. Soc. Aust.*, 18, 83.

Harris, L. J. (1933), Chemical test for vitamin C, and the reducing substances present in tumour and other tissues, *Nature (London)*, 132, 27.

Hawley, E. E. and Pearson, O. (1938), Vitamin C and its relation to cataract, *Arch. Ophthalmol.*, 19, 959.

Heath, H., Beck, T. C., Rutter, A. C., and Greaves, D. P. (1961), Biochemical changes in aphakia, *Vision Res.*, 1, 274

Heath, H. (1962), Distribution and possible functions of ascorbic acid in the eye, *Exp. Eye Res.*, 1, 362.

Hess, A. F. (1935), Scurvy, in *A Textbook of Medicine*, 3rd ed., Cecil, R. L., Ed., W. B. Saunders, Philadelphia.

Hood, J., Burns, C. A., and Hodges, R. E. (1970), Sjögren's syndrome in scurvy, *N. Engl. J. Med.*, 282, 1120.

Hood, J. and Hodges, R. E. (1969), Ocular lesions in scurvy, *Am. J. Clin. Nutr.*, 22, 559.

Hughes, R. E. and Maton, S. C. (1968), The passage of vitamin C across the erythrocyte membrane, *Br. J. Haematol.*, 14, 247.

Hughes, R. E., Hurley, R. J., and Jones, P. R. (1971), The retention of ascorbic acid by guinea pig tissues, *Br. J. Nutr.*, 26, 433.

Hughes, R. E. and Jones, P. R. (1971), The influence of sex and age on the deposition of L-xyloascorbic acid in tissues of guinea pigs, *Br. J Nutr.*, 25, 77.

Josephson, E. M. (1935), Ascorbic acid in cataract with special reference to dinitrophenol cataracts, *Science*, 82, 222.

Kahan, Á. (1958), Über die den Zustand des Glaskörpers beinflussenden Faktoren, *Acta Physiol. Acad. Sci. Hung*, Suppl., 14, 48.

Kato, K. (1932), A critique of the Roentgen signs of infantile scurvy; with report of 12 cases, *Radiology*, 18, 1096.

Killey, F. P., Edelhauser, H. F., and Aaberg, T. A. (1980), Intraocular fluid dynamics: measurements following vitrectomy and intraocular sulfur hexafluoride administration, *Arch. Ophthalmol.*, 98, 1448.

Kronfeld, P. C. (1946), Contribution to the chemistry of the aqueous humor in the glaucomas, *Trans. Am. Ophthalmol. Soc.*, 44, 134.

Kronfeld, P. C. (1952), The ascorbic acid content of the aqueous of surgically aphakic human eyes, *Trans. Am. Ophthalmol. Soc.*, 50, 347.

Kurz, D. and Eyring, E. J. (1974), Effects of vitamin C on osteogenesis imperfecta, *Pediatrics*, 54, 56.

Lam, K.-W., Lee, P.-F., and Fox, R. (1976), Aqueous ascorbate concentration in hereditary buphthalmic rabbits, *Arch. Ophthalmol.*, 94, 1565.

Lee, P.-F., Fox, R., Henrick, I., and Lam, W. K. W. (1978), Correlation of aqueous humor ascorbate with intraocular pressure and outflow facility in hereditary buphthalmic rabbits, *Invest. Ophthalmol. Visual Sci.*, 17, 799.

Lee, P.-F., Lam, K.-W., and Lai, M.-M. (1977), Aqueous humor ascorbate concentration and open-angle glaucoma, *Arch. Ophthalmol.*, 95, 308

Levinson, R. A., Paterson, C. A., and Pfister, R. R. (1976), Ascorbic acid prevents corneal ulceration and perforation following experimental alkali burns, *Invest. Ophthalmol. Visual Sci.*, 15, 986; as cited in *Am J Ophthalmol.*, 83, 609, 1977

Lin, Y.-H., Lam, K.-W., Shihab, Z., and Lee, P. (1982), Biochemical changes in aqueous humor of eyes with neovascular glaucoma, *Glaucoma*, 4, 210.

Lindahl, C. (1941), Remarks on origin of senile cataract and some other forms of cataract, *Upsala Laekarefoeren Foerh.*, 47, 107, as cited in *JAMA*, 120, 162, 1942.

Linner, E. (1969), The pressure lowering effect of ascorbic acid in ocular hypertension, *Acta Ophthalmol.*, 147, 685, as cited in *Am. J. Ophthalmol.*, 69, 529, 1970.

Loewe, W. R. (1955), Physiological control of certain retinopathies, *Eye Ear Nose Throat Mon.*, 34, 108.

Lohmann, W. (1987), Ascorbic acid and cateract, *Ann. N.Y Acad. Sci.*, 498, 307.

Lyle, T. K. and McLean, D. W. (1941), Vitamin "C" (ascorbic acid), its therapeutic value in inflammatory conditions of the cornea, *Br J. Ophthalmol.*, 25, 286.

Mann, G. V. (1974), Hypothesis: the role of vitamin C in diabetic angiopathy, *Perspect Biol. Med.*, 17, 210

Maurice, D. M. (1962), in *The Eye*, Vol. 1, Davson, H., Ed , Academic Press, New York, 238

Mitsukawa, K. (1963), Production of vitamin C in the plasmoid aqueous, *Acta Soc Ophthalmol. Jpn* , 67, 839; as cited in *Am J. Ophthalmol.*, 58, 155, 1964.

Mitsukawa, K. (1964), Outflow of vitamin C into the vortex vein, *Acta Soc Ophthalmol. Jpn.*, 68, 69; as cited in *Am. J Ophthalmol.*, 58, 902, 1964.

Monjukowa, N. K. and Fradkin, M. J. (1935), Neue experimentelle Befunde über die Pathogenese der Katarakt, *Arch. Ophthalmol.*, 133, 328

Müller, H. K. and Buschke, W. (1934), Vitamin C in Linse, Kammerwasser und Blut bei normalem und pathologischem Linsenstoffwechsel, *Arch. Augenheilkd.*, 108, 368.

Nirankari, V. S., Varma, S. D., Lakhanpal, V., and Richards, R. D. (1981), Superoxide scavenging agents in treatment of alkali burns, *Arch. Ophthalmol.*, 99, 886.

Ogielska, E. (1963), Experimental iritis and the level of ascorbic acid in the aqueous, *Klin. Oczna*, 33, 113; as cited in *Am J. Ophthalmol.*, 59, 743.

Ogino, S., Ishihara, T., and Ito, A. (1954), Vitamin C metabolism in patients with senile cataract, *Acta Soc. Ophthalmol Jpn* , 58, 466; as cited in *Am. J. Ophthalmol* , 38, 446.

Pfister, R. R., Paterson, C. A., Spiers, J. W., and Hayes, S. A. (1980), The efficacy of ascorbate treatment after severe experimental alkali burns depends upon the route of administration, *Invest. Ophthalmol. Visual Sci.*, 19, 1526; as cited in *Am. J. Ophthalmol.*, 91, 425, 1981.

Purcell, E. F., Lerner, L. H., Kinsey, V. E. (1954), Ascorbic acid in aqueous humor and serum of patients with and without cataract, *AMA Arch. Ophthalmol.*, 51, 1.

Sabatine, P. L., Rosen, H., Geever, E. F., and Levenson, S. M. (1961), Scurvy, ascorbic acid concentration and collagen formation in the guinea pig eye, *AMA Arch. Ophthalmol.*, 65, 32.

Sarji, K. E., Kleinfelder, J., Brewington, P., Gonzalez, J., Hempling, H., and Colwell, J. A. (1979), Decreased platelet vitamin C in diabetes mellitus; possible role in hyperaggregation, *Thromb. Res.*, 15, 639.

Shimizu, H. (1964), Intake and distribution of ascorbic acid within the eyeball, *Acta Soc. Ophthalmol. Jpn.*, 68, 1447, as cited in *Am. J. Ophthalmol* , 60, 951, 1965.

Summers, T. C. (1946), Penicillin and vitamin C in the treatment of hypopyon ulcer, *Br. J. Ophthalmol.*, 30, 129.

Tahata, S. (1937), Vitamin C in the eye of various mammals, *Acta. Soc. Ophthalmol. Jpn.*, 41, 2004; as cited in Heath, H., Beck, T. C., Rutter, A. C., and Greaves, D. P. (1961), *Vision Res.*, 1, 274.

Talbot, F. B., Dodd, W. J., and Peterson, H. O. (1913), Experimental scorbutus and the roentgen ray diagnosis of scorbutus, *Boston Med. Surg. J.*, 169, 232.

Verma, S. D. (1987), Ascorbic acid and the eye with special reference to the lens, *Ann. N.Y. Acad. Sci.*, 498, 280.

Varma, S. D., Chand, D., Sharma, Y. R., Kuck, J. F., Jr., and Richards, R. D. (1984), Oxidative stress on lens and cataract formation: role of light and oxygen, *Curr. Eye Res.*, 3, 35.

Varma, S. D., Kumar, S., and Richards, R. D. (1979), Light-induced damage to ocular lens cation pump. prevention by vitamin C, *Proc. Natl. Acad. Sci. U.S.A.*, 76, 3504.

Verlangieri, A. J. and Sestito, J. (1981), Effect of insulin on ascorbic acid uptake by heart endothelial cells: possible relationship to retinal atherogenesis, *Life Sci.*, 29, 5.

Virno, M., Bucci, M. G., Pecori-Giraldi, J., and Cantore, G. (1966), Intravenous glycerol-vitamin C (sodium salt) as osmotic agents to reduce intraocular pressure, *Am. J. Ophthalmol.*, 62, 824.

Virno, M., Bucci, M. G., Pecori-Giraldi, J., and Missiroli, A. (1967), The effect of the oral administration of ascorbic acid on the intra-ocular pressure, *Boll. Ocul.*, 46, 259; English abstr. in *Am. J. Ophthalmol.*, 65, 952, 1968.

von Euler, H. and Martius, C. (1933), Über den Gehalt der Augenlinsen an Sulphydrylverbindungen und an Ascorbinsäure, *Z. Physiol. Chem.*, 222, 65.

Wichard, A., van Heuven, J., Lam, K.-W., and Ray, G. S. (1982), Source of subretinal fluid on the basis of ascorbate analyses, *Arch. Ophthalmol.*, 100, 976.

Young, P., Briggs, H. H., and Fry, J. (1968), Spontaneous remission and successful therapy in chloroquine retinopathy, *Arthritis Rheum.*, 11, 122.

Yudkin, A. M. (1951), Ocular disturbances associated with malnutrition, *Am. J. Ophthalmol.*, 34, 901

Chapter 19

CEREBRAL HEMORRHAGE AND THROMBOSIS

I. VASCULAR FRAGILITY DUE TO ASCORBIC ACID DEFICIENCY

In a treatise on scurvy, the famous English physician Willis (1668) mentioned the occurrence of intracranial hemorrhage in the course of this disease. Likewise, Hayem (1871), describing his findings at autopsy in eight patients who died of scurvy during the siege of Paris, mentioned one case of hemorrhagic pachymeningitis.

Sutherland (1894) described the findings of recent and old subdural hemorrhages, respectively, in two young children who died at 24 and 14 months of age. Both had classical signs of scurvy, with subcutaneous petechiae, ecchymoses, and subperiosteal hemorrhages, but not the spongy bleeding gums which are said to be rare in infancy and early childhood. On opening the dura mater of the older child there was found a large amount of blood, covering the vertex on both sides, from the middle of the frontal to the posterior part of the parietal regions. The surface of the brain was covered with a layer of blood clot and the cortex was deeply stained. The blood also extended between the hemispheres. The brain contained two hemorrhagic cavities under the cortex, situated in the posterior frontal region, one on each side. That on the right side was the size of a small tangerine orange. On microscopic examination the basilar artery was found to be occupied by a thrombus which had undergone some degree of organization. In the other child there was in the subdural space, a dense fleshy, purplish red, smooth and soft deposit which completely covered the surface of the brain. It formed a false membrane from 1/16 to nearly 3/4 in. thick, being thickest towards the base of the skull in the anterior and posterior fossae. Beneath this layer, and easily separable from it, was a second dense white membrane (of fibrin), and on the under surface of this was a third layer of soft, yellowish, coagulated lymph, somewhat closely adherent to it and to the pia mater beneath; under it the brain was generally shrunken and atrophied. This old layered subdural hemorrhage had undoubtedly been responsible for a fit accompanied by strabismus, which had occurred when the child was 10 months old and with which she had survived for 4 months. Clearly, this child would have been permanently brain damaged if she had survived. It is therefore very important to think of ascorbate deficiency and to diagnose it early, so that it can be treated before permanent harm can occur.

In the same year, at a meeting of The Pathological Society of London, Ord (1894) reported another instance of subdural hemorrhage in an infant with scurvy who died of bronchopneumonia. A large clot occupied the whole vault of the cranium, but there was no sign of hemorrhage at the base. At that meeting, Sutherland commented that he had seen a third infant with subdural hemorrhage in association with scurvy.

Another case of pachymeningitis hemorrhagica interna was described by Meyer (1896) who found a fibrinous hemorrhagic exudate under the dura mater on post-mortem examination of an infant who died with scurvy at 9 months of age.

A report on 379 cases of infantile scurvy by the American Pediatric Society (1898) ascribed 3 out of 29 deaths to cerebral hemorrhage. Sato and Nambu (1908) also reported a case of subdural hemorrhage in scurvy.

Ellis (1909), in a detailed study of 31 brains from men and women who had died of cerebral hemorrhage, at a mean age of 60 years, found mild, moderate, or severe atheroma in the great vessels at the base of the brain in 29, and macroscopic aneurysms of the intracerebral arteries in 20. The incidence of aneurysms was found to be related to the degree of atheroma:

Advanced, 14 cases   Macroscopic aneurysms in 11, or 79%
Moderate, 11 cases   Macroscopic aneurysms in 8, or 73%
Slight, 4 cases   Macroscopic aneurysms in 1, or 25%
None, 2 cases   Macroscopic aneurysms in 0, or 0%

Microscopic examination revealed that, "the so-called miliary aneurysms of cerebral hemorrhage are false aneurysms, due primarily to vascular disease and are preceded by simple rupture or by dissecting aneurysm, probably more often the latter, of the affected artery."

Though having the gross appearance of true aneurysms, the primary pathology was found to be degeneration of the endothelium, leading to rupture and elevation of the intima, and effusion of blood beneath it. In places there was perforation of the media and dissection of the adventitia by effusion of blood between its layers.

True aneurysms are "balloons" and are lined by an attenuated blood vessel wall. These false aneurysms, on the other hand, were "blow outs" and had none of the coats of the blood vessel with which they communicated, except the adventitia which extended a short distance in some instances; they were largely filled with coagulum and were lined by a layer of fibrin. These observations by Ellis are relevant because the endothelium of the artery is the same tissue as the wall of the capillary, which is the tissue most affected by ascorbic acid deficiency, or by a disturbance of ascorbate metabolism.

Scherer (1913) reported 13 autopsy records obtained within the course of a few months on young adults with scurvy whose ages ranged from 20 to 25 years. Of the examinations, 25% revealed intracranial hemorrhage of one type or another. In none of the cases was there any evidence of trauma or of inflammatory change that might have caused the hemorrhage.

A case of subdural hemorrhage in the spinal cord was described by Feigenbaum (1917). Following typical signs of scurvy for a month, a German soldier developed the signs of a transverse cord lesion of severe degree. With appropriate dietary treatment, the external bleeding ceased, but the neurological signs showed no tendency to regress and he eventually died of sepsis. At autopsy a flat, partially organized subdural hemorrhage was found anterior to the cord, extending from the sixth to the tenth thoracic segment. There was also an extensive hematoma lying between the nerve roots in the subarachnoid space.

Sammis (1919) reported the occurrence of subdural hemorrhage in a malnourished 1-year-old infant with classical signs of rickets and scurvy. A general clonic convulsion was followed by death 4 h later. Post-mortem examination revealed a blood clot over the left postcentral gyrus, lying between the dura and the arachnoid.

McClelland (1921), working in Glamorganshire, described his observations of a 40-year-old coal miner who developed frank scurvy as a result of his dietary prejudices. The patient referred to vegetables as "pig's food" and appears to have lived principally on bacon, bread and tea. Presenting with a headache and weakness, he was found to have tachycardia, bleeding gums, foul breath and ecchymoses of the arms and legs. A few days later he became unconscious and developed a deep coma. The pupils were equal, but soon his right arm was noted to have become flaccid and he died, so we may conclude that he had a cerebral hemorrhage or thrombosis due to scurvy.

Vitamin B_1 deficiency is well recognized as an accompaniment of hyperemesis gravidarum, but a concomitant vitamin C deficiency is very common and must be treated. Indeed, Banhidi (1960) has shown that the thiamine disulfide of food may be unavailable until it has been activated by the reducing action of ascorbic acid or cysteine. Swanson (1927) described a patient with hyperemesis gravidarum complicated by scurvy, which was manifested by epistaxis, black vomitus, and a subperiosteal hematoma. She survived following hydration, blood transfusion, and treatment with lemon and orange juice, but might well have died with hemorrhagic encephalopathy if the vitamin C deficiency had not been recognized and treated. Hyperemesis gravidarum is hardly ever fatal nowadays, but death used to be a real danger. The brain in patients with terminal encephalopathy used to show typical changes in the corpora mammilaria, the thalami, and the floor of the fourth ventricle. These consisted of endothelial swelling and bead-like dilation of the capillaries with surrounding petechial hemorrhages of symmetrical distribution.

Gilman and Tanzer (1932), working in Rochester, NY, described the development of neurological signs, including twitching of the right side of the face, rolling of the eyes, internal strabismus, dilation of the left pupil, and vomiting, due to cerebral hemorrhage in a 16-month-old boy, 10 d after admission to hospital, and while under treatment for classical scurvy. It is noteworthy that he received both ferrous carbonate and copper sulfate as treatment for his anemia; the copper may have been counterproductive.

Stokes and Campbell (1932) working at the University of Pennsylvania Hospital, reported three cases of infantile scurvy with severe hyperpyrexia suggestive of pontine hemorrhage, their temperatures reaching 106.8, 107, and 108°F (41.6, 41.7, and 42.2°C) before death, in spite of appropriate dietary treatment with orange juice. Autopsy consent for the first two infants was not obtained; post-mortem examination of the third child showed subdural hemorrhages.

Sherwood (1930), working at the Infants and Children's Hospital in Boston, wrote an extensive review of "chronic subdural hematoma". He preferred this name to the older term "pachymeningitis interna haemorrhagica", as in most instances there was no evidence of infection. Statistics were cited which showed the disease to be present in 2 to 7% of all infants coming to autopsy when less than 1 year of age. After 1 year, there was a falling off of the incidence up to the age of 30 to 40 years, when it began to increase. The largest percentage of cases occurred in persons between 70 and 80 years of age. He described nine infants with this condition and remarked on the fact that five of them came from foster homes or institutions, so they might not have received the loving care that a mother would have provided. Convulsions and enlargement of the head were the two most consistent signs in infants. Retinal hemorrhages were frequently seen, and a bulging fontanelle and vomiting were also frequent accompaniments. The bleeding time was increased from 4.5 to 8 min in four of the eight in whom it was measured. The cerebrospinal fluid was usually blood tinged or xanthochromic at first, but later became clear as the subdural hematoma became encysted and walled off; then a subdural tap would reveal bloody or xanthochromic fluid.

Sherwood provided an illustration of the gross pathology of the subdural membrane formation in chronic subdural hematoma (Figure 1). He cited Putnam's histopathological differentiation between the traumatic and the spontaneous varieties of this condition. The spontaneous type showed fibrinous tissue which was well demarcated from the dura, but which differed from the traumatic type in that giant capillaries filled with red blood cells were present. The smallest capillaries in this type were comparable to the largest ones found in the traumatic type.

On follow-up, three of Sherwood's nine patients seemed to have made a full recovery and were apparently normal children. One still had an active process; one was feeble-minded; two showed retarded development, and one of these had strabismus; one was dead; one was too young for a definite prognosis to be made, but was reported to be doing well. He cited the sequelae on follow-up of ten cases of chronic subdural hematoma by Rosenberg as two idiots, three imbeciles, one person with a neuropathic constitution, one stutterer, one bedwetter, and two normal subjects.

While Sherwood did not think that infection was the direct cause of chronic subdural hematoma, he felt that in some instances the patient's resistance had been lowered by previous infection and in others he thought that the subdural hematoma had predisposed to subsequent infection. He did not find clinical manifestations of scurvy in any of his nine cases.

Ingalls (1936), also working at Harvard University, continued these studies of chronic subdural hematoma and reached the conclusion that subclinical scurvy was an important

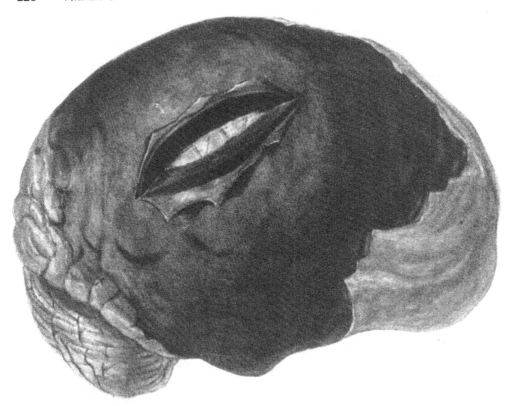

FIGURE 1.   Sketch made at autopsy, showing dura, subdural membrane, clot beneath subdural membrane, and brain substance under clot (From Sherwood, D [1930], *Am. J. Dis. Child.*, 39, 980. ©1930 American Medical Association With permission.)

cause of this condition even though convulsions and blood-tinged cerebrospinal fluid were the most characteristic and often the only findings. He asserted that subclinical scurvy may exist before clinical manifestations and X-ray changes become apparent. He gave reasons for believing that chronic subdural hematoma, as encountered at autopsy, is the end stage of chronic low-grade scurvy in infants, and that a similar picture may be present in alcoholics and in the insane. While he conceded that infants, alcoholics, and the insane are all subject to head trauma, they are often vitamin C deficient and a smaller increment of trauma is necessary to produce bleeding when added to the "haemorrhagic diathesis". While in chronic subdural hemorrhage the source of the bleeding is usually regarded as seepage following rupture of "bridging veins" between the pia and the dura mater, Ingalls suggested that the hemorrhage could be from minute bleeding points where the arachnoid villi were torn away at their invaginations into the dural sinuses following a minor injury such as would not have harmed a nonscorbutic individual.

Alexander et al. (1938), also working in Boston, confirmed Ingalls' suspicion of ascorbic acid deficiency in many alcoholics. They felt that subclinical scurvy was a contributing cause of Wernicke's polioencephalitis hemorrhagica and felt justified in considering, "vitamin C deficiency as a predisposing etiologic factor in subdural hematoma in patients with chronic alcoholism."

Levrat and Ballivet (1939), using a suprasystolic tourniquet test of capillary strength, found capillary fragility in 31 out of 33 patients with hypertension and in only 1 out of 7 nonhypertensive control subjects of similar age (over 60 years) with normal blood pressure. Moreover, the one control subject with a positive tourniquet test had chronic rheumatism which may have accounted for the abnormal test result. They concluded that nearly all hypertensive people have capillary fragility and wondered whether it might not predispose to cerebral or retinal hemorrhage. They also pondered the question as to whether the capillary lesions were the cause or the result of the hypertension.

Paterson (1940) described capillary rupture and hemorrhage from within the intima of the cerebral arteries as being the apparent initial cause of cerebral thrombosis. The hemorrhages were usually confined to the diseased and thickened intimal layer of sclerotic cerebral arteries, but in a transverse section, one such hemorrhage was seen to have broken through the media and involved the adventitia of the artery, so cerebral hemorrhage and thrombosis may have a common etiology. Paterson stated that the intimal hemorrhages in sclerotic cerebral arteries are similar in structure to those seen in sclerotic coronary arteries and result from rupture of capillaries derived from the main arterial lumen. He noted that hypertension, atheromatous degeneration and capillary fragility associated with advancing age and vitamin deficiency were all contributing factors leading to cerebral hemorrhage and cerebral thrombosis, and that similar changes were often seen in the coronary arteries of the same patients, most of whom were hypertensive and some diabetic.

An observation by Lund (1942) showed the possibility of ascorbic acid deficiency leading to cerebral hemorrhage to be very real. Patient Number 1 in his paper entitled "Ascorbic Acid Deficiency Associated with Gastric Lesions", was recorded as having had an ascorbic acid intake of 1 mg/d during the month before admission to hospital. He was found to have a plasma ascorbic acid level of 0.15 mg/100 ml before surgery (normal range 0.4 to 1.3 mg/100 ml), but received no supplement and died of a cerebral hemorrhage following a posterior gastroenterostomy.

Moreover, Sanford et al. (1942) drew attention to the low blood vitamin C values obtained in 6 newborn infants with cerebral hemorrhage (0.03 to 0.25 mg/100 ml). They emphasized the need for vitamin C and discussed the value of vitamin K.

Griffith and Lindauer (1944) observed that hypertensive persons who have increased capillary fragility are especially predisposed to apoplexy, retinal hemorrhage, and death. Beaser et al. (1944) demonstrated increased capillary fragility in 54% of hypertensive patients and in 100% of diabetic hypertensives, who are especially prone to cerebrovascular accidents.

Vilter et al. (1946) recorded the development of hemiplegia, due to thrombosis of the right middle cerebral artery, in a 54-year-old man admitted to hospital with scurvy. Lamy et al. (1946) observed thrombosis of the large left pial veins leading to the longitudinal sinus; also, hemorrhagic softening of the parietal and occipital regions in a 35-year-old man who died of malnutrition following release from a prison camp.

Alvarez (1946) observed that, "one of the commonest diseases of man is a slow petering out toward the end of life, and one of the commonest ways of petering out is that in which the brain is slowly destroyed by the repeated thrombosis of small sclerotic blood vessels." He noted that these thrombotic episodes can start even as early as age 30 and that many small apoplexies affecting silent areas of the brain may go unnoticed or may produce minor attacks of dizziness, fainting, or confusion. The step becomes more hesitant, the memory less trustworthy, and falls more frequent. He cited sayings by elderly patients including, "death is taking little bites of me," and another, "I don't mind dying of a stroke, but I dread not dying with the first big one. I don't want to take ten years to peter out, a nuisance to myself and to my family."

Minor cerebral embolic, thrombotic, or hemorrhagic incidents may be recognized principally by the suddenness of onset of a change in cerebral function, but may otherwise be difficult to recognize for what they are. They may take the form of sudden changes in temperament, personality, or mortality. Nervousness, depression, or even psychotic behavior may develop following a fall, or the change may be more subtle with only loss of memory, clumsiness, worsening handwriting, or loss of grooming as evidence of recurrent episodes.

Griffith and Lindauer (1947) confirmed their earlier finding that, "apoplexy and retinal haemorrhage occur more frequently in hypertensive subjects with increased capillary fragility and permeability." In a series of 1200 hypertensive subjects it was found that 20% showed increased capillary fragility, while an additional 10% showed increased capillary permeability with normal fragility. A history of apoplexy and retinal hemorrhage was obtained in 9% of those with increased capillary fragility or permeability, and in 2% of the remainder.

Thewlis and Gale (1947) found that one of the most frequent dietary deficiencies in older people is vitamin C. They recommended dietary supplements of ascorbic acid and the bioflavonoid rutin to decrease capillary fragility in the hope of preventing cerebral hemorrhage and thrombosis.

Soloff and Bello (1948), using the same capillary fragility test as Griffith and Lindauer, found that only 2 out of 50 hypertensive patients showed capillary fragility after treatment with ascorbic acid, 150 mg, niacinamide, 150 mg, thiamine chloride, 15 mg, and riboflavin, 9 mg daily for over a month.

Gale and Thewlis (1953) reported 32 patients who took vitamin C and rutin for 1 to 4 years. Their ages varied from 43 to 92 years with 66 as the average. Every patient had suffered one or more vascular disturbances. "There were six deaths among the 32 patients, and four were direclty due to a heart attack or a cerebral episode. Of these four not one had taken more than 100 mg of Vitamin C daily. In each of five who died, the final attack was severe enough to cause death within a short period. The 26 patients who remained alive were active and had few or no symptoms referable to the vascular system."

Unfortunately, it was not possible to compare the results with patients who had similar symptoms and illnesses without such vitamin supplements. These authors cited statistics showing that heart attacks and cerebral accidents tend to occur within a few weeks after upper respiratory tract infections and advised that much benefit may be derived from the administration of 180 mg of rutin and 900 mg of ascorbic acid daily at such times.

In a survey of 50 patients in general practice in Wakefield, RI, Gale and Thewlis found that 36 or 72% of their patients stated that they bruised easily. Gingivitis, epistaxis, loss of weight, anorexia, and muscle and joint pain were other common symptoms in the aged, which they thought of as possibly being due to vitamin C deficiency.

Griffith et al. (1955) used the bioflavonoid rutin alone in the treatment of a large series of hypertensive patients with capillary fragility and reported a marked reduction in the subsequent incidence of apoplexy as shown in Table 1. This is pertinent because of the beneficial effect of this flavonoid on ascorbic acid metabolism, both *in vivo* and *in vitro*, as described in Chapter 11, Volume I.

Transient ischemic attacks are believed to be due to platelet emboli from the carotid arteries or to relative perfusion deficiencies in watershed areas, but Saelhof et al. (1959) studied "little stroke", with minor neurological symptoms, which is one of the most common diseases of aging man, and gave reasons for believing that the vast majority of such episodes are due to hemorrhage from small capillaries and venules in the brain, associated with generalized capillary fragility and are not associated with thrombosis of a cerebral blood vessel. They treated 21 cases of little stroke with citrus bioflavonoids and ascorbic acid with encouraging results; 14 cases, some of whom received anticoagulants, served as controls. In the bioflavonoid-treated group there was one episode of little stroke, without fatality. Seven episodes of little stroke and two fatalities were recorded in the control group.

Not only do ascorbic acid and bioflavnonoids seem to be important in reducing the risk of cerebral hemorrhage and thrombosis, but Kubala and Katz (1960) have reported that orange juice supplements were even of value in increasing mental alertness. They found a significant improvement in mean "intelligence quotient" test performance by students from kindergarten through college ($p < 0.05$).

In X-ray studies of barium and gelatin-infused arteries of brain slices from normotensive

Table 1
EFFECT OF RUTIN THERAPY ON INCIDENCE OF APOPLEXY

Capillary fault	No. in group	Rutin therapy	Incidence of apoplexy	
			Number	%
Absent	346	No	22	6.3
Present	199	Yes	16	8.0
Present	217	No	46	21.2

EFFECT OF RUTIN THERAPY ON MORTALITY ASSOCIATED WITH APOPLEXY

				Observed deaths from apoplexy				
							Ratio to expected deaths in	
Initial capillary fault	Subjects in group (number)	Rutin therapy	Total deaths (number)	Incidence in group		Ratio to total deaths (%)	General population (%)	Hypertensive population (%)
				Number	%			
Absent	346	No	59	13	3.7	22	66	21
Present	199	Yes	23	7	4.0	30	46	13
Present	217	No	93	37	17.0	40	324	106

Note: Of hypertensive patients with capillary fragility, 21% developed apoplexy, while only 6% of hypertensive patients without capillary fragility developed apoplexy. Treatment with the chelating flavonoid rutin reduced the incidence of apoplexy in hypertensive patients with capillary fragility to a level similar to that of those without capillary fragility The effects of rutin treatment on the mortality of hypertensive patients with capillary fragility were equally impressive.

From Griffith, J. Q., Krewson, C. F., and Naghski, J. (1955), in *Rutin and Related Flavonoids*, Mack, Easton, PA, 173. With permission.

and hypertensive individuals, RossRussell (1963) found multiple ''aneurysms'' to be common in hypertensive and rare in normotensive individuals. They were definitely associated with cerebral hemorrhage, and were most frequently seen in the putamen, pallidum, and thalamus, but also occurred in the caudate nucleus, internal capsule, centrum semiovale, and cortical grey matter. The ''aneurysms'' were clearly acquired lesions due to weakness of the arterial wall and high intraluminal pressure. They are the points of least resistance in a diseased arterial wall. In histological studies the muscular tissue of the parent vessel was seen to terminate abruptly at the point of origin of the aneurysm and remnants of the elastica could be seen to extend for a short distance into the aneurysm before disappearing. ''The wall of the aneurysm was composed of connective tissue only, an inner hyaline layer derived from the intima fusing with an outer collagenous layer continuous with the adventitia of the parent vessel.''

Cole and Yates (1967) made very similar studies of brains from 100 hypertensive and 100 normotensive subjects. They confirmed the association of ''aneurysms'' with hypertension and with cerebral hemorrhage. Indeed, 86% of patients with cerebral hemorrhage were found to have these ''aneurysms''. They noted that, ''at points on the parent vessel for a short distance on either side of the aneurysm the intima shows hyaline thickening and this is directly continuous with similar material lining the lumen of the sac. At this latter site, however, the hyaline material is often arranged in loose bands or laminations between which pockets of red cells are sometimes seen . . . Collections of fresh blood, lymphocytes

Table 2
MORTALITY[a] DUE TO ALL CAUSES, CEREBRAL VASCULAR LESIONS, NEOPLASM, AND HEART DISEASE (1958—1962)

Area	Total	Cerebral vascular lesions	Neoplasm	Heart disease
Japan	745.0	159.5	99.9	70.6
Iwate	788 0	212.0	74.2	72.0
Village A	954.4	272.6	85.8	75.4
Village B	844.0	104.0	4.4	80 4

Note: Comparison of death rates from three diseases in Japan as a whole, in Iwate Province of northern Honshu, and in two villages of that province. The mortality due to cerebral vascular lesions in Village A was nearly three times that in Village B

[a]   Per 100,000 population.

From Kimura, T. and Ota, M. (1965), *Am. J. Clin. Nutr.*, 17, 381. © American Society for Clinical Nutrition. With permission.

and plasma cells are frequently seen around the aneurysms both in the cleared specimens and in histological sections. Older haemosiderin pigment is also found, indicating a previous leak.''

Fisher (1971) preferred to use the term ''fibrin globe'' for the sites where small cerebral arteries had ruptured. ''The larger hemorrhage contained one large fibrin globe and three smaller ones, all consisting of whorls of fibrin enclosing masses of red blood cells with a central clump of platelets lying adjacent to a break in the artery.'' He also saw evidence of smaller old hemorrhages at other sites in the brain, but there was no evidence of an aneurysm at the site of hemorrhage. He conceded that the vessel might have bulged before rupturing. There was a subintimal accumulation of a lipid-rich hyaline material that finally weakens and destroys the artery. ''This process which is strongly correlated with the occurrence of hypertensive hemorrhages has been given various names — fibrinoid necrosis, hyaline arterionecrosis, atherosclerosis of small arteries, hyaline fatty change, plasmatic vascular destruction, hyalinosis, angionecrosis, fibrinoid arteritis and its chronic healed stage segmental arterial disorganization.'' Fisher chose to call it ''lipohyalinosis''. In fact, it is almost identical to the intimal lesion observed by Sulkin and Sulkin (1975) in the aortas of guinea pigs fed on a marginal vitamin C-deficient diet for 104 d (Figures 1 to 4, Chapter 8, Volume II).

A study of the death rates from various diseases in Japan showed the mortality from cerebral vascular lesions to be very high in Iwate Prefecture in northern Honshu, the main island of Japan. Between the years 1958 and 1962, it was 212 per 100,000 of population. However, the mortality from this cause was found to vary from area to area within this region (Table 2), being 272 per 100,000 in one village (A) and 104 per 100,000 in another village (B). An epidemiological study of cerebrovascular disease by Kimura and Ota (1965) was therefore devoted to the differences between the dietary, social, and other factors affecting these two village populations.

''Village A is a typical rice-producing community situated on an open field in a basin of the Kitakami River. In 1961 the total population was 15,501 and the total death rate was 952 per 100,000 population. Village B, 35 miles from A, is situated at the foot of the Kitakami Mountain Range. It is an oceanside mountain village producing dairy products and cereals, excluding rice, and is known as one of the most underdeveloped villages in Japan. In 1961 the total population was 4,440 and the total death rate 855 per 100,000. The elevation above sea level of Villages A and B are 100 and 105 meters respectively.''

The high mortality from cerebrovascular lesions in Village A was associated with an increased incidence of hypertension. Among male subjects 50 to 59 years of age, 33.6% of the villagers in A had blood pressure levels of 160 mmHg (systolic) and/or 95 mmHg (diastolic) and higher, as opposed to only 22.7% in Village B. Salted fish and other salted foods were commonly eaten in both villages and this had been suspected as contributing to the hypertension in northern Honshu, but dietary histories revealed that less salt was actually consumed in Village A, where hypertension and death from cerebrovascular lesions were so much more common. The consumption of animal protein was low, especially in Village B, but total protein intake was lower in Village A.

There was no significant difference between the climatic conditions of the two villages; nor was there any significant difference between the serum lipid levels of the villagers (Table 3). There was, however, a difference in the consumption of certain nutrients; the intakes of vitamin A, riboflavin, ascorbic acid, and pantothenic acid were found to be lower in Village A (Table 4).

''Most common vegetables such as Chinese cabbage, cucumber, egg plant and radish and its leaves were eaten in the preferred 'salted' form which is called 'tsukemono'. Radish, common in the rural area, was consumed as 'takuan' which is the product of fermentation with a large amount of salt for some weeks or even months. 'Takuan' and 'Tsukemono' are stock vegetables and are served as the most common and inexpensive side dishes in the area. Few fresh vegetables are available during the long winter period which starts in December and ends in April. Consequently, the ascorbic acid intake varies from season to season.''

The intake of ascorbic acid in village A in winter and early spring was assessed as being far below the minimum need. On the other hand, potatoes, consumed in large quantities in B, supplied a large amount of ascorbic acid in winter.

The mean serum ascorbic acid levels were studied in Village A and were found to be 1.08 mg/100 ml in August and 0.52 mg/100 ml in February. One would not expect an ascorbate level of 0.52 to have any serious consequences, but one must remember that these are mean figures. The full range of values undoubtedly included many more individuals with seriously low ascorbate levels in winter in Village A than in Village B.

The incidence of deaths from cerebral vascular lesions was much greater during the late winter and spring (Figure 2). This certainly suggests a correlation with dietary ascorbic acid deficiency.

When compared with the U.S., the mortality from heart disease in both villages was very low, as is that of all Japan; moreover, it was similar in the two villages.

These data strongly suggest that ascorbic acid deficiency plays a significant role in the etiology of death from cerebrovascular lesions in northern Japan, just as ascorbate deficiency seems to be a factor in the etiology of coronary heart disease in Finland. We may perhaps suppose that it is the low meat consumption and the low serum cholesterol levels of northern Honshu that protect them from coronary atheroma, but we do not know what other factors predispose them to cerebrovascular lesions. The vitamin A consumption in both villages was far lower than desirable; it was as low as one fifth of the required amount for Japanese adults according to the Japan Ministry of Health and Welfare.

It may be that a combination of low vitamin C, low vitamin A, and possibly riboflavin or thiamine deficiency conspire to injure the endothelium of the cerebral blood vessels and

Table 3
MEAN SERUM LIPID LEVEL BY AGE AND SEX IN VILLAGES A AND B

Age group (years)	Village A							Village B						
	No. of subjects	Cholesterol (mg/dl)		Phospholipid (mg/dl)		β + γ/α Lipoprotein (mg/dl)		No. of subjects	Cholesterol (mg/dl)		Phospholipid (mg/dl)		β + γ/α Lipoprotein (mg/dl)	
		Mean	SD	Mean	SD	Mean	SD		Mean	SD	Mean	SD	Mean	SD
Male Subjects														
20—39	88	152.4	30.7	278	75	2.5	0.4	25	156.6	42.3	275	84	2.4	0.4
40—59	102	160.7	33.2	302	42	2.6	0.3	22	149.1	50.9	288	85	2.6	0.7
Female Subjects														
20—39	132	154.2	31.3	310	46	2.4	0.3	25	152.9	45.1	288	74	2.2	0.4
40—59	88	168.5	41.2	320	60	2.8	0.5	20	158.7	38.8	305	67	2.5	0.8

Note: The concentration of serum phospholipid is calculated by multiplying the concentration of lipid phosphorus by 25. The mean serum cholesterol levels of male and female subjects in both villages were quite similar. The levels were remarkably low in comparison with those of many Western populations and even in the rest of Japan.

From Kimura, T. and Ota, M. (1965). *Am. J. Clin. Nutr.*, 17, 381. © American Society for Clinical Nutrition. With permission.

Table 4
PER CAPITA INTAKE OF NUTRIENTS
PER DAY IN VILLAGES A AND B

Nutrient	Village A	Village B
Total calories	2222	2119
Protein		
Vegetable (g)	44 2	68.3
Animal (g)	27 7	18.5
Total (g)	71.9	86.8
Fat		
Vegetable (g)	13.6	18.1
Animal (g)	8 3	4.3
Total (g)	21 9	22 4
Calories (%)	8.8	9.5
Carbohydrate (g)	434	443
Calcium (mg)	349	563
Iron (mg)	19.8	50.8
Vitamins		
A (carotene) (IU)	1235	3270
Thiamine (mg)	1.02	1.32
Riboflavin (mg)	0 63	1 22
Ascorbic acid (mg)	54 3	83 4
Pantothenic acid (mg)	4.9	12.0
Salt (g)	34.7	40.5

Note: Vitamin A consumption in both villages was far
lower than desirable. It was especially low in Village A Ascorbic acid consumption was also lower
in Village A, especially in winter and early spring.

From Kimura, T. and Ota, M (1965), *Am. J. Clin. Nutr*,
17, 381. © American Society for Clinical Nutrition. With
permission.

lead to cerebral hemorrhage, but the seasonal incidence of deaths from this condition suggests that ascorbic acid deficiency plays a major role in the conspiracy.

Certainly vitamin A deficiency alone can cause clinically significant intracranial hypertension in the absence of any signs of xerophthalmia or kerotomalacia, as reported by Kasarkis and Bass (1982), but this is not associated with any cerebral hemorrhage or thrombosis, so the seasonal vitamin C deficiency that occurs in some people in northern Honshu is most likely to be an essential factor in the etiology of the cerebrovascular lesions which are so common in that part of Japan.

Saelhof and McConnell (1962) cited evidence that little stroke was affecting over 1 million people a year in the U.S. In the majority of little strokes there may be personality changes, a growing irritability, and inability to concentrate. In some cases, the episodes follow one after another, within months, bringing death in 2 or 3 years or earlier. In other instances the disease drags on for many years. They found that cerebral arteriography revealed no occlusions in the majority of patients with little stroke, but generalized capillary fragility was very evident. They treated 112 patients who had suffered one or more little strokes by oral medication with capsules containing citrus bioflavonoids and ascorbic acid, 200 mg of each, 4 to 6 times a day for periods ranging from 1 to 5 years. Other medications were administered to 73 patients who served as controls. The capillary fragility index was significantly reduced in the bioflavonoid-treated group, but remained unchanged in the control group. One severe and six mild episodes occurred in the group of patients treated with citrus

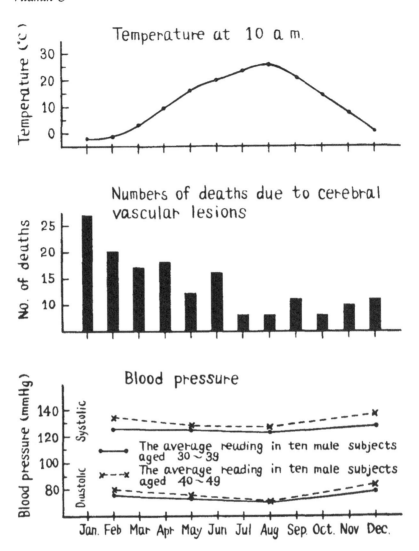

FIGURE 2.   The heights of the vertical black bars indicate the number of deaths due to cerebral vascular lesions in each month of the year in Village A. Mean environmental temperatures and examples of average blood pressures in men and women in that village are also shown. It is clearly evident that deaths from cerebrovascular lesions are most frequent in the late winter and spring. (From Kimura, T. and Ota, M. [1965], *Am. J. Clin. Nutr.*, 17, 381. © American Society for Clinical Nutrition With permission.)

bioflavonoids. There occurred 19 severe episodes of little stroke (including 6 fatalities) and 17 mild episodes in the control group.

Chazan and Mistilis (1963) reported seven patients admitted to the Boston City Hospital with frank scurvy in 1 year. Case number 1, a 69-year-old single white man is of particular interest here, for, "During the patient's period in the hospital he became disoriented and confused, with neurological signs consistent with thrombosis of the middle cerebral artery. He was given ascorbic acid and 2 units of whole blood and was transferred to the neurological service twenty-four hours after admission. The patient made an uneventful recovery after being treated with a "house diet" and supplemental multiple vitamins. He was discharged six weeks later, much improved with only a slight hemiparesis."

Acheson (1966) analyzed the age-adjusted death rates for cerebrovascular disease in the U.S. for the years 1914 to 1960 (Figure 3). In an extensive discussion of the shortcomings

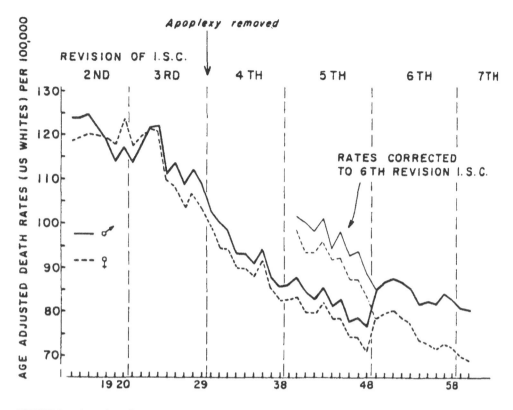

FIGURE 3. Age-adjusted death rates of U.S. white population, both sexes, for vascular lesions affecting the central nervous system, 1914 to 1960. Continuous line, males; discontinuous line, females. Figure shows the duration of each revision of the International Statistical Classification of Diseases, and the effect of 6th revision. (From Acheson, R. M. [1966], in Cerebrovascular Disease Epidemiology: A Workshop, Public Health Service Monogr. 76, U.S. Government Printing Office, Washington, D.C., 23.)

of mortality statistics in general, he pointed out that the accuracy of differentiation between the ICD* rubrics — 330 for subarachnoid haemorrhage, 331 for cerebral hemorrhage, 332 for cerebral thrombosis and embolism, 333 for spasm of the cerebral arteries, and 334 for other vascular diseases of the central nervous system, was of necessity uncertain. He also discussed changes in the coding that had occurred during those years. Nevertheless, he stated that, "if these trends, both for the whole group and the individual rubrics are genuine, the overall decrease must be a consequence of a reduction in the incidence of cerebral hemorrhage." While season appeared to be a major factor (Figure 4), Acheson was inclined to attribute this to deaths from secondary causes such as pneumonia in winter. He cited several studies showing a highly significant negative correlation between rubrics 330 to 334 and almost all constituents of "hard water" in municipal water supplies in the U.S., the U.K., and Japan. He discussed the suggestion of Schroeder (1966) that soft water will release heavy metals from corroded water pipes. This is of particular interest as copper is a potent catalyst for the oxidation of ascorbic acid which soon becomes hydrolyzed and loses its vitamin activity. Moreover, excessive iron stores are known to deplete ascorbic acid stores (Chapter 10, Volume I).

These data certainly suggest that ascorbic acid alone may not relieve the problem. We need either to remove copper water pipes from our homes or increase our intake of insoluble heavy metal-chelating food fiber, such as bioflavonoids or catechins, to remove heavy metals from the blood stream into the lumen of the bowel for excretion.

* International Classification of Diseases (6th and 7th revisions.)

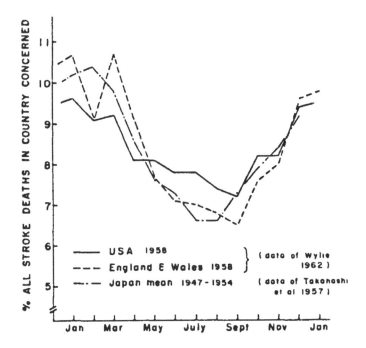

FIGURE 4.   Seasonal variation in deaths from all forms of cerebrovascular disease (Rubrics 330 to 334) in England and Wales and the U.S. in 1958, and from cerebral hemorrhage in Japan 1947 to 1954. (From Acheson, R. M. [1966], in Cerebrovascular Disease Epidemiology: A Workshop, Public Health Service Monogr. 76, U.S. Government Printing Office, Washington, D.C., 23.)

In a detailed study of 117 consecutive patients with nonembolic cerebral infarction at Kyushu University in Fakuoka, Japan, Omae et al. (1969) reported finding significantly increased plasma thromboplastic activity and fibrinogen content. They concluded that both of these changes were secondary to brain damage.

In a survey of 159 patients admitted to an acute geriatric unit in southwestern England, Wilson et al. (1972) found that those with low leukocyte total ascorbic acid levels ($<$12 μg/10^8 cells) had a significantly higher mortality within 4 weeks of admission than those with a high level ($>$25 μg/10^8 cells), as shown in Table 5, Chapter 5, Volume I. However, a subsequent trial of ascorbic acid supplementation (200 mg daily) for half the patients on admission, which was reported by Andrews and Wilson (1973), showed no benefit. There were 13 cerebrovascular episodes among 98 men and women who received ascorbic acid and 14 among 96 receiving placebo. Coronary episodes were actually more frequent among those receiving ascorbic acid than among the controls.

Perhaps it is too late to hope for much benefit when treatment is started so late in the progress of the disease, but a trial of ascorbic acid, 200 mg three times a day, in the form of catechin-coated tablets would certainly be worthwhile, for the mean leukocyte ascorbic acid levels of these elderly patients had not risen to 25 μg/10^8 cells following treatment with ascorbic acid alone.

Studies by Acheson (1964), Knox (1973), Acheson and Sanderson (1978), Acheson and Williams (1980), Garraway et al. (1983), and Editorials in *The Lancet* and the *New England Journal of Medicine* (1983) have revealed a steady fall in the incidence and mortality rates from cerebrovascular disease over the last 3 decades in Britain and even more so in the U.S. (Table 5). Various theories have been advanced in attempts to explain this fortunate decrease;

Table 5
AVERAGE ANNUAL AGE-SPECIFIC INCIDENCE RATES PER 100,000
POPULATION FOR STROKE IN VARIOUS PERIODS

Age group (years)	1945—1949		1950—1954		1955—1959		1960—1964		1965—1969		1970—1974		1975—1979	
	Rate	No.	Rate	No.	Rate	No.	Rate	No.	Rate	No.	Rate	No.	Rate	No.
<55	25	27	33	40	24	34	18	31	21	42	13	30	17	40
55—64	439	53	291	40	444	67	308	51	330	61	233	48	170	37
65—74	1079	77	1063	92	851	88	841	101	658	89	589	87	510	80
75—84	2156	66	2228	83	1720	79	1791	105	1662	126	1186	112	957	102
≥85	3925	25	3457	27	3700	39	2817	42	1628	34	1503	42	1026	36

Note: The incidence of stroke in the population of Rochester; MN; it has continued to decline, up to the end of the last decade.

From Garraway, W. M., Whisnant, J. P., and Drury, I. (1983), *Mayo Clin. Proc*, 58, 520. With permission.

hypertension is clearly important, but it does not seem to have shown any tendency to decline in incidence or severity. Knox (1973), in a correlation analysis between nutrient intakes and various causes of death in different regions of England and Wales, observed that the strongest negative association was between cerebrovascular disease and vitamin C intake (see Table 1, in Chapter 20 of this volume).

Taylor (1976) has observed capillary fragility in patients with cerebrovascular disease and has suggested chronic or intermittent vitamin C deficiency as the major cause of arterial disease and stroke. It is quite possible that improved transportation and distribution of fresh fruits and vegetables, in all seasons, may account for the progressive reduction in both the incidence and the mortality rates of stroke in recent years.

A very erudite article by Greer (1977), on the uncommon causes of stroke, listed and described 10 infective, 14 inflammatory noninfective, 5 metabolic, and 9 miscellaneous causes of cerebrovascular accidents in addition to neoplastic disease, but dismissed vitamin C deficiency with the following words: "In well-nourished populations, the chance that Vitamin C deficiency will be so severe as to cause cerebral haemorrhage is extremely remote."

This has in fact been the usual teaching throughout the English-speaking part of the world, but it takes no account of disturbances of ascorbic acid metabolism which are so common. If a disturbance of ascorbic acid metabolism were not the commonest cause of capillary fragility, it would not be possible to increase the capillary strength of so many patients by the use of rutin or rutin and ascorbic acid. Vitamin C deficiency is only thought to be rare because plasma ascorbate levels are so rarely measured. If all hospitals were set up to analyze plasma ascorbic acid and whole blood histamine levels routinely, then ascorbate-responsive histaminemia would be recognized as being very common indeed (Chapter 1 of this volume).

Acheson and Williams (1980) observed a strong and statistically significant negative correlation between National Food Survey estimates of vitamin C intakes and standardized death rates from cerebrovascular disease in the nine statistical regions of the U.K. (Figure 4). They also found a significant positive correlation between per capita expenditure on tobacco and cerebrovascular disease (Figure 5), which is certainly understandable. Smoking is known to reduce vitamin C levels (Chapter 4, Volume I); these same workers, however,

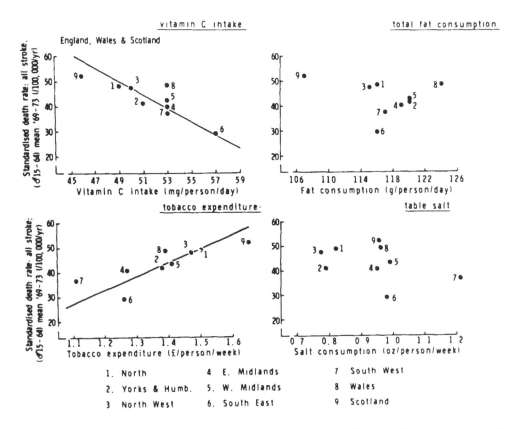

FIGURE 5.   Relationships between mortality from cerebrovascular disease (CVD) and indices of consumption of four factors in the general population of the nine statistical regions of the U.K. An inverse relationship between CVD and vitamin C intake and a direct relationship between CVD and tobacco expenditure are clearly evident. Table salt and fat consumption did not show any evident relationship to cerebrovascular disease. (From Acheson, R. M. and Williams, D. R. R. [1980], in *Clinical Neuroepidemiology*, Rose, F. C., Ed., Pitman Medical, London, 88 With permission.)

found no evident correlation between table salt or total fat consumption and death from cerebrovascular disease.

The mortality from cerebrovascular disease is lower in the south of England than the north (Table 6) and also lower in the more wealthy (Table 7), whether from the south or the north. Studies by Acheson and Williams (1983) showed certain significant negative correlations between cerebrovascular disease (CVD) mortality and estimates of fresh vegetable and fresh fruit consumption, both for geographical areas of England and for population density (Table 8). When calculations were made with sex-specific mortality rates, the correlations with female rates were higher than when classification was by geographical region, but higher with male rates when it was by population density. With the odd exception of fresh potatoes, correlations were consistently negative: some were of high order and some were statistically significant ($p < 0.05$). The absence of any significant correlation with potato intake suggests that not only ascorbic acid, but some other factor such as the bioflavonoids of fresh fruits and vegetables may be essential for the chelation of heavy metals in Britain.

Acheson and Williams pointed out that the patterns of correlations were in general similar when the population was divided and classified in two different ways. They suggested that this implies that the association is not likely to be due to chance (Table 8). While these authors urged caution in the interpretation of their results, they felt justified in concluding that, "Eating fresh fruit and fresh green vegetables does indeed seem to be associated with

Table 6
STANDARDIZED MORTALITY RATIOS FOR ALL CEREBROVASCULAR DISEASE AND CEREBRAL THROMBOSIS IN THE ENGLISH REGIONAL HEALTH AUTHORITIES AND IN WALES FOR MALES AND FEMALES SEPARATELY (1979)

Region	All cerebrovascular disease		Cerebral thrombosis	
	Males	Females	Males	Females
Northwest	124	117	115	117
Northern	121	112	151	135
Wales	116	114	144	141
Yorkshire	110	101	127	113
Merseyside	105	102	95	106
Trent	102	101	118	114
West midlands	105	101	114	105
Oxford	83	95	72	78
East Anglia	95	100	100	109
Southwest	98	104	85	88
Wessex	96	102	85	99
NW Thames	75	77	63	60
SW Thames	91	98	84	86
NE Thames	85	83	57	57
SE Thames	85	91	79	87

Note: The Registrar General's statistical reports for England and Wales, in recent decades, have shown a geographical gradient in the mortality rates for cerebrovascular disease and for cerebrovascular thrombosis, with a rise in mortality from southeast to northwest.

From Acheson, R. M. and Williams, D. R. R. (1983), *Lancet*, 1, 1191. With permission

Table 7
STANDARDIZED MORTALITY RATIOS FOR ALL CEREBROVASCULAR DISEASE AND CEREBRAL THROMBOSIS BY SOCIAL CLASS FOR DEATHS IN MALES AGED 15 TO 64 (1970—1971)

Cause of death	Social class					
	I	II	IIIN	IIIM	IV	V
All cerebrovascular disease	80	86	98	106	111	136
Cerebral thrombosis only	75	80	87	108	110	150

Note: Standardized mortality rates for all cerebrovascular diseases and for cerebral thrombosis have been found to be lower in the higher social classes, both in the north and in the south of England.

From Acheson, R. M. and Williams, D. R. R. (1983), *Lancet*, 1, 1191. With permission.

Table 8
COEFFICIENTS OF CORRELATION (r) BETWEEN MORTALITY FOR ALL CEREBROVASCULAR DISEASE AT AGES 45 TO 64, FRUIT AND VEGETABLE CONSUMPTION, AND RESIDENCE AT DEATH CLASSIFIED BY GEOGRAPHICAL AREA AND BY POPULATION AGGREGATION IN 1970 (BY SEX)

	Scotland, Wales, and regions in England (n = 11)				Population aggregation, England and Wales (n = 9)			
	Males		Females		Males		Females	
National Food Survey	r	p	r	p	r	p	r	p
Fresh green vegetables	−0.5200	0.15	−0.7708	0.01	−0.8917	0.02	−0.6306	0.18
Other fresh vegetables	−0.4556	0.22	−0.6051	0.08	−0.8087	0.05	−0.7930	0.06
All fresh vegetables	−0.5476	0.12	−0.8029	0.01	−0.9184	0.01	−0.6955	0.12
Fresh fruit	−0.6777	0.04	−0.7393	0.02	−0.8528	0.03	−0.6614	0.15
Fresh potatoes	0.5022	0.17	0.3350	0.38	0.1326	0.8	0.1786	0.73
All fresh fruit and vegetables	−0.2144	0 58	−0.5085	0.16	−0.3979	0.43	−0.0657	0 90

Note: Correlation analyses of cerebrovascular disease mortality and National Food Survey estimates of fresh vegetable and fresh fruit consumption are shown both by region and by population aggregation. With the exception of fresh potatoes, correlations are consistently negative; some are of a high order, and several are statistically significant.

From Acheson, R. M. and Williams, D. R. R. (1983), *Lancet*, 1, 1191. With permission.

reduced risk of early death from CVD, possibly through the vitamin C they contain. Nevertheless, it would be wise to keep an open mind about mechanism and simply assert that consumption of fruit and vegetables may offer a safe and agreeable form of preventive medicine against a disease which is much more important to modern society than scurvy.''

Vollset and Bjelke (1983), reporting a prospective nutritional study of 16,713 Norwegian postal survey respondents, stated their agreement with the conclusion of Acheson and Williams. Not only did they find that those with the highest consumption of vegetables, fruits, and berries had the lowest incidence of deaths from cerebrovascular disease (CVD), they also observed a similar inverse relationship between the intake of potatoes and deaths from CVD. The association was stronger for an index obtained by weighing the intakes of 20 items in these food groups by their vitamin C contents (Table 9).

Arad and Eyal (1983) studied premature newborn human infants with and without cerebral hemorrhage, proven by ultrasound examination. They concluded that there is an appreciable efflux of ascorbic acid from the brain into the blood plasma following intraventricular hemorrhage, so even the finding of a normal plasma ascorbate level following a cerebrovascular accident does not guarantee a normal intracerebral level. It behooves us to supplement all our patients.

Ohmoto et al. (1979) studying the exposed basilar arteries of cats, have concluded that the heme components of hemolyzed blood may be responsible for the delayed cerebral vasospasm which is so often seen following subarachnoid hemorrhage in man. These workers found that the vasospasm caused by hemolyzed blood could be prevented by topical application of *O*-phenanthroline, which chelates iron, by methyl prednisolone or by ascorbic acid. It seems that all patients with cerebral hemorrhage, whether or not they had cerebrovascular ascorbate deficiency before the hemorrhage, will subsequently have local ascorbate deficiency due to hemolysis (Chapter 15, Volume I). One can try to correct the local increase in the oxidation-reduction potential by chelation of iron (Chapter 10, Volume I), by administration of cortisone (Chapter 13, Volume I), or by administration of ascorbic acid. Indeed,

Table 9
OBSERVED/EXPECTED DEATHS (1969—81) FROM CEREBROVASCULAR DISEASE BEFORE AGE 80 BY LEVELS OF VITAMIN C INDEX

Age (December 31, 1967)	Deaths	O/E with vitamin C index			p^a	RR[b]
		<15	15—21	≥22		
45—54	13	6/2.7	5/5.0	2/5.2	0.02	0.17
55—64	86	23/21 1	37/33.8	26/31.1	0.32	0.75
65—74	124	46/38.6	40/46.9	38/38.5	0.36	0 87
Total	223	75/62.5	82/85.8	66/74.8	0.06	0.75
Ages 45—64 by specific diagnosis (ICD-8)						
Subarachnoid hemorrhage (430)	7	2/1.6	3/2 9	2/2.6	0.63	0.62
Cerebral hemorrhage (431)	15	6/3.9	6/5.8	3/5.4	0.13	0.35
Cerebral infarction (432—433)	13	1/3 2	9/5.0	3/4.8	0.89	1.1[c]
Other and ill-defined CVD (436—438)	64	20/15.3	24/25.3	20/23.4	0.18	0.64
Total (430—438)	99	29/23.9	42/38.8	28/36.3	0.08	0.62

Note. In this prospective study of 16,713 Norwegian respondents to a postal dietary survey, subsequent mortality from cerebrovascular disease (CVD) showed a significant decreasing trend with increasing vitamin C intake. This was especially notable in the 45- to 54-year age group, where the ratio of observed to expected deaths from CVD showed a significant inverse relationship to dietary ascorbic acid index ($p < 0.02$).

[a]   Two sided, for linear trend.
[b]   Relative risk for ≥22 vs. <15.
[c]   Departure from linear trend ($p = 0.02$)

From Vollset, S. F. and Bjelke, E. (1983), *Lancet,* September 24, 742. With permission.

Ohmoto et al. reported encouraging results in the treatment of delayed vasospasm following subarachnoid hemorrhage in two out of five patients by intrathecal administration of ascorbic acid (200 to 1000 mg).

A statistical analysis of the incidence of coronary heart disease and stroke among men in eastern Finland by Salonen and Puska (1983) concluded that both serum triglycerides and cholesterol levels were equally strong predictors of stroke, but only cholesterol was predictive of coronary heart disease when their data were corrected for other known risk factors such as age, tobacco consumption, and blood pressure. This study made no mention of ascorbic acid levels, even though there must be a considerable risk of ascorbic acid deficiency during the long winters in such a northern region. In fact, the incidences of coronary heart disease and stroke in northern Karelia, Finland, seem to be greater than anywhere else in the world.

Recent observations by Enwonwu and Okolie (1983) have revealed that the feeding of a low-protein (2% casein) diet to infant nonhuman primates *(M. nemestrina)* caused a highly significant reduction in the blood and brain ascorbate levels of these animals, and also a significant increase in their brain histamine levels. A similar elevation of the brain histamine level was observed when these monkeys were fed an ascorbate-deficient, 20% protein diet. These findings are in keeping with our knowledge of the effects of protein deficiency on ascorbate metabolism (Chapter 12, Volume I), and of ascorbate deficiency on histamine metabolism (Chapter 1 of this volume). Perhaps the high incidence of cerebrovascular accidents in northern Honshu results when the effects of low protein, low ascorbate, and perhaps low vitamin A intake are combined.

Certainly we have to consider cerebrovascular accident as a multifactoral disease. If we are to reduce the risk of cerebrovascular disease we should start early in life with a proper diet and good living habits; but even later in life it may be possible to reduce the risk of stroke. Not only must we control excesses of hypertension, perhaps by medication, we should also provide plenty of fresh fruit and vegetables or a dietary supplement of catechin-ascorbic acid pills, as well as sufficient first-class protein, vitamin A, and other vitamins in the diet. We must remove any focus of sepsis and insist on the importance of adequate sleep, as well as the avoidance of stress. A change from cigarette to pipe smoking may be advisable if abstinence is too miserable. Excessive alcohol consumption must be avoided. Most of these factors either affect ascorbic acid metabolism and thereby histamine metabolism or else affect histamine metabolism directly.

II. HYPERTENSION

All would agree that hypertension plays a very important role in the development of cerebral hemorrhage, so studies relating ascorbic acid status and hypertension are very pertinent here.

Koh and Stewart (1974), studying a rural black population in southwest Mississippi, reported a negative correlation between plasma ascorbic acid (TAA)* levels and blood pressure and suggested that ascorbic acid may in some way help to prevent the development of hypertension. However, their 304 subjects ranged in age from 5 to 80 years, so one wonders whether the results may not simply reflect the known increase in blood pressure and fall in ascorbate levels with increasing age.

A subsequent study by Koh and Chi (1978) was restricted to 439 persons aged 34 and over and it still showed a negative correlation between vitamin C (serum TAA) levels and systolic blood pressure ($p < 0.001$) in each of four groups, comprised by black and white men and women, and with diastolic blood pressure in black men ($p < 0.01$) and white men ($p < 0.05$). Oddly enough, in all four groups combined, the diastolic blood pressure was reported to be almost twice as useful as systolic blood pressure in predicting vitamin C levels. These studies showed no relationship between serum ascorbic acid and total cholesterol levels, but did show reciprocal relationships between serum ascorbic acid and two other pathological conditions, namely hypertension and obesity, which were consistently observed in both studies. These observations were impressive, but they needed to be confirmed in comparable 10-year age groups, or after statistical correction for age.

Rouse et al. (1983a), working at the University of Western Australia, observed a significant reduction (5 to 6 mmHg) in the systolic blood pressure of a group of omnivorous subjects within 4 weeks after changing to a lacto-ovovegetarian diet ($p < 0.01$). There was also a modest fall (2 to 3 mmHg) in the diastolic pressure, but this was not statistically significant ($p < 0.1$). Moreover, the blood pressure trends were reversed when the diets of the two groups were reversed. Further analysis of the data by Rouse et al. (1983b) revealed that the reduction of blood pressure was related to an increased intake of polyunsaturated fat, dietary fiber, vitamin C, vitamin E, magnesium and calcium, and a decreased protein intake, but it was not possible to know which factor or factors played major roles in reducing the blood pressure.

The recent finding by Yoshioka (1984) of a highly significant inverse correlation between serum total ascorbic acid and both systolic ($p < 0.001$) and diastolic ($p < 0.01$) blood pressure in healthy Japanese men aged 30 to 39 years suggests that chronic subclinical ascorbic acid deficiency not only damages the vascular endothelium, but may also cause hypertension.

Recent observations by Yoshioka et al. (1985) have demonstrated that spontaneously hypertensive (SH) rats have significantly lower tissue ascorbate (TAA) levels in the liver, lungs, adrenals, and kidneys than normal rats, suggesting some defect in the synthesis or

---

* TAA — total ascorbic acid, reduced and oxidized forms.

metabolism of ascorbic acid in these hypertensive animals. Moreover, these workers have reported that dietary supplementation with ascorbic acid prevented the blood pressure elevation in this breed of rats. This certainly seems to be a discovery of major importance and led these authors to conclude that: "The abnormalities of ascorbic acid metabolism in the SH rats may be associated with, in part, their high blood pressure, because exogenous ascorbic acid prevented the blood pressure elevation of SH rats, but some other mechanism may also be involved in the effect of ascorbic acid on blood pressure."

If indeed it is confirmed that low tissue ascorbate levels do cause hypertension, as well as weakening the walls of the blood vessels, the reasons for an association between low serum ascorbic acid levels and a high incidence of cerebrovascular accidents will be quite evident.

REFERENCES

Acheson, R. M. (1966), Mortality from cerebrovascular disease in the United States, in Cerebrovascular Disease Epidemiology: A Workshop, Public Health Service Monogr. 76, U.S. Government Printing Office, Washington, D.C., 23.

Acheson, R. M. and Sanderson, C. (1978), Strokes: social class and geography, *Popul. Trends,* 12, 13.

Acheson, R. M. and Williams, D. R. R. (1980), Epidemiology of cerebrovascular disease; some unanswered questions, in *Clinical Neuroepidemiology,* Rose, F. C., Ed., Pitman Medical, London, 88.

Acheson, R. M. and Williams, D. R. R. (1983), Does consumption of fruit and vegetables protect against stroke?, *Lancet,* 1, 1191.

Alexander, L., Pijoan, M., Schube, P. G., and Moore, M. (1938), Cevitamic acid content of blood plasma in alcoholic psychoses, *Arch. Neurol. Psychiatry,* 40, 58.

Alvarez, W. C. (1946), Cerebral arteriosclerosis; with small, commonly unrecognized apoplexies, *Geriatrics,* 1, 189.

American Pediatric Society (1898), American Pediatric Society's collective investigation on infantile scurvy in North America, *Arch. Pediatr.,* 15, 481.

Andrews, C. T. and Wilson, T. S. (1973), Vitamin C and thrombotic episodes, *Lancet,* 2, 39.

Arad, I. D. and Eyal, F. G. (1983), High plasma ascorbic acid levels in premature neonates with intraventricular hemorrhage, *Am. J. Dis. Child.,* 137, 949.

Bánhidi, Z. G. (1960), Weight increase in the rat due to thiamine disulfide activation by ascorbic acid, *Int. Z. Vitaminforsch.,* 30, 305.

Beaser, S. B., Rudy, A., and Seligman, A. M. (1944), Capillary fragility in relation to diabetes mellitus, hypertension and age, *Arch. Intern. Med.,* 73, 18.

Chazan, J. A. and Mistilis, S. P. (1963), The pathophysiology of scurvy. A report of six cases, *Am. J. Med.,* 34, 350.

Cole, F. M. and Yates, P. O. (1967), The occurrence and significance of intracerebral micro-aneurysms, *J. Pathol. Bacteriol.,* 93, 393.

Editorial, Walker, W. J. (1983), Changing U.S. life style and declining vascular mortality — a retrospective, *N. Engl. J. Med.,* 308, 649.

Editorial (1983), Why has stroke mortality declined?, *Lancet,* 1, 1195.

Ellis, A. G. (1909), The pathogenesis of spontaneous cerebral haemorrhage, *Int. Clin.,* Ser. 19, 2, 271.

Enwonwu, C. O. and Okolie, E. (1983), Differential effects of protein malnutrition and ascorbic acid deficiency on histidine metabolism in the brains of infant nonhuman primates, *J. Neurochem.,* 41, 230.

Feigenbaum, D. (1917), Ein Beitrag zur Kenntniss der Ruckenmarksblutungen beim Skorbut, *Wien. Klin. Wochenschr.,* 30, 1455; as cited by **Gilman, B. B. and Tanzer, R. C. (1932),** *JAMA,* 99, 989.

Fisher, C. M. (1971), Pathological observations in hypertensive cerebral haemorrhage, *J. Neuropathol. Exp. Neurol.,* 30, 536.

Gale, E. T. and Thewlis, M. W. (1953), Vitamins C and P in cardiovascular and cerebrovascular disease, *Geriatrics,* 8, 80.

Garraway, W. M., Whisnant, J. P., and Drury, I. (1983), The continuing decline in the incidence of stroke, *Mayo Clin. Proc.,* 58, 520.

Gilman, B. B. and Tanzer, R. C. (1932), Subdural hematoma in infantile scurvy, *JAMA,* 99, 989.

Greer, M. (1977), Uncommon causes of stroke. I. Diseases of the vessel wall, *Geriatrics,* 32, 28 and 39.

Griffith, J. Q., Krewson, C. F., and Naghski, J. (1955), Effect of rutin therapy on mortality, in *Rutin and Related Flavonoids*, Mack, Easton, PA, 173.

Griffith, J. Q., Jr. and Lindauer, M. A. (1944), Increased capillary fragility in hypertension: incidence, complications and treatment, *Am. Heart J.*, 28, 758.

Griffith, J. Q. and Lindauer, M. A. (1947), Rutin: therapy for capillary abnormality in hypertension, *Ohio State Med. J.*, 43, 1136.

Hayem, M. G. (1871), Note sur l'anatomie pathologique du scorbut, *C. R. Seances Mem. Soc. Biol.*, Ser. 5, Vol. 3, Sect. 23, Part 2, 3.

Ingalls, T. H. (1936), The role of scurvy in the etiology of chronic subdural haematoma, *N. Engl. J. Med.*, 215, 1279

Kasarkis, E. J. and Bass, N. H. (1982), Benign intracranial hypertension induced by deficiency of vitamin A during infancy, *Neurology*, 32, 1292.

Kimura, T. and Ota, M. (1965), Epidemiologic study of hypertensive: comparative results of hypertensive surveys in two areas in Northern Japan, *Am. J. Clin. Nutr.*, 17, 381

Knox, E. G. (1973), Ischaemic heart disease mortality and dietary intake of calcium, *Lancet*, 1, 1465.

Koh, E. T. and Chi, M. S. (1980), Relationship of serum vitamin C and globulin fractions with anthropometric measurements in adults, *Nutr. Rep. Int.*, 21, 1537.

Koh, E. T. and Stewart, T. (1978), Interrelationship among blood components and anthropometric measurements, *Nutr. Rep. Int.*, 18, 539.

Kubala, A. L. and Katz, M. M. (1960), Nutritional factors in psychological test behavior, *J. Genet. Psychol.*, 96, 343.

Lamy, M., Lamotte, M., and Lamotte, S. B. (1946), Etude anatomique des états de dénutrition, *Bull. Mem. Soc. Med. Hop. Paris*, 435.

Levrat, M. and Ballivet, J. (1939), Le purpura provoqué chez les hypertendus, *Lyon Med.*, 162, 243.

Lund, C. C. (1942), Ascorbic acid deficiency associated with gastric lesions, *N. Engl. J. Med.*, 227, 247.

McClelland, R. S. (1921), A case of scurvy, *Lancet*, 2, 608.

Meyer, E. (1896), Ueber Barlowsche Krankheit, *Arch. Kinderheilkd.*, 20, 202; as cited by Gilman, B. B. and Tanzer, R. C. (1932), *JAMA*, 99, 989.

Ohmoto, T., Yoshioka, J., Shibata K., Morooka, H., Matsumoto, Y., and Nishimoto, A. (1979), Cerebral vasospasm following subarachnoid hemorrhage. Experimental and clinical studies (in English), *Neurol. Med. Chir.*, 19, 73.

Omae, T., Katsuki, S., Nishimaru, K., Yamaguchi, T., Takeya, Y., Fujishima, M., and Kato, M. (1969), Clinical features of cerebral infarction in the Japanese, *J. Chron. Dis.*, 21, 585.

Ord, W. (1894), Subdural haemorrhage in scurvy, *Br. Med. J.*, 2, 1430.

Paterson, J. C. (1940), Capillary rupture with intimal hemorrhage in the causation of cerebral vascular lesions, *Arch. Pathol.*, 29, 345.

RossRussell, R. W. (1963), Observations on intracerebral aneurysms, *Brain*, 86, 425.

Rouse, I. L., Beilin, L. J., Armstrong, B. K., and Vandongen, R. (1983a), Blood-pressure-lowering effect of a vegetarian diet: controlled trial in normotensive subjects, *Lancet*, 1, 5.

Rouse, I. L., Beilin, L. J., Mahoney, D. P., Margetts, B. M., Armstrong, B. K., and Vandongen, R. (1983b), Vegetarian diet and blood pressure, *Lancet*, September 24, 742.

Saelhof, C. C., Martin, W. C., Sokoloff, B., and McConnell, B. (1959), Bioflavonoid therapy in the little stroke: a report of 21 cases, *Clin. Med.*, 6, 207.

Saelhof, C. C. and McConnell, B. H. (1962), Clinical study of little stroke in 112 cases, *Clin. Med.*, 69, 901.

Salonen, J. T. and Puska, P. (1983), Relation of serum cholesterol and triglycerides to the risk of acute myocardial infarction, cerebral stroke and death in Eastern Finnish male population, *Int. J. Epidemiol.*, 12, 26.

Sammis, J. F. (1919), A case of scurvy with cerebral hemorrhage, *Arch. Pediatr.*, 36, 274.

Sanford, H. N., Shmigelsky, I., and Chapin, J. M. (1942), Is administration of vitamin K to the newborn of clinical value?, *JAMA*, 118, 697.

Sato, T., and Nambu, K. (1908), Zur Pathologie und Anatomie des Skorbuts, *Virchows Arch. Pathol. Anat.*, 194, 151; as cited by Gilman, B. B. and Tanzer, R. C. (1932), *JAMA*, 99, 989.

Scherer (1913), Ueber Skorbut in Deutsch Südwest Africa, *Arch. Schiffs Trop. Hyg.*, 17, 191; as cited by Gilman, B. B. and Tanzer, R. C. (1932), *JAMA*, 99, 989.

Schroeder, H. A. (1966), Municipal drinking water and cardiovascular death rates, *JAMA*, 195, 81.

Sherwood, D. (1930), Chronic subdural hematoma in infants, *Am. J. Dis. Child.*, 39, 980.

Soloff, L. A. and Bello, C. T. (1948), "Capillary fragility" in hypertension: the effect of antiscorbutic therapy on the results of tests for "capillary fragility", *Am. J. Med. Sci.*, 215, 655.

Stokes, J. Jr. and Campbell, F. C. (1932), Three fatal cases of infantile scurvy with a discussion of febrile reactions, possibly secondary to antiscorbutic therapy, *Med. Clin. North Am.*, 16, 219.

Sulkin, N. M. and Sulkin, D. F. (1975), Tissue changes induced by marginal vitamin C deficiency, *Ann. N.Y. Acad. Sci.*, 258, 317.

Sutherland, G. A. (1894), On haematoma of the dura mater associated with scurvy in children, *Brain*, 17, 27.

Swanson, C. N. (1927), Scurvy complicating vomiting of pregnancy, *JAMA*, 88, 26.

Taylor, G. (1976), Vitamin C and stroke, *Lancet*, 1, 247.

Thewlis, M. W. and Gale, E. T. (1947), Nutrition in the aged, *J Am Inst. Homeopathy*, 40, 266.

Vilter, R. W., Wolford, R. M., and Spies, T. D. (1946), Severe scurvy; a clinical and hematologic study, *J. Lab. Clin. Med* , 31, 609.

Vollset, S. F. and Bjelke, E. (1983), Does consumption of fruit and vegetables protect against stroke?, *Lancet*, September 24, 742.

Willis, T. (1668), In quo agitur de morbis convulsivis et de scorbuto, pathologia cerebri et nervosi generis specimen; as cited by **Gilman, B. B. and Tanzer, R. C. (1932),** *JAMA*, 99, 989

Wilson, T. S., Weeks, M. M., Mukherjee, S. K., Murrell, J. S., and Andrews, C. T. (1972), A study of vitamin C levels in the aged and subsequent mortality, *Gerontol. Clin* , 14, 17.

Yoshioka, M., Aoyama, K., and Matsushita, T. (1985), Effects of ascorbic acid on blood pressure and ascorbic acid metabolism in spontaneously hypertensive (SH) rats, *Int. J. Vitam. Nutr. Res* , 55, 301.

Yoshioka, M., Matsushita, T., and Chuman, Y. (1984), Inverse association of serum ascorbic acid level and blood pressure or rate of hypertension in male adults aged 30-39 years, *Int J. Vitam. Nutr. Res.*, 54, 343.

Chapter 20

CORONARY THROMBOSIS AND MYOCARDIAL INFARCTION

I. CLINICAL AND PATHOLOGICAL OBSERVATIONS

The older records of scurvy contain many references to sudden death. James Lind (1753) wrote, "Persons that appear to be but slightly scorbutic, are apt to be suddenly and unexpectedly seized with some of its worse symptoms. Their dropping down dead upon exertion of their strength, or change of air, is not easily foretold." Such sudden deaths suggest heart attacks or pulmonary emboli, but we cannot now know the causes of these deaths which occurred so long ago.

O'Shea (1918) reported praecordial pain and bradycardia (40 beats per minute) associated with cardiac irregularity lasting for 10 d in one patient with scurvy and tachycardia (120 to 140 beats per minute), in three other patients.

Ralli and Friedman (1938) recorded the occurrence of coronary occlusion in a 45-year-old man with piled up, bleeding, ulcerated gums and petechiae over both arms due to scurvy. His blood vitamin C level was recorded as 0.37 mg/100 ml, but he had to receive 6700 mg of ascorbic acid by mouth before any appreciable quantity of ascorbic acid began to appear in his urine.

Bronte-Stewart (1953) recorded 32 adults admitted to the Groote Schuur Hospital from 1946 to 1950 with the diagnosis of scurvy. One who died soon after admission was found to have had a coronary thrombosis.

Thomson (1954) recorded 2 deaths from myocardial infarction among 100 patients with scurvy admitted to Stobhill General Hospital in Glasgow. Three other deaths were attributed to pneumonia, to chronic bleeding duodenal ulcer, and to inanition.

Ascorbic acid deficiency is normally considered as a cause of hemorrhage rather than thrombosis. However, the coagulation time has been found to be normal in guinea pigs with scurvy by Findlay (1921) and in guinea pigs with hypovitaminosis C by Ginter et al. (1968). It has also been found to be normal in human scurvy by Bronte-Stewart (1953) and by Cutforth (1958). Moreover, blood coagulation is the normal mechanism for the arrest of hemorrhage, so subendothelial hemorrhages could well be a precipitating cause of thrombosis. Indeed, Paterson (1936a, b) observed some degree of hemorrhage into the intimal tissue in all of ten recently thrombosed coronary arteries and noted collections of hemosiderin in the deep layers of the coronary arteries in most older thromboses. All showed vascularization of the intima from the lumen, and this process seemed to be confined almost exclusively to arteries with atherosclerosis, either with or without thrombosis. The portion of the intima affected by the hemorrhage was sometimes quite small, but it was a consistent finding in recent thromboses and almost certainly initiated the thrombus formation. Wartman (1938) reported seven deaths from coronary occlusion due to hemorrhage in the vessel wall, without any evidence of thrombosis. He suggested that intramural coronary arterial hemorrhage may lead to thrombosis or may itself cause coronary occlusion. Subsequent work by Paterson (1938, 1940, 1941) confirmed the existence of intimal hemorrhage adjacent to the thrombus in 52 out of 58 thrombosed coronary arteries (see Figure 1).

Serial sections of more than 100 thrombosed coronary arteries by Horn and Finkelstein (1940) at Mt. Sinai Hospital, New York, confirmed that intramural hemorrhage is the most frequent underlying mechanism in coronary occlusion. All of the thrombosed coronary arteries showed sclerotic changes, and hemorrhages into the walls of these diseased vessels occurred with great frequency. Intramural hemorrhage was considered to have caused 62.5% of the coronary occlusions, whereas thrombosis on an arteriosclerotic plaque was observed

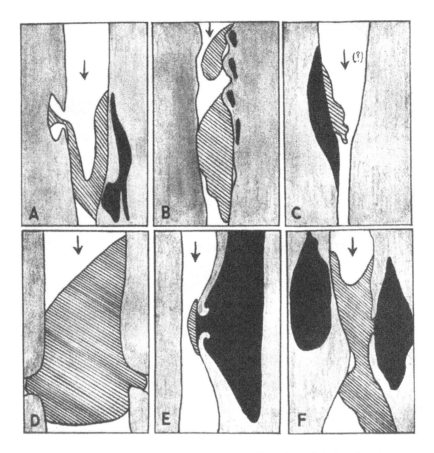

FIGURE 1.   Longitudinal diagrams constructed from serial sections of six thrombosed coronary arteries, showing intimal hemorrhages (black) beneath the thrombi (diagonally shaded) in five of the six specimens. (From Paterson, J.C. [1938], *Arch. Pathol.*, 25, 474. ©1940 American Medical Association. With permission.)

in 37.5%. Iron pigment deposition in the walls of the coronary arteries provided evidence of old hemorrhage and made it difficult to ascertain with any certainty the pathogenesis of the severe narrowing which affected these arteries.

Seymour and Sowton (1964) reported that the administration of ascorbic acid restored upright T-waves in the electrocardiograms of two "healthy" human volunteers treated with large doses of digoxin. Shafar (1967) reported rapid reversion of electrocardiographic changes, like those of myocardial infarction, after treatment in two cases of scurvy. One wonders what would have happened to these patients if they had not been diagnosed as having scurvy and had not been treated with ascorbic acid, for Hume et al. (1972) reported a further fall in leukocyte ascorbic acid to scorbutic levels after myocardial infarction. Follis, in 1942, reported sudden deaths in three infants with scurvy and concluded on autopsy that they died of heart failure.

Sament (1970) reported sudden death to be a well-recognized hazard among Africans with scurvy in Johannesburg. He therefore treated scurvy as a medical emergency, requiring immediate intravenous ascorbic acid and found that the electrocardiographic changes were reversed in minutes after intravenous ascorbic acid. This is clearly too rapid a response for an impending coronary thrombosis, so it must be attributed to reversal of scorbutic cardiac myopathy or to an effect on the fibers of Purkinje.

Spodick (1970) drew attention to another cardiac disorder in scurvy, namely hemopericardium, which he noted as having been recorded over a century earlier.

Singh and Chan (1974) described a woman in Singapore who presented with a 1-week history of breathlessness on exertion and swelling of the ankles. She was found to be lethargic, with spongy septic bleeding gums and multiple skin petechiae. There was massive edema of the legs, ascites, pleural effusion, and cardiac enlargement, yet her neck veins were not distended and her electrocardiogram was normal. She was completely restored to normal health within 1 week by oral treatment with ascorbic acid, 400 mg daily.

Clearly, sudden deaths in scorbutic individuals can be due to cardiac myopathy, to cardiac conduction defects, to pericardial hemorrhage, or to pulmonary embolism, but they can be due to coronary occlusion.

II. ANIMAL EXPERIMENTS

In an experiment involving control, scorbutic, infected, and scorbutic-infected guinea pigs, Taylor (1937) made detailed examinations of the hearts. They were opened, fixed, and embedded and approximately every 30th and 31st sections were taken and stained with hematoxylin and eosin and by the Gram-Weigert method for bacteria. Although he was most interested in searching for evidence of rheumatic carditis, his diligent search revealed that, "the coronary arteries of the scorbutic and scorbutic-infected pigs frequently showed organized thrombi, but without evidence of infarction." In order to study chronic rather than acute scurvy, some of the guinea pigs had been allowed 2 ml and some 4 ml of orange juice on alternate days. This allowed them to live longer and it seems to have been mostly those dying later that developed coronary thrombosis. Indeed, partial deficiency simulates a more likely human condition than does a completely deficient diet.

Human coronary thrombosis usually occurs in blood vessels which have already been damaged by atherosclerosis, so the guinea pig studies of Willis (1953) and of Ginter (1974), already cited in the section on atherosclerosis, are very pertinent here. Willis showed that ascorbic acid deficiency damages the intima of the arteries and markedly accelerates the subendothelial deposition of cholesterol. This inevitably leads to the development of atherosclerosis.

Moreover, Chambers and Zweifach (1940), in experiments on frog mesenteric capillaries, showed that injury to the vascular endothelium results in the development of a stickiness of the interendothelial cement, and Samuels and Webster (1952), studying the endothelium of the veins of dogs, showed that platelets adhere along the lines of interendothelial cement when a vessel is injured.

III. HUMAN EXPERIMENTS

Krebs (1953) reported a study of ten healthy volunteers, aged 21 to 24 years, on a vitamin C-free diet. The following is a quotation from his report.

Some important abnormalities were observed in single cases. One man developed effusions into both knee joints and ecchymoses of the leg during the 30th week of deprivation after a long walk. Another was taken ill 4 weeks later, 19 h after heavy physical exercise. He had severe pain in the lower sternal region, and became dyspnoeic and cyanosed. The pulse was rapid and the blood pressure low. The clinical picture was that of an acute cardiac emergency. He was immediately admitted to hospital and dosed with vitamin C. The lower sternal pain, which at first had increased in intensity, passed off after 9 h. The electrocardiogram showed high ST levels in leads I and II. A radiogram of the chest showed no abnormality. Eighteen days later another deprived volunteer complained of a sudden constrictive pain in the chest. Physical examination revealed a systolic murmur which had not been heard before, and the electrocardiogram showed a partial heart-block, the P-R interval being 0.32 sec. Before the experiment the electrocardiogram had been normal with a P-R interval of 0.20 sec. It was thought necessary to treat this volunteer immediately with large doses of vitamin C. The chest pain and the systolic murmur disappeared with 24 h, but during the following months the P-R interval showed variable periods between 0.13 and 0.32 sec depending on posture, breathing, administration of drugs, and other factors.

These incidents occurred at a stage when occasional skin petechiae were the main clinical manifestations of scurvy and general fitness appeared to be fairly good. We can only conjecture as to whether it was subendothelial hemorrhage, or hemorrhage followed by thrombosis, in one of the coronary arteries that caused these heart attacks.

Griffith et al. (1955) reported a reduction in the incidence of coronary occlusion in patients with capillary fragility, from 5 to 3% per annum as a result of treatment with vitamin C and rutin.

Simonson and Keys (1961) reviewed the extensive Russian literature on the relationship between vitamin C and atherosclerosis. Miasnikova in 1947, and later, Miasnikov in 1954, reported that vitamin C supplements retarded the development of atherosclerosis in cholesterol-fed rabbits. This led to clinical trials of ascorbic acid therapy in patients with coronary heart disease by Tiapina, by Sedov, and by others. Apparently ascorbic acid is now accepted as a standard and essential part of the treatment of coronary thrombosis in Russia.

Sokoloff et al. (1966) of Lakeland, FL, conducted a trial of ascorbic acid supplementation in 60 patients with high blood cholesterol levels and/or coronary heart disease. "In 10 cases the administration of ascorbic acid (2.0 to 3.0 gm daily for twelve to thirty months had no effect. In the remaining 50 cases there was definite and often marked improvement; on the average, lipoprotein lipase activity increased by 100 per cent and the level of triglycerides declined by 50—70 percent. In contrast to the results in animals, the total cholesterol level did not change significantly."

IV. DEMOGRAPHIC AND NUTRITIONAL SURVEYS

Death rates from ischemic heart disease and other forms of cardiovascular disease have been found to be higher in soft-water areas than in hard-water areas throughout the U.S., the U.K., and Japan, where they have been studied, as reported by Schroeder (1966). This has been discussed in Chapter 10, Volume I, where the potentially harmful effects of soft water and certain heavy metals on ascorbic acid metabolism were outlined. It is of particular interest that Knox (1973) found a positive correlation between dietary iron and ischemic heart disease and a negative correlation between this form of heart disease and vitamin C intake, as shown in Table 1.

Armstrong et al. (1975) of Oxford University conducted multiple regression and partial correlation analyses between ischemic heart disease and the consumption of various food substances in the British Isles. They found a strong negative relationship between ischemic heart disease mortality and the consumption of total fresh green vegetables.

Phillips et al. (1978) reported a 6-year prospective study of 24,000 members of the Seventh-Day Adventist religious order in California. The standard mortality rate of these people for coronary heart disease was found to be half that for the California population as a whole. No doubt, this was partly due to the quieter life of these people and their total abstinence from tobacco and alcohol. Approximately half of them were lacto-ovovegetarians, and these authors found the vegetarians to have a coronary death rate which was one third of that for the nonvegetarians ($p < 0.01$). Lower total or saturated fat intake may have played a role in protecting the vegetarians from coronary heart disease, but the high intakes of ascorbic acid and chelating fiber which are characteristic of a vegetarian diet may also have been important protective factors.

Diabetics are particularly prone to vascular disease, with a greatly increased risk of coronary thrombosis, so it is interesting to note that Sarji et al. (1979) found the mean platelet ascorbic acid level of male insulin-dependent diabetics (25.5 μg/10^{10} platelets) to be significantly lower than that of normal nonsmoking control men (45.2 μg/10^8 cells; $p < 0.001$). Moreover, these authors reported that ascorbic acid decreased platelet aggregation, both *in vivo* and *in vitro*, in normal volunteers, but not in a diabetic patient. They suggested

Table 1
MATRIX OF CORRELATIONS BETWEEN S.M.R.S AND NUTRIENT INTAKES

	Causes of death						
Nutrients	**All causes**	**Ischemic heart disease**	**Cancer of stomach**	**Diabetes**	**Cerebrovascular disease**	**Hypertension**	**Bronchitis**
k cal	+0.54	+0.44	+0.57	+0.27	+0.53	+0.63	+0.34
Protein	+0.29	+0.28	+0. 29	+0.17	+0.29	+0.31	+0.16
Animal protein	−0.40	−0.33	−0 45	−0.34	−0.45	−0 29	−0.31
Fat	+0.25	+0.33	+0.28	+0.04	+0.23	+0.38	+0.06
Carbohydrate	+0.56	+0.38	+0 58	+0.33	+0.55	+0.61	+0.40
Calcium	−0 61	−0.67	−0.56	−0.24	−0.59	−0 02	−0.52
Iron	+0.37	+0.49	+0.37	+0.19	+0.39	+0 12	+0.17
Vitamin A	−0.04	+0.04	+0.09	−0.30	−0.16	+0.04	−0.08
Thiamine	+0.09	+0.07	+0 17	+0.17	+0.13	+0.35	+0.04
Riboflavin	−0.04	−0.04	−0.47	−0.33	−0.47	−0.19	−0.38
Nicotinic acid	+0.22	+0.28	+0.20	+0.01	+0.16	+0.09	+0.16
Vitamin C	−0 63	−0.49	−0.45	−0.19	−0.68	−0.13	−0.52
Vitamin D	+0.57	+0.58	+0.43	+0.12	+0.49	−0 16	+0.56

Note: Results of a correlation analysis carried out between standardized mortality ratios (S.M.R.S.) for ischemic heart disease and other disease states in different regions of England and Wales in different years and dietary intakes of a number of nutrients. It is of interest to note that all the correlations of S.M.R.S. with calcium and vitamin C were negative All those with iron and thiamine and all but one of those with vitamin D were positive.

From Knox, E. G. (1973), *Lancet,* 1, 1465. With permission

that there may be a decreased transfer of ascorbic acid from plasma to platelets in diabetes mellitus, as postulated by Mann (1974).

There is little doubt that a high-cholesterol diet is a major contributor to the development of atherosclerosis and coronary heart disease in man and in other primates, as stressed by Stamler (1979), but dietary vitamin C intake is also very important, as ascorbic acid aids the conversion of cholesterol to bile acids (Chapter 5 of this volume), and the histaminemia of ascorbic acid insufficiency (Chapter 1 of this volume) predisposes to subendothelial hemorrhages which favor subendothelial cholesterol deposition, an aberrant form of arterial wound healing.

Walker (1983) discussed the 18 consecutive annual declines in age-specific coronary mortality reported for the U.S. in 1977 and suggested that a reduced consumption of tobacco and a general change in the national diet had been largely responsible. No doubt, the Surgeon General's warning about the risks of smoking had a beneficial effect, and so did a reduced intake of cholesterol and saturated fat, but it seems likely that an increased intake of ascorbic acid, and the year-round availability of fresh fruits and vegetables may also have played important roles in reducing the mortality from this multifactorial disease.

It is surely pertinent that coronary heart disease is common in Finland, where the harvesting of garden produce is restricted to a short summer season, and where vitamin C deficiency is known to exist among those who cannot afford enough imported fruits and vegetables in winter and spring. Indeed, an editorial in the *British Medical Journal* (1977) noted that the highest incidence of coronary heart disease in the world is reported from eastern Finland.

Salonen and Puska (1983), at the University of Kuopio, have demonstrated a positive correlation between elevated serum total cholesterol levels and an increased risk of death from acute myocardial infarction in eastern Finnish men. These high cholesterol levels may in part be due to a high cholesterol intake, but decreased conversion of endogenous and exogenous cholesterol to bile acids may be an even more important factor during periods of ascorbic acid deficiency (see Chapter 5 of this volume).

Gey et al. (1987) analyzed blood samples from groups of European men aged 40 to 49 years, living in regions of high coronary heart disease (CHD) mortality (North Karelia, Finland, and Edinburgh, Scotland), medium CHD mortality (Belfast, Northern Ireland), and low CHD mortality (Thun, prealpine, Switzerland, and Sapri, southern Italy). Not only did they find that plasma total cholesterol levels correlated with high incidence of CHD, they also found low plasma ascorbic acid levels (near marginal vitamin C deficiency between January and April), and a low vitamin E status in people of areas with high CHD mortality.

V. ASCORBIC ACID LEVELS

Paterson (1941), in a study of 455 consecutive adult admissions to the public wards of the Ottowa Civic Hospital, recorded fasting plasma vitamin C levels of less than 0.5 mg/100 ml in 56% of all patients and in 81% of patients with proven coronary thrombosis on the morning after admission to hospital. It was noted that this high incidence of vitamin C deficiency was not reached by any other type of disease in the series. Even gastrointestinal diseases, including peptic ulcer, showed a higher vitamin C level on the average than did coronary occlusion. It was suggested that hypertension combines with atheroma and capillary fragility due to vitamin C deficiency to cause intimal hemorrhages which lead to coronary thrombosis. Paterson also noted that coronary thrombosis occurred most frequently during the months of February, March, April, and May, when the vitamin C levels were lowest. Cases of cerebral thrombosis also showed a pronounced degree of vitamin C deficiency, although somewhat less than in the coronary occlusion group.

Trimmer and Lundy (1948) observed that 15 out of 23, or 65% of patients with coronary thrombosis had vitamin C (AA)* levels of 0.35 mg/100 ml or less, as compared with 87 out of 411, or 21%, of miscellaneous patients in their practice.

Sokoloff (1966) reported that, "in coronary and related diseases the concentration of ascorbone (dehydroascorbic acid) is about three times higher than in healthy persons of the same age groups;" these data suggest the existence of a profound disturbance of ascorbate metabolism, with a markedly reduced AA:DHAA ratio in people with coronary heart disease.

Hume et al. (1972) found the mean ascorbic acid concentration of heart muscle from patients who had died of coronary occlusion (47.9 ± 26.7 μg/g) to be significantly higher than that of patients dying from other causes (mean 35.7 ± 19.2 μg/g; $p < 0.05$). They believe that this is because leukocytes have migrated to the damaged myometrium, carrying ascorbic acid to the site where it is needed for the process of healing.

Andrews and Wilson (1973) found no significant differences between the leukocyte ascorbic acid levels of geriatric patients who subsequently developed coronary thrombosis and those who did not. They also reported that ascorbic acid supplements, 200 mg daily for 3 or 4 weeks after admission to hospital, did not reduce, but seemed to increase the incidence of coronary episodes. Indeed, there were 19 coronary episodes among 271 patients receiving vitamin C supplements and only 12 among 267 receiving placebo tablets. This presents a paradox, because Spittle (1973), studying thrombosis in another site, namely the deep veins of the calf, observed marked benefit from the administration of an ascorbic acid supplement. In a double-blind study of venous thrombosis diagnosed by the ^{125}I-fibrinogen-scanning technique, she found that ascorbic acid, 1 g daily, almost halved the incidence of deep-vein thrombosis, from 60 to 33%. There was also a striking reduction in the clinical evidence of thrombosis in the treated patients.

Vallance et al. (1978) studied the serum and leukocyte total ascorbic acid levels of seven

---

* AA — ascorbic acid, reduced form.

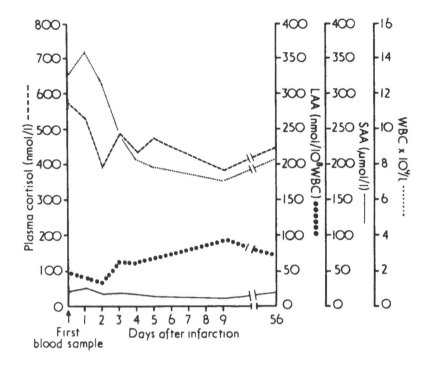

FIGURE 2. Mean values for six patients following acute myocardial infarction. LAA = leukocyte total ascorbate; SAA = serum total ascorbate; WBC = white blood cell count. The fall in the leukocyte ascorbic acid level results from the granulocytosis which occurs after the infarction; it shows some recovery as the leukocytosis subsides. In contrast, the serum total ascorbate level starts much lower than normal and remains low in unsupplemented patients. (From Vallance, B. D., Hume, R., and Weyers, E. [1978], *Br. Heart J.*, 40, 64. With permission.)

men (mean age 56) and five women (mean age 61) who had sustained acute myocardial infarctions. In six who received no ascorbic acid supplement, the leukocyte ascorbic acid and the serum ascorbic acid levels remained subnormal throughout the period of the 56 d of the study, as shown in Figure 2. In some instances the serum ascorbic acid was so low as to be unrecordable. They found that granulocytes contain about half as much ascorbic acid as lymphocytes, so they attributed a fall in the leukocyte ascorbic acid levels to the leukocytosis which is associated with an increase in the proportion of granulocytes. These authors state that, "While the initial apparent fall in leucocyte ascorbic acid after an acute myocardial infarction seems to be a reflection of the granulocytosis, the eventual return of leucocyte ascorbic acid to normal when the acute stress has subsided, is dependent on an 'adequate' supply of ascorbic acid." They believe that the very low serum ascorbic acid levels which persisted in unsupplemented patients were the result of myocardial damage. This is probably true, but inspection of Table 2 suggests that the ascorbic acid levels were probably low before the occurrence of infarction.

Ramirez and Flowers (1980) found the leukocyte ascorbic acid level to be markedly reduced in patients with radiological evidence of coronary atherosclerosis ($p < 0.001$), suggesting that ascorbic acid deficiency or a disturbance of ascorbic acid metabolism (as by smoking, etc.) may play an important role in the etiology of coronary artery diseases.

VI. BLOOD CHOLESTEROL LEVELS

Ginter, working at the Institute of Human Nutrition in Bratislava, Czechoslovakia, has shown that ascorbic acid has a major effect on the rate of conversion of cholesterol to bile

Table 2ª

	Serum total ascorbate μmol/ℓ	Leukocyte total ascorbate nmol/10⁸ WBC
6 normal volunteers	52 ± 23	138 ± 49
31 professional footballers before exercise	80 ± 28	
The same footballers after exercise	97 ± 34	
6 acute myocardial infarct patients; initial values (Group S)	13 ± 12	42 ± 18
6 acute myocardial infarct patients after 24 h (Group U)	19 ± 17	35 ± 11

ª Data from the work of Vallance et al. (1978).

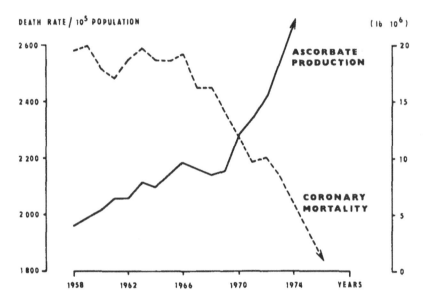

FIGURE 3. Age-adjusted (35 to 74) death rates from ischemic heart disease per 100,000 population and production of ascorbic acid during the past 20 years in the U.S. Sources: Vital Statistics of the U.S. and Census of Manufacturers, U.S. Department of Commerce. (From Ginter, E. [1979], *Am. J Clin. Nutr.*, 32, 511. © American Society for Clinical Nutrition. With permission.)

acids (Chapter 5 of this volume). He believes that ascorbic acid deficiency predisposes to atherosclerosis not only in guinea pigs, but also in man. Discussing the decline in the total death rate from coronary heart disease in the U.S. in the last 20 years, Ginter (1979) stated that the use of beta blockers (like propranolol) must have played a part, but gave reasons for believing that the declining coronary death rate may be related to the increased production of ascorbic acid, as shown in Figure 3. It would indeed be ironic if the populace were taking ascorbic acid in the hope of preventing the common cold and were actually reducing their chances of death from coronary thrombosis.

A study of elderly men and women by Horsey et al. (1981) confirmed that men with ischemic heart disease tend to have elevated total cholesterol (not significant) and very low-density-lipoprotein (VLDL) cholesterol levels ($p < 0.02$), but a lower than normal percentage of cholesterol in the high-density-lipoprotein (HDL) fraction ($p < 0.05$). Moreover, these authors reported finding a significant positive correlation between the initial leukocyte ascorbic acid levels and HDL cholesterol in men ($p < 0.02$), but not in women. Oral administration of ascorbic acid, 1 g daily for 6 weeks, resulted in a rise in the HDL cholesterol concentration, both in men ($p < 0.02$) and in women ($p < 0.05$) with ischemic heart disease. Ascorbic acid treatment also reduced the low-density-lipoprotein (LDL) cholesterol concentration in the men ($p < 0.05$) and reduced the total and VLDL triglyceride concentration ($p < 0.01$; $p < 0.05$) in women.

VII. AGING

Adams et al. (1974), at Guy's Hospital Medical School, have shown by autopsy studies that the connective tissue of the Achilles tendon and the coronary artery age along parallel lines. Bassler (1977) has pointed out that ascorbic acid may be the common link, being essential both in the formation of collagen for the tendon and for stabilization of the cement between the endothelial cells of the coronary arteries.

It certainly seems that our inability to synthesize ascorbic acid may be our Achilles heel. Daily exercise and weight reduction should be of great benefit to many of us, but will be of no avail if we neglect our dietary vitamin C intake.

Meacham et al. (1969) have shown that ascorbic acid can be of value as an antisludging agent for blood in the extracorporeal circulation machines which are used for cardiac surgery. It would seem wise to pay attention to the needs for this vitamin long before surgery becomes necessary.

Catechin coating of ascorbic acid tablets (Chapter 11, Volume I) may prove to be especially important in geriatric patients, so as to prevent gastric distress due to release of ascorbate free radical and the loss of vitamin activity in the achlorhydric stomach which is so prevalent among elderly men and women.

REFERENCES

Adams, C. W. M., Bayliss, B., Baker, R. W. R., Abdulla, Y. H., and Hunter-Craig, C. J. (1974), Lipid deposits in ageing human arteries, tendons and fascia, *Atherosclerosis,* 19, 429.

Andrews, C. T. and Wilson, T. S. (1973), Vitamin C and thrombotic episodes, *Lancet,* 2, 39.

Armstrong, B. K., Mann, J. I., Adelstein, A. M., and Eskin, F. (1975), Commodity consumption and ischemic heart disease mortality, with special reference to dietary practices, *J. Chronic Dis.,* 28, 455.

Bassler, T. J. (1977), Parallel aging of Achilles tendon and coronary artery, *Br. Med. J.,* July 23, 262.

Bronte-Stewart, B. (1953), The anaemia of adult scurvy, *Q. J. Med.,* 22, 309.

Chambers, R. and Zweifach, B. W. (1940), Capillary endothelial cement in relation to permeability, *J. Cell. Comp. Physiol.,* 15, 255.

Cutforth, R. H. (1958), Adult scurvy, *Lancet,* 1, 454.

Editorial (1977), Why does coronary heart disease run in families?, *Br. Med. J.,* August 13, 415.

Findlay, G. M. (1921), The blood and blood vessels in guinea-pig scurvy, *J. Pathol. Bacteriol.,* 24, 446.

Follis, R. H. (1942), Sudden death in infants with scurvy, *J. Pediatr.,* 20, 347.

Gey, K. F., Stähelin, H. B., Puska, P., and Evans, A. (1987), Relationship of plasma level of vitamin C to mortality from ischemic heart disease, *Ann. N.Y. Acad. Sci.,* 498, 110.

Ginter, E. (1979), Decline in coronary mortality in the United States and vitamin C, *Am. J. Clin. Nutr.,* 32, 511.

Ginter, E. (1974), Vitamin C in lipid metabolism and atherosclerosis, in *Vitamin C. Technological and Nutritional Aspects of Ascorbic Acid,* Applied Science, London, 179.

Ginter, E., Bobek, P., and Ovecka, M. (1968), Model of chronic hypovitaminosis C in guinea-pigs, *Int. Z. Vitaminforschung.,* 38, 104.

Griffith, J. Q., Krewson, C. F., and Naghski, J. (1955), *Rutin and Related Flavonoids,* Mack, Easton, PA.

Horn, H. and Finkelstein, L. E. (1940), Arteriosclerosis of the coronary arteries and the mechanism of their occlusion, *Am. Heart J.,* 19, 655.

Horsey, J., Livesley, B., and Dickerson, J. W. T. (1981), Ischaemic heart disease and aged patients: effects of ascorbic acid on lipoproteins, *J. Hum. Nutr.,* 35, 53.

Hume, R., Weyers, E., Rowan, T., Reid, D. S., and Hillis, W. S. (1972), Leucocyte ascorbic acid levels after acute myocardial infarction, *Br Heart J.,* 34, 238.

Knox, E. G. (1973), Ischaemic-heart-disease mortality and dietary intake of calcium, *Lancet,* 1, 1465

Krebs, H. A. (1953), The Sheffield experiment on the vitamin C requirement of human adults, *Proc. Nutr. Soc.,* 12, 237

Lind, J. (1757), *A Treatise of the Scurvy,* 2nd ed., Millar, London.

Mann, G. V. (1974), Hypothesis: the role of vitamin C in diabetic angiopathy, *Perspect Biol. Med.,* 17, 210.

Meacham, E. J., Sachs, D., and Circotti, J. J. (1969), Rationale for ascorbic acid as an anti-sludging agent in extracorporeal circulation, *J Cardiovasc. Surg.,* 10, 152.

O'Shea, H. V. (1918), Scurvy, *Practitioner,* 101, 283.

Paterson, J. C. (1936a), Vascularization and hemorrhage of the intima of arteriosclerotic coronary arteries, *Arch. Pathol.,* 22, 313.

Paterson, J. C. (1936b), Capillary rupture with intimal hemorrhage as a causative factor in coronary thrombosis, *Arch. Pathol.,* 22, 474

Paterson, J. C. (1938), Capillary rupture with intimal hemorrhage as a causative factor in coronary thrombosis, *Arch. Pathol.,* 25, 474.

Paterson, J. C. (1940), Capillary rupture with intimal hemorrhage in the causation of cerebral vascular lesions, *Arch. Pathol.,* 29, 345

Paterson, J. C. (1941), Some factors in the causation of intimal haemorrhages and in the precipitation of coronary thrombi, *Can. Med. Assoc. J.,* 44, 114.

Phillips, R. L., Lemon, F. R., Beeson, W. L., and Kuzma, J. W. (1978), Coronary heart disease mortality among Seventh-Day Adventists with differing dietary habits: a preliminary report, *Am. J Clin. Nutr.,* 31, S191.

Ralli, E. P. and Friedman, G. J. (1938), The response to the feeding of cevitamic acid in normal and deficient subjects as measured by a vitamin C excretory test, *Ann. Intern. Med.* 11, 1996.

Ramirez, J. and Flowers, N. C. (1980), Leukocyte ascorbic acid and its relationship to coronary artery disease in man, *Am. J. Clin. Nutr.,* 33, 2079

Salonen, J. T. and Puska, P. (1982), Relation of serum cholesterol and triglycerides to the risk of acute myocardial infarction, cerebral stroke and death in Eastern Finnish male population, *Int. J. Epidemiol.,* 12, 26.

Sament, S. (1970), Cardiac disorders in scurvy, *N. Engl. J. Med.,* 282, 282.

Samuels, P. B. and Webster, D. R. (1952), The role of venous endothelium in the inception of thrombosis, *Ann. Surg.,* 136, 422.

Sarji, K. E., Kleinfelder, J., Brewington, P., Gonzalez, J., Hempling, H., and Colwell, J. A. (1979), Decreased platelet vitamin C in diabetes mellitus: possible role in hyperaggregation, *Thromb. Res.,* 15, 639

Schroeder, H. A. (1966), Municipal drinking water and cardiovascular death rates, *JAMA,* 195, 81.

Seymour, J. and Sowton, E. (1964), Action of ascorbic acid on digitalis effects in the cardiogram, *Br. Med. J.,* June 13, 1551.

Shafar, J. (1967), Rapid reversion of electrocardiographic abnormalities after treatment in two cases of scurvy, *Lancet,* 2, 176.

Simonson, E. and Keys, A. (1961), Research in Russia on vitamins and atherosclerosis, *Circulation,* 24, 1239.

Singh, D. and Chan, W. (1974), Cardiomegaly and generalized oedema due to vitamin C deficiency, *Singapore Med. J.,* 15, 60.

Sokoloff, B., Hori, M., Saelhof, C. C., Wrzolek, T., and Imai, T. (1966), Aging, atherosclerosis and ascorbic acid metabolism, *J. Am. Geriatr. Soc.,* 14, 1239.

Spittle, C. R. (1973), Vitamin C and deep vein thrombosis, *Lancet,* 2, 199.

Spodick, D. H. (1970), Another cardiac disorder in scurvy, *N. Engl. J. Med.,* 282, 686.

Stamler, J. (1979), Population studies, in *Nutrition, Lipids and Coronary Heart Disease,* Levy, R., Rifkind, B., Dennis, B., and Ernst, N., Eds., Raven Press, New York, 25.

Taylor, S. (1937), Scurvy and carditis, *Lancet,* 1, 973.

Thomson, T. J. (1954), Scurvy — a rare disease?, *Glasgow Med. J.,* 35, 363.

Trimmer, R. W. and Lundy, C. J. (1948), A nutrition survey in heart disease, *Am. Pract.,* 2, 448.

Vallance, B. D., Hume, R., and Weyers, E. (1978), Reassessment of changes in leucocyte and serum ascorbic acid after acute myocardial infarction, *Br. Heart J.,* 40, 64.

Walker, W. J. (1983), Changing U.S. life style and declining vascular mortality — a retrospective, *N. Engl. J. Med.,* 308, 649.

Wartman, W. B. (1938), Occlusion of the coronary arteries by hemorrhage into their walls, *Am. Heart J.,* 15, 459.

Willis, G. C. (1953), An experimental study of the intimal ground substance in atherosclerosis, *Can. Med. Assoc. J.,* 69, 17.

Index

INDEX

Printed and bound by CPI Group (UK) Ltd, Croydon, CR0 4YY

22/10/2024

01777630-0013